Optical Wireless Communication Theory and Technology

Series Editor

Xizheng Ke, School of Automation and Information Engineering, Xi'an University of Technology, Xi'an, Shaanxi, China

The book series Optical Wireless Communication Theory and Technology aims to introduce the key technologies and applications adopted in optical wireless communication to researchers of communication engineering, optical engineering and other related majors. The individual book volumes in the series are thematic. The goal of each volume is to give readers a comprehensive overview of how the theory and technology in a certain optical wireless communication area can be known. As a collection, the series provides valuable resources to a wide audience in academia, the communication engineering research community and anyone else who are looking to expand their knowledge of optical communication.

Xizheng Ke · Chenghu Ke

Noise Models in Optical-Wireless Communication Systems

Science Press
Beijing

Springer

Xizheng Ke
School of Automation and Information Engineering
Xi'an University of Technology
Xi'an, Shaanxi, China

Chenghu Ke
School of Information Engineering
Xi'an University
Xi'an, Shaanxi, China

ISSN 2731-5967 ISSN 2731-5975 (electronic)
Optical Wireless Communication Theory and Technology
ISBN 978-981-97-7549-1 ISBN 978-981-97-7550-7 (eBook)
https://doi.org/10.1007/978-981-97-7550-7

Jointly published with Science Press
The print edition is not for sale in China mainland. Customers from China mainland please order the print book from: Science Press.

This Springer imprint is published by the registered company Springer Nature Singapore Pte Ltd.
The registered company address is: 152 Beach Road, #21-01/04 Gateway East, Singapore 189721, Singapore

Preface

Optical-wireless communication has characteristics of radio-frequency and optical-fiber communication. Unlike optical-fiber communication, optical-wireless communication does not require optical fibers, and the system is flexible and convenient. Compared with microwave communication, optical-wireless communication does not require spectrum permits, and the receiving device is very light. The transmission signal in optical-wireless communication is not subject to the physical restrictions of electromagnetic-wave transmissions. It can transmit over a greater distance at a faster rate. Noise in optical-wireless communication systems limits further improvements.

This book analyzes the noise mechanism in optical-wireless communication systems, explores noise-suppression methods, and seeks a way to approach the channel-capacity limit. Moreover, it improves the quality of optical-wireless communication systems, lays a theoretical foundation for their research and system design, and provides supporting experiments.

Chapter 1 introduces noise classification in optical-wireless communication systems, as well as the current research progress at home and abroad. Source, channel, and sink noises in optical-wireless communication systems are introduced. Chapter 2 introduces the detector noise mechanism. The outlook on domestic and foreign research progress helps readers gain a comprehensive understanding of the field. Methods for applying various types of detectors are introduced.

Chapter 3 begins with the atmospheric-turbulence mechanism and introduces the characteristics of optical-wireless communication channels, including the classification of atmospheric turbulence and its mechanism. Chapter 4 introduces the measurement of noise caused by atmospheric turbulence, and various results are displayed. Through an analysis of actual measurement data, a general guiding conclusion is obtained. Chapter 5 introduces atmospheric-turbulence suppression methods and their research progress. These include diversity technology, large-pore receiving technology, adaptive optics, and partial coherent optical-transmission theory.

Chapter 6 introduces noise in visible-light communication systems and its impact on the system. Chapter 7 introduces the noise mechanism in underwater optical-wireless communication, as well as noise models and cross-agency noise models.

Chapter 8 introduces the noise and measurement results encountered in ultraviolet (UV) communications. Chapter 9 establishes the noise model of an optical-wireless communication system and analyzes various noises, providing readers with a comprehensive understanding of the noise in the system.

Master's graduate students Liao Zhiwen and Xu Dongsheng assisted in the writing of this book. The experiments in this book are part of their academic theses. Dr. Qin Huanhuan participated in the study on visible-light noise models. The author expresses his heartfelt respect for these graduate students and thanks them for their youth and enthusiasm for scientific research.

This book represents part of the author's time as an Honorary Professor at Xi'an University. It is also the research results of the "Academician Workstation of Ke Xizheng," Xi'an University.

This research was supported by the National Natural Science Foundation of China and the major industrial chain fund of Shaanxi Province.

The author expresses his thanks to the authors of the literature that was referenced, as well as the large number of authors who were not referenced. Their work has provided him with many opportunities for in-depth study.

Qinling Old House
Midsummer 2023

Xizheng Ke
Chenghu Ke

Contents

Chapter 1
Introduction

1.1 Research Background and Significance

In optical-wireless communication, data information is loaded onto a laser beam, which is similar to a carrier wave, for data transmission. It has clear advantages in information transmission, confidentiality, transmission speed, etc. [1]. In addition, it can effectively overcome the "last mile" bottleneck of information-transmission technology in broadband networks, and provides more convenient services for users [2].

Optical-wireless communication includes interplanetary optical communication for satellite-to-satellite communications; satellite-to-ground optical communication; fixed-ground optical-wireless communication between stations; onboard, shipborne, and airborne optical-wireless communication between mobile devices and mobile stations or fixed points; and underwater optical communication [2].

Optical-wireless communication has characteristics of both radio-frequency and fiber-optic communication, with unique advantages. Compared to fiber-optic communication, optical-wireless communication systems do not need auxiliary optical fibers, and system installations are very flexible. Compared with microwave communication, optical-wireless communication devices do not require frequency licensing, have higher transmission rates, and the transceiver device is very lightweight. Transmission signals in optical-wireless communication are not subject to the physical limitations of electromagnetic-wave propagation; they can cover a longer distance and propagate faster.

The atmospheric refractive index is the ratio of the propagation speed of light in a vacuum to the propagation speed of light in atmosphere. When light propagates in atmosphere, owing to the uneven distribution of the atmospheric refractive index, its propagation path is not always along a straight line. The light refracts, affects the transmission of optical signals in an atmospheric channel. Physically, the air density changes with changes in other factors, such as temperature, humidity, and pressure,

X. Ke and C. Ke, *Noise Models in Optical-Wireless Communication Systems*,
Optical Wireless Communication Theory and Technology,
https://doi.org/10.1007/978-981-97-7550-7_1

which vary at different times and locations. This can lead to random fluctuations in the atmospheric refractive index.

When a laser is transmitted through the atmosphere, random fluctuations in the atmospheric refractive index affects it, causing such things as beam expansion, spot drift, intensity scintillation, and fluctuations in the arrival angle. This leads to a significant decrease in the accuracy and transmission distance of the laser communication and its tracking and aiming system; in severe cases, this may even lead to the interruption of the laser-transmission link [3, 4].

1.2 Current Research Status of Optical-Wireless Communication Noise Models

The function of a communication system is to transmit information from a source to one or more destinations. In optical-wireless communications, the information generated by a source is first modulated onto an optical carrier, and the carrier (beam or field) is emitted through an optical channel called a transmitter. Optical transmitters include source and channel coding, modulation, optical-signal amplification, and transmission antennas. Subsequently, optical methods are used to collect and process optical signals at the receiver end. The optical receiver includes an optical-signal collection antenna, spatial optical-fiber coupling unit, preamplifier, detector, and demodulator.

In the optical-field transmission process, various forms of noise can significantly affect the communication-system performance. Therefore, modeling and analyzing these noises are crucial. This section will analyze each type of noise introduced from the transmitter, channel, and receiver perspectives. A typical optical-wireless communication system and the noise introduced therein are shown in Fig. 1.2.

Since the discovery of lasers, researchers have actively attempted to use them for communication. In 1967, Brookner began simulating and modeling wireless laser communications from Earth to Venus in space [5], and measured the laser transmission in the atmosphere [6]. The development of communication theory provided a theoretical basis for the advancement of optical-wireless communication technologies. In 1970, Gagliadi conducted preliminary research on the theory of optical-wireless communication [7], and in 1976, he systematically expounded the principle of optical-wireless communication in his book, *Optical-Wireless Communication.*

The emergence of semiconductor lasers solved the problem of light-source miniaturization. In 1989, Katz summarized the application of new lasers in wireless laser communication and further elaborated on a system model [8]. In 1994, Hinton observed that the development of integrated circuits brought wireless laser communication into a rapid development stage [9]. Large-scale integrated circuits enabled complex algorithms to be applied to optical-wireless communication. In 2010, Wolf indicated that subcarrier modulation would become mainstream in the integration of wireless laser communication with telecommunication networks [10].

During the same period, optical-wireless communication research in China increased from a specialized-technology research stage to a system research stage. In 2004, Ke systematically studied the mathematical models of the source, channel, and sink components in wireless laser communication systems and highlighted the importance of communication coding in improving system performance [11]. In 2010, Jiang summarized a space laser-communication system and elaborated on the acquisition, pointing, and tracking (APT) system of a flight platform [12]. In 2015, Ma summarized the development of satellite-to-ground laser communication in China and provided a detailed introduction to technologies, such as the design of satellite laser-communication terminals, preview and acquisition of space beams, and satellite-orbit prediction [13].

These studies show that research in the wireless laser-communication field in China is advancing. Although the wireless laser-communication noise model is quite in-depth, it is still in the isolated research stage and lacks a unified-system noise model. Therefore, to study the noise model of wireless laser communication systems, this section begins with three aspects: source, channel, and sink, and elaborates on the research progress of related noise models.

1.2.1 Channel Noise

A laser beam propagating through an atmospheric channel is affected by the atmospheric environment. In wireless laser communication, the reception of optical signals is often affected by weather conditions. The influence of atmospheric channels on the propagation of light beams includes phenomena such as intensity attenuation, flickering, and beam-spread drift. In addition, noise disturbances are caused by other environmental factors, such as background-light noise and aiming-error noise, which are unfavorable factors in optical-wireless communication systems.

An early physical explanation of beam propagation in the atmosphere was provided by Kolmogorov [14]. The beam propagates in weakly turbulent channels. Two approximate theories, the Born and Rytov theories, can derive similar results. A lognormal model with the logarithmic variance of light intensity as a parameter can describe the probability distribution of the received light intensity [15]. The Rytov approximation uses the logarithmic light intensity and phase to describe the light field [16]. When the propagation distance is small, the theoretical and experimental data match well.

The description of beam behavior in medium-to-strong turbulence is very important. In 1979, Parry discovered experimentally that the K distribution describes the distribution of light intensity under certain harsh conditions [17]. In 1982, Andrews used a complex Gaussian model with a Rayleigh intensity distribution and a uniform phase distribution to describe the propagation of light in medium and strong turbulence [18]. Subsequent researchers further developed the model into the modified Rice–Nakagami distribution [19]. However, the premise of such models is the Born approximation, which leads to discrepancies between the numerical model and experimental results.

In 1985, Andrews developed the I-K distribution [20], which had better universality than the K distribution. Subsequently, researchers conducted extensive testing on the propagation of lasers in various environments and accumulated a large amount of data. In 1988, Miyazaki measured the propagation of lasers in smoke [21]. In 1993, Vostretsov et al. measured the intensity distribution of laser propagation in snow [22]. The results indicated that the intensity distribution followed a certain gamma distribution.

In 1997, Valadas et al. proposed the use of angle diversity in indoor optical-wireless communication systems to combat environmental noise [23]. The results showed that the optical gain increased with the relative weight of the directional noise within the unit and the sharpness of the beamwidth of the directional noise source. In 2000, Wang et al. [24] discussed the baseband digital-modulation technology of an additive Gaussian white-noise channel, and conducted a more in-depth study on the optimal receiver performance of the channel. They focused on the relationship between the bit-error rate of the signal transmission and the signal-to-noise ratio, and found an effective method to determine the best receiver.

In 2001, Andrews derived the gamma–gamma model, which was consistent with experimental data under moderate-to-strong turbulence conditions [25]. In 2002, Rollins et al. [26] proposed that the detector sensitivity was affected by the level of background light. They proposed that harmful environmental effects, including interference from sunlight in the instantaneous field of view of the receiver, must be considered when designing and testing optical transceivers. In 2005, Razavi et al. [27] investigated the effects of atmospheric turbulence, background light, and receiver thermal noise. Numerical results were obtained using the conditional Gaussian approximation method, and the power losses generated under various operating conditions were measured.

In 2005, Kiasaleh et al. [28] considered weak-turbulence (clear air) scenarios and a negative exponential distribution of the received signal intensity, and provided numerical results under pulse-position modulation to elucidate the impact of turbulence on the overall performance. In 2007, Hranilovic provided a mathematical model of the aiming error and calculated the system-capacity limit under its influence [29]. In 2008, Sermsak et al. [30] used vector radiation-transmission theory to model underwater optical-wireless communication channels and then quantified the scattering effect as a function of distance and bit rate through numerical Monte Carlo simulations.

In 2008, Zhu [31] conducted in-depth research on the performance of various modulation methods in indoor optical-wireless communication and studied the root-mean-square delay generated by multipath channels during high-speed transmission and the impact of the intersymbol interference it caused. In 2008, Ma et al. [32] explored the characteristics of laser-propagation attenuation in foggy media and analyzed the effect of changes in the field-of-view angle on the signal-to-noise ratio. In 2009, Mullen et al. [33] investigated the effects of underwater scattering on the optical communication signals, received optical power, and link ranges.

In 2009, Zwick et al. [34] studied the effectiveness and limitations of the additive Gaussian white-noise model in quasi-linear, long-distance, zeroing, and direct-detection optical-fiber communication systems. In 2010, Meng [35] used the classic quadratic generalized Lagrange method to analyze a optical-wireless communication model under equation constraints and established a optical-wireless communication model affected by the atmospheric-transmission window, as well as atmospheric turbulence. In 2011, Jurado-Navas further proposed the Malaga distribution model [36]. In 2011, Wang [37] constructed an efficient and accurate mathematical-calculation model to evaluate the error performance of optical-wireless communication systems under weak turbulence conditions.

In 2012, Tu [38] proposed a binary amplitude-keying modulation method based on a mixed-noise model to calculate the bit-error rate. The simulation results showed that the free-space optical communication performance was significantly affected by atmospheric turbulence and noise. In 2012, Abd et al. introduced the impact of adverse weather conditions, such as rain, fog, and snow, and the resulting scattering loss on the transmission performance of optical-wireless communication systems [39]. In 2013, Yao et al. [40] analyzed the channel model of underwater optical-wireless communication and derived an expression for the false packet rate of fixed-length digital pulse-spacing modulation. In 2013, Ansari [41] investigated the aiming-error problem in optical communication systems using the Meijer G-function and integrated the obtained results into a unified expression.

In 2013, Rao studied the proportion of each component in receiver noise under different conditions, providing guidance for suppressing system noise [42]. In 2014, H.E. Nistazakis et al. studied the bit-error rate performance of atmospheric-turbulence channels modeled using gamma–gamma or negative exponential distributions [43]. In 2014, Varotsos [44] analyzed the effects of factors, such as time dispersion and atmospheric turbulence, on free-space optical communication systems. In 2014, Zhu et al. [45] proposed a phase-noise measurement scheme based on the delay-line method, which employed a multifunctional microwave photonic processor that simultaneously achieved an electro-optical conversion, provided a time delay, and controlled the phase of the output microwave signal.

In 2015, Ke [46] proposed a spatial-reception diversity-system model of optical-wireless communication under a strong-turbulence model and verified its applicability and anti-interference ability in atmospheric environments through simulation. In 2015, Yang [47] derived a theoretical expression for the symbol-error rate of optical-wireless communication systems under multi-pulse position modulation conditions using the exponential Weber fading model. In 2015, according to the spatial-correlation characteristics of strong turbulence, Ke et al. fitted the correlation function based on the 1-norm, obtained the correlating K-distribution random field using a spherical invariant stochastic process, and established a two-dimensional correlation K-distribution turbulence-channel simulation model [48].

In 2015, Tian et al. [49] proposed a model for evaluating the performance of coherent optical transmission systems under the influence of a balanced enhanced phase noise. In 2016, Han et al. derived an expression for the intensity-scintillation index of array beams propagating in a non-Kolmogorov turbulent atmosphere using the Rytov method, based on the generalized Huygens–Fresnel principle [50]. In 2016, Ke et al. used logarithmic-normal and gamma–gamma distributions to model the channel. The analysis showed that the former was not suitable for the light-intensity fluctuation behavior under medium and strong turbulence, while the latter had a wider application range [51].

In 2017, Fan et al. [52] established a sky background-light noise model and conducted theoretical analyses and simulation research. The simulation results showed that the sky background light caused a decrease in the signal-to-noise ratio of the detectors used in the coarse and fine tracking units, resulting in a decrease in tracking accuracy and an increase in the communication error rate. In 2017, Zhang [53] described the specific classification of noise in free-space optical communication systems and the time-varying and fixed fading models, owing to turbulence.

In 2018, Ke [54] investigated the effect of phase noise from the demodulation module on the error performance of optical-wireless communication systems for optical-wireless subcarrier multivariate phase-shift keying modulation systems. They also simulated and analyzed the effect of the variance of the light-intensity rise and fall, phase noise, and modulation order on the system-error performance.

In 2019, Wang [55] derived the Fourier-series expression for the probability-density function of the phase-noise model with a Tikhonov distribution. In 2020, Nafkha [56] derived the traversal-capacity expression for multiple-input multiple-output (MIMO) Jacobi and MIMO Rayleigh-fading channels and used Monte-Carlo method validation.

In 2020, Xing [57] analyzed the effects of absorption, scattering, and solar-radiation noise on the performance of underwater optical-communication systems and power-allocation performance. In 2020, Liu [58] analyzed the effect of absorption and scattering of seawater channels on light attenuation and compared the time-domain expansion phenomenon of optical pulses under different water-quality conditions by simulation. In 2020, Wang [59] modeled the underwater optical-propagation characteristics and analyzed the attenuation due to light absorption and scattering in the ocean, background noise, and the energy-consumption model when transmitting data.

In 2020, Fu et al. [60] conducted a simulation study of an underwater optical-wireless communication (UOWC) system, exploring the beamwidth, receiving aperture, ocean-turbulence parameters, and targeting error on the average BER and outage-probability performance. In 2020, Zhang et al. [61] proposed an asymptotic growth-element learning adaptive-multirate scheme based on orthogonal frequency-division multiplexing and analyzed it over UOWC turbulent channels. In 2021, Lee-Leon et al. [62] designed a receiver system based on a deep-confidence network. The simulation results showed that the performance improved 13.2 dB at a 10^{-3} BER.

In 2021, Behnaz [63] introduced an analytical model for scattering noise and investigated its impact on the BER performance of UOWC systems over three types of water and a range of transmission-link spans. In 2021, Han [64] analyzed the channel characteristics and noise model for indoor visible-optical communication systems, focusing on scattering-noise and thermal-noise effects on the system performance. In

2021, Ke et al. [65] proposed a background-light noise model of the nighttime visible-light communication system of the Internet of Vehicles, based on the superposition of a double-Gaussian function.

In 2021, Ke et al. [66] conducted in-depth research on the impact of atmospheric turbulence on optical-wireless communication and explored phenomena such as beam diffusion, drift, and scintillation. In 2021, Yang et al. used the double-generalized gamma distribution and temporal-covariance function of the atmospheric-turbulence channel to establish time-series models that could reflect the dynamic characteristics of the receiving light intensity [67].

In 2021, Shen et al. deduced an analytical expression for light-intensity scintillation based on the classic weak-fluctuation turbulence theory and simulated the light-intensity scintillation of plane and spherical waves under different turbulence intensities and Fresnel numbers. The results showed that when the Rytov variance was less than 4.8, there was an intersection point in the light-intensity scintillation curve of the plane and spherical waves, which corresponded to the same scintillation intensity of the plane and spherical waves under the Fresnel number. When the Rytov variance was greater than 4.8, the scintillation intensity of the plane wave was always less than that of the spherical wave [68].

In 2022, Miao et al. used a series representation method to study the traversal capacity and interrupt performance of free-space optical (FSO) communication systems on lognormal turbulent channels with generalized pointing errors. They derived approximate analytical expressions for lognormal Rician channels with generalized pointing errors. The results indicated that the pointing errors significantly reduced the performance of the FSO links. By selecting the optimal beamwidth on the transmitter side, the effect of the pointing error on the traversal capacity could be significantly reduced [69]. In 2022, Li et al. [70] conducted systematic research, compared the theory and construction of various atmospheric-turbulence spectral models, and detailed the atmospheric-turbulence models used in practical research.

In 2022, Zhang et al. [71] analyzed the receiving noise of underwater optical-wireless communication systems and built a UOWC experimental platform based on 852-nm wavelengths under environmental-light noise interference. The results showed that the use of filters could effectively improve the communication performance and link length of UOWC under environmental-light noise interference. In 2022, Chen et al. [72] proposed an architecture for wavelength-division multiplexing (WDM) coherent data-center links and found that, in each frequency comb, the phase noise in all combs was completely determined by the optical phase noise of the seed laser and the phase noise of the microwave oscillator.

In 2023, Li et al. [73] discussed optical design, driver circuits, and channel modeling in visible-light full-duplex communication. In 2023, Jin et al. [74] discussed the performance of broadband long-haul nonlinear fiber-optic communication systems in the presence of balanced enhanced phase noise and showed that an increase in the symbol rate and transmission distance exacerbated the distortion of C-band systems.

1.2.2 Signal and Other Noise

The carrier in a wireless laser communication system primarily refers to a photodetector. Commonly used photodetectors in communications include PIN-type photovoltaic detectors and photoresistor-type photoconductive detectors. Among these detectors, thermal noise, dark-current noise, shot noise, and 1/f low-frequency noise are common.

In 1976, Boisrobert studied the circuit characteristics of PIN detectors when receiving sine-wave modulated signals, provided an equivalent circuit of PIN detectors, and analyzed the scattering noise of the detectors [75]. In 1986, Bowers used the heterodyne method to measure the received signal-to-noise ratio of PIN detectors for amplitude- and frequency-modulation signals [76], observing that the modulation depth was an important parameter affecting system performance.

In 1995, Wieczorek conducted a detailed analysis and model of the PIN scattering noise, dark-current noise, and 1/f noise. The study showed that dark-current noise was independent of the detection signal and that 1/f noise originated from the fluctuation of the injection current [77]. In 1997, Fang et al. demonstrated that the position of the input impedance in the equivalent-noise model of a photoreception amplifier was different from that in a typical noise model; therefore, the equivalent-noise model was not suitable for application [78].

In 2003, Haas et al. [79] conducted a shot-noise-limited operation, where the detector output was a double-random Poisson process; its rate was proportional to the sum of the transmission power through lognormal random-fading scaling. They proved that the ergodic capacity of this fading channel was equal to or exceeded the ergodic capacity of the channel with the determined path gain. In 2004, Guo et al. conducted a detailed analysis and research on the noise model of a photomultiplier tube by summarizing previous experience, and provided the corresponding equivalent circuit [80].

In 2005, Bao et al. [81] conducted experimental and theoretical research on the flicker noise (1/f noise) of optoelectronic-coupling devices before and after electrical stress, and established a quantitative-analysis model based on the mechanism of the carrier-number and mobility fluctuations. The experimental results were in good agreement with the model. In 2007, Gao et al. [82] conducted a numerical simulation of a mathematical model of avalanche-photodiode characteristics and obtained numerical results for the internal electric-field distribution, dark-current characteristics, excess noise, and gain-bandwidth characteristics of avalanche photodiodes.

In 2008, Wang [83] analyzed and modeled the granular noise, thermal noise, and background noise in a optical-wireless communication system, and conducted a qualitative analysis of the transmission performance of each part of the system. In 2010, Hu et al. [84] elaborated on the optoelectronic-conversion circuit, amplification-circuit noise, amplification-circuit stability, and other issues that needed attention, and designed an amplification circuit that effectively reduced the noise and temperature drift, with a large dynamic input range.

In 2012, Wang et al. [85] theoretically analyzed the impact of shot noise induced by background light on a communication system, proposed a theoretical model that allowed receiver sensitivity and the corresponding power penalty, and verified the model through experiments. The experimental results showed that when the working bit rate was low, the background light led to a power loss of several dB. As the bit rate increased, the noise caused by the preamplifier increased, ultimately dominating the noise process.

In 2012, Yan analyzed the random noise of photodetectors from the perspective of statistics, indicating that normal and logarithmic-normal distributions could describe the random-noise characteristics of photodetectors [86]. In 2013, Sakaharas et al. [87] analyzed the performance of PIN and APD photodetectors commonly used in optical communication systems in terms of bandwidth, gain, extinction ratio, and shot noise.

In 2019, Yan et al. [88] proposed a method for directly recovering the clock and data from a discrete random pulse-sequence output using a single-photon avalanche diode. Through experiments, the effects of the system setting parameters, such as the number of gating signals, synchronous character length, and baud rate, on the average bit-error rate were verified. In 2022, Mostafapour et al. [89] discussed the situation of multiplicative noise in visible-light communication, modeled and implemented mobile adaptive networks, and analyzed their performance.

1.3 Current Research Status of the Atmospheric-Turbulence Model

1.3.1 Atmospheric-Turbulence Effect

When a light beam carrying a signal passes through a turbulent atmospheric channel, the atmospheric-turbulence effect causes the optical power within the beam cross-section to undulate randomly. The resulting phenomena of light-intensity scintillation, beam drift, and phase undulation cause the communication performance of the optical communication system to deteriorate.

As shown in Fig. 1.1, the impact of a turbulent atmospheric channel on optical communications, including light-intensity flicker, beam drift, beam expansion, and phase undulation, causes the optical signal to randomly undulate, so that the optical power attenuates. Moreover, it decreases the communication system's signal-to-noise ratio, reducing the communication quality. Considering the series of effects caused by atmospheric turbulence on optical-wireless communication systems, the study of methods to inhibit atmospheric turbulence is of great significance for the development of optical-wireless communication.

When a transmission-signal beam passes through an turbulent atmospheric channel, the optical signal is subject to turbulence interference so that the optical power is changed or even reduced, resulting in a decline in the communication quality. The atmospheric-turbulence scale of the different optical signals and the light-beam

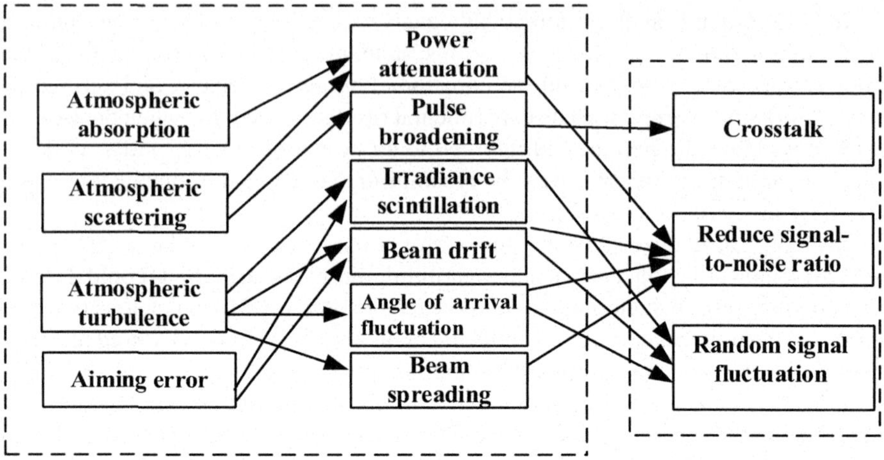

Fig. 1.1 Atmospheric turbulence and its impact on optical communication

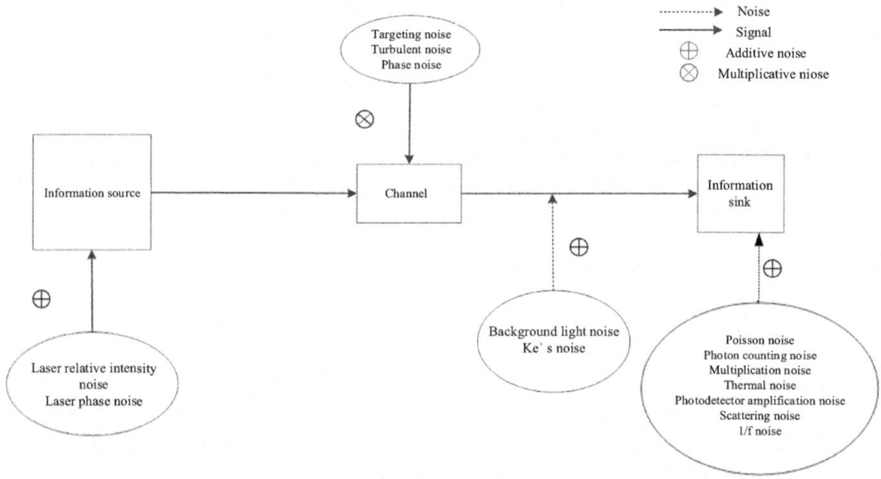

Fig. 1.2 Noise distribution in optical-wireless communication systems

quality will change the optical signal and beam quality in different manners with differing impacts. Studies have shown that changes in the atmospheric-turbulence scale can have a significant impact on optical-wireless communication systems, including changes in the optical signal and beam quality, which, in turn, depend on the diameter of the laser beam and geometry of the turbulence.

1.3.1.1 Light-Intensity Scintillation

When the cross-sectional diameter of the launch beam is much larger than the turbulence geometry, the beam cross section includes many small-scale turbulence pairs; each small-scale turbulence incidence will affect the beam differently, and can cause beam scattering and diffraction. At this time, the light intensity and phase will undulate randomly, affecting the optical communication system performance and quality.

1.3.1.2 Beam Drift

When a beam passes through atmospheric turbulence, it will experience multiple refractions, causing the beam to drift. This means that the beam cannot be vertically irradiated on the receiving surface. Beam drift causes the transmission of optical signals with random phase variations, thus reducing the communication quality of the optical-wireless communication system.

1.3.1.3 Beam Expansion

Atmospheric turbulence changes the refractive index of the atmosphere. These random changes cause the transmitted beam to expand, greatly reducing the energy in the beam cross-section. When the beam energy is sufficiently reduced, the beam cannot be successfully transmitted through the receiving lens to reach the photodetector, which reduces the communication quality of the optical-wireless communication.

1.3.1.4 Phase Undulation

In optical-wireless communication through an atmospheric channel, the inhomogeneity of the transmission medium causes wavefront distortion, which can change the arrival angle of the beam, causing it to undulate. If the optical signal is transmitted in a uniform medium, the wavefront will not be aberrant. The atmospheric turbulence distorts the beam wavefront such that its phase undulates.

1.3.2 Research Progress on Atmospheric-Turbulence Models

Theoretical studies of the transmission characteristics of light waves in atmospheric turbulence rely on a power-spectrum model of atmospheric turbulence. Kolmogorov proposed the concept of an atmospheric-turbulence power-spectral model and established it in the well-known Kolmogorov turbulence theory published in 1941 [90].

Subsequently, significant research has been conducted on the atmospheric-turbulence power spectrum, and different models of atmospheric-turbulence spectra have been successively established. The most commonly used models include the Kolmogorov, von Karman, Hill, and non-Kolmogorov spectra.

In 1941, Kolmogorov et al. established the theory of locally uniform isotropic turbulence and proposed approximating the turbulence as a Kolmogorov spectrum of local isotropy. Within the inertial subregion, the relationship between the atmospheric refractive-index structure function and turbulence structure constant follows the famous two-thirds law, with a power-spectral power of $-5/3$ [91]. In 1948, von Karman et al. proposed the von Karman spectrum to describe turbulence generation in circular tubes. This spectrum included both internal and external scale effects, and avoided singularities in the power spectrum at the origin, making it suitable for the full wavenumber range [92].

In 1961, Tatarskii et al. introduced an internal-scale influence factor as the truncation function of the power-spectrum model, established the Tatarskii spectrum, and extended its applicability to the dissipative region (i.e., the region where the energy dissipation exceeds the kinetic energy) [93]. In 1963, von Karman et al. considered the influence of the external and internal atmospheric-turbulence scales on the Kolmogorov spectrum and proposed a modified von Karman spectrum. This power spectrum effectively compensated for the deficiency of the Kolmogorov and Tatarskii spectra, which could not integrate singularities [94].

In 1974, Greenwood et al. conducted experimental measurements on turbulence spectra in a rough underlying environment and proposed the Greenwood–Tarazano spectrum based on experimental data [95]. In 1978, Hill considered the influence of sudden changes in the power spectrum near high-wavenumber regions and proposed the Hill spectrum, a precise numerical model for describing the inertial and dissipative regions [96]. In 1992, Andrews et al. proposed the Andrews spectral model based on the Hill spectral model, which described the protrusions of the power spectrum near high wavenumbers and had a mathematical form that was easy to analyze and calculate [97].

In 2004, Lipatack et al. defined the generalized Tatarskii spectrum that extended the structural function to smaller power-law values with a singularity at $K = 0$, which was valid for all beams [98]. In 2008, Zilberman et al. studied the turbulence characteristics of communication links under logarithmic-amplitude changes based on a three-layer model of the refractive-index fluctuation spectrum in a free atmosphere. Based on experimental and theoretical data of turbulence profiles in the troposphere and lower troposphere, they proposed a turbulence-spectrum model that varied with height [99]. In 2008, Toselli et al. proposed a non-Kolmogorov turbulence power-spectrum model that explained the nonuniformity and anisotropy of atmospheric turbulence [100].

To study the turbulence effects, more research has focused on beam-propagation and intensity-scintillation characteristics. In 1970, Whitman et al. [101] derived a beam-propagation formula using the wave-scalar equation and conducted a specific analysis using the Kolmogorov spectrum. The results indicated that as the laser beam propagated further away, the beam-propagation radius correspondingly increased.

In 1972, Poitier et al. [102] used the Born approximation method to study the broadening effects of focused beams under atmospheric turbulence. In 1975, Fante [103] introduced the modified von Karman spectrum to derive a long-term beam-expansion formula and used the cross-correlation function to derive a short-term beam-expansion formula.

In 1999, Fan [104] used a Mellin transform to study the beam spread of a Gaussian collimated beam and calculated the beam spread at different target heights in combination with Kunming atmospheric-turbulence intensity data. In 2001, Andrews et al. [105] discussed the influence of internal and external turbulence scales on the long-term beam radius and used numerical-simulation methods for beam calibration. In 2005, Pan et al. [106] analyzed the intensity-fluctuation characteristics of laser-beam propagation in satellite-to-ground links. The results showed that in weak, medium, and strong turbulent regions, multiple beam propagations could effectively suppress the intensity fluctuation.

In 2007, Zhang et al. [107] used the modified von Karman turbulence spectrum to study and obtain the relationship between the average intensity of a Gaussian beam propagating in a turbulent atmosphere. This included the turbulence-scale effect, short-term beam–expansion factor including the outer turbulence-scale effect, and the relationship between the equivalent radius of the Gaussian beam and the propagation distance, initial beam radius, and wavelength of the light wave.

In 2007, Guo et al. [108] used the modified Rytov method to extend the classic Rytov method to the medium and strong turbulence areas, and extended the horizontal propagation to slanted propagation according to the atmospheric-turbulence structure-constant model of the International Telecommunications Union Radio Communications Section (ITU-R), obtaining a zero internal and external scale plane wave, and the variation law of a spherical-wave scintillation index with modified Rytov variance. In 2008, Qian et al. [109] used a multi-layer phase-screen simulation method to numerically simulate the slanted propagation of ground-to-air lasers in atmospheric turbulence. They analyzed the changes in the beam-extension radius, spot-centroid drift variance, and scintillation index under different zenith angles, laser wavelengths, and initial radii.

In 2008, Wei et al. [110] derived the extended radius of a laser beam under a slanted-range transmission path. In the same year, Wei further discussed the scintillation-index characteristics of slanted-range Gaussian beams on the scintillation models of spherical and plane waves in slanted-range paths [111]. In 2010, Chu et al. derived an expression for the intensity distribution of partially coherent Gaussian–Schell beams during oblique propagation under atmospheric turbulence. They studied both relay and direct transmissions, and concluded that the intensity distribution during relay transmissions was better than that during direct transmissions [112].

In 2012, Zhang et al. derived an expression for the beamwidth of partially coherent light propagating in atmospheric turbulence by using an arbitrary refractive-index fluctuation power-spectrum model. Then, using a partially coherent flat-topped beam as an example, the effects of the beam order, spatial coherence, zenith angle, and other parameters on the beamwidth during oblique propagation were analyzed

[113]. In 2013, Duan et al. conducted a study on the average intensity and mean-square beamwidth of Gaussian–Schell-model (GSM) beams propagating through three different paths in atmospheric turbulence: upstream, downstream, and horizontal. They found that the beam was the least affected when transmitted through a downstream path during atmospheric turbulence [114].

In 2014, Zhang et al. studied the change in the intensity distribution of a nonparaxial polychromatic partially coherent Laguerre–Gaussian beam when propagating in free space and found that the intensity-distribution characteristics of the beam were determined by its initial parameters. Moreover, they found that the spectral shift of the beam when propagating in free space was related to the initial beam parameters and the propagation distance [115]. In 2016, Asmaa et al. [116] used the Hufnagel–Valley model to derive a closed-form expression for the scintillation index of spherical waves in atmospheric-turbulent oblique propagation, and studied the effects of the signal-to-noise ratio and bit-error rate on the system performance.

In 2018, Leng et al. [117] used multilayer phase screens to simulate and analyze laser transmissions in marine atmospheric environments. The results showed that as the laser emission power at the transmitting end gradually increased, the thermal-halo effect gradually strengthened, and the light-spot distortion gradually became severe, leading to poor beam quality. In 2020, Lin et al. [118] studied the beam-drift characteristics of partially coherent elliptical vortex beams in anisotropic non-Kolmogorov turbulence, based on the generalized Huygens–Fresnel theorem. The results indicated that both twisted-phase modulation and coherent modulation had a saturation zone, and outside the saturation zone, twisted-phase modulation effectively suppressed beam drift.

In 2022, Yang et al. [119] conducted a study on the attenuation of light intensity in atmospheric-turbulence channels under dusty weather conditions. The results showed that when the weather visibility decreased, the multiple-scattering effect increased, resulting in a corresponding reduction in the light-intensity attenuation and bit-error rate. In 2022, Baykal et al. [120] used the Rytov method to study the relationship between the scintillation index of multimode laser beams in weak atmospheric-turbulence regions and the size of the light source, propagation distance, and zenith angle.

In 2023, Xu et al. [121] deduced analytical expressions for the scintillation index and bit-error rate of a partially coherent twisted Gaussian beam propagating in a turbulent atmosphere. Numerical simulation results showed that the partially coherent twisted Gaussian beam had an advantage over the partially coherent Gaussian beam in reducing the scintillation caused by turbulence in short-distance transmissions. This advantage was further enhanced by an increase in the distortion factor.

Theoretical research results have certain limitations, owing to the complexity of atmospheric turbulence. Therefore, while conducting theoretical research on laser atmospheric transmission, researchers have also conducted corresponding experimental research.

In 1995, the Phillips Laboratory in the United States conducted an experiment to measure the atmospheric characteristics of airborne laser transmissions. A laser transceiver was installed on two aircraft flying in the stratosphere, and the slope-structure function of the laser wavefront was measured in the experiment. The measurement results were consistent with the theoretical results [122].

In 2000, Rao et al. experimentally measured the spot drift of a focused laser beam. The research results indicated that the spot drift in the vertical direction was slightly higher than that in the horizontal direction; under clear weather and a high turbulence intensity, the measured results were quantitatively consistent with the theoretical results, and their variation trend was approximately the same as that of scintillation [123]. In 2002, Byron et al. built a 48.5-km laser-transmission link platform and conducted experimental research on diversity technology. Their results showed that as the number of emitted laser beams increased, the scintillation index at the receiving end decreased [124].

In 2004, Ke et al. designed a laser-spot measurement system and used it to measure the light-intensity flicker and spot jitter under different meteorological conditions [125]. In 2005, Shi et al. experimentally measured the horizontal propagation of focused beams in atmospheric turbulence and obtained a calibration relationship between the beam expansion and turbulence-effect characteristic parameters of a focused beam under different obstruction ratios [126].

In 2005, Phillips et al. established three experimental platforms for laser-transmission links at distances of 1, 2, and 5 km. The results showed that when the emission aperture of the transmitting end remained unchanged, the drift of the beam emitted by a multimode laser was essentially the same as that of a single-mode laser [127]. In 2015, Wu et al. conducted laser-transmission experiments with distances of 1.4 and 0.9 km, and compared and analyzed the laser-transmission characteristics under water and on road surfaces. The results indicated that the atmospheric refractive-index structure constant in highway environments had a large order-of-magnitude span and strong fluctuations in light intensity, and its turbulence intensity and scintillation index were greater than those of the sea surface [128].

In 2016, Wang et al. [129] built a laser-transmission platform in three different underlying environments—near the sea, beach, and highway—and conducted experimental measurements and studies on the beam-drift and spot-expansion characteristics. The research results indicated that the fluctuation degree of the spot expansion and beam drift was the strongest in offshore environments, with fluctuations of 1.5 cm and 0.6 cm, respectively.

In 2018, Wang et al. conducted an experimental study on laser transmissions with a 610-m link length in the Lanzhou area. They measured the atmospheric refractive-index structure constant using the scintillation method under clear, cloudy, sandstorm, and sleet weather conditions. The results showed that atmospheric turbulence under various weather conditions was medium intensity [130]. In 2021, Zhao et al. established non-Kolmogorov turbulence spectra with anisotropic tilt angles on horizontal and satellite links and derived analytical expressions for the scintillation index of the anisotropic non-Kolmogorov horizontal and satellite links [131].

1.4 Development Trends

The noise models in an optical-wireless communication system can be discussed from the following perspectives:

1. More accurate noise models: Currently, the noise models used in optical-wireless communication systems are mainly thermal and quantum noise models. The influence of atmospheric-turbulence noise should be considered.
2. Improved adaptability to new devices: With the continuous development of technology, the devices used in optical-wireless communication systems are constantly being updated. Therefore, the development trend in noise models includes improving the adaptability of new devices to ensure their reliability and accuracy.
3. Attention to system noise and human-interference suppression: In addition to the noise of the device itself, other noise and human interference affect the system. Future development trends will focus on the impacts of these factors and propose targeted-noise and human-interference suppression methods.

Overall, with the continuous development of optical-wireless communication technology, the development trend of noise models in optical-wireless communication systems is expected to become more comprehensive, accurate, and practical.

References

1. Goodwin FE (1970) A review of operational laser communication systems. Proc IEEE 58(10):1746–1752
2. Zhang WT, Zhu BH (2007) Research on the influence of atmospheric turbulence on laser signal transmission. J Univ Electron Sci Technol China 36(4):784–787
3. Jiao Y (2012) Current status and future development trends of laser communication technology. Inf Commun 10(5):206–207
4. Ke XZ, Wu JL, Yang SJ (2021) Research progress and prospects of atmospheric turbulence for optical-wireless communication. J Radio Sci 36(03):323–339
5. Brookner E, Kolker M, Wilmotte RM (1967) Deep-space optical communications. Spectrum IEEE 4(1):75–82
6. Brookner E (1970) Atmosphere propagation and communication channel model for laser wavelengths. IEEE Trans Commun Technol 18(4):396–416
7. Karp S, O'Neill EL, Gagliardi RM (1970) Communication theory for the free-space optical channel. Proc IEEE 58(10):1611–1626
8. Katz J (1989) Free-space optical communication systems. Springer, Netherlands
9. Hinton HS (1994) Progress in free-space digital optics. Springer, US
10. Wolf M, Li J, Grobe L et al (2010) Challenges in Gbps optical-wireless transmission. Springer, Berlin Heidelberg
11. Jiang HL, Tong SF, Zhang LZ et al (2010) Space laser communication technology and systems. National Defense Industry Press
12. Yu S, Ma Z (2015) Overview and trend of steady tracking in free-space optical communication links. Proc SPIE—Int Soc Opt Eng 9521(8):2390–2398
13. Pauwels H (1966) Phase and amplitude fluctuations of the laser oscillator. IEEE J Quantum Electron 2(3):54–62

14. Barabanenkov YuN, Kravtsov et al (1970) Status of theory of wave propagation in randomly inhomogeneous media. Appl Phys Lett 7(3):67–69
15. Lin FC, Fiddy MA (1991) The Born–Rytov controversy: I. Comparing analytical and approximate expressions for the one-dimensional deterministic case. J Opt Soc Am A 9(7):1102–1110
16. Ho TL (1969) Log-amplitude fluctuations of laser beam in a turbulent atmosphere. J Opt Soc Am 59(4):385–388
17. Parry G, Puaey PN (1979) K distributions in atmospheric propagation of laser light. J Opt Soc Am 69(5):796–798
18. Phillips RL, Andrews LC (1982) Universal statistical model for irradiance fluctuations in a turbulent medium. J Opt Soc Am 72(7):864–870
19. Churnside JH, Clifford SF (1987) Log-normal Rician probability-density function of optical scintillations in the turbulent atmosphere. J Opt Soc Am A 4(10):1923–1930
20. Andrews LC, Phillips RL (1985) I-K distribution as a universal propagation model of laser beams in atmospheric turbulence. J Opt Soc Am A 2(2):160–163
21. Ooyama N, Miyazaki Y (1988) Light scattering measurements of Gaussian laser beam in random media of fog and smoke. Jpn J Appl Phys 27(8):1504–1507
22. Vostretsov NA, Zhukov AF, Kabanov MV et al (1993) Probability distribution of intensity fluctuations of laser radiation during a snowfall. Optika Atmosfery I Okeana 6(1):37–41
23. Valadas RT, Tavares AR, Duarte AM (1997) Angle diversity to combat the ambient noise in indoor optical-wireless communication systems. Int J Wireless Inf Networks 4(4):275–288
24. Wang YH, Yuan XX (2000) Research on optimal receiver performance of additive Gaussian white noise channel J Changsha Jiaotong Univ 16(04):19–22
25. Alhabash A, Andrews LC (2001) New mathematical model for the intensity PDF of a laser beam propagating through turbulent media. Int Soc Opt Eng 40(8):1554–1562
26. Rollins D, Baars J, Bajorins DP et al (2002) Background light environment for free-space optical terrestrial communication links. Int Soc Opt Eng 48(73):99–110
27. Razavi M, Shapiro JH (2005) Optical-wireless communications via diversity reception and optical preamplification. IEEE Trans Wireless Commun 4(3):975–983
28. Kiasaleh K (2005) Performance of APD-based, PPM free-space optical communication systems in atmospheric turbulence. IEEE Trans Commun 53(09):1455–1461
29. Farid AA, Hranilovic S (2007) Outage capacity optimization for free-space optical links with pointing errors. J Lightwave Technol 25(7):1702–1710
30. Jaruwatanadilok S (2008) Underwater optical-wireless communication channel modeling and performance evaluation using vector radiative transfer theory. IEEE J Sel Areas Commun 26(09):1620–1627
31. Zhu J (2008) Research on indoor optical-wireless communication uplink and downlink based on white LED. Jiangsu University, Jiangsu
32. Ma BK, Guo LX (2008) The impact of fog on the signal-to-noise ratio of optical-wireless communication systems. J Xi'an Univ Posts Telecommun 67(01):25–27
33. Cochenour BM, Mullen LJ, Laux AE (2009) Characterization of the beam-spread function for underwater optical-wireless communications Links. IEEE J Oceanic Eng 33(4):513–521
34. Zweck J, Menyuk CR (2009) Validity of the additive white Gaussian noise model for quasi-linear long-haul return-to-zero optical fiber communications systems. J Lightwave Technol 27(16):3324–3335
35. Meng HC (2010) Augmented Lagrange algorithm and its application in optimization design of optical-wireless communication systems. Wuhan University of Technology, Wuhan, Hubei
36. Jurado-Navas A, Garrido-Balsells JMA, Paris JF et al (2011) A unifying statistical model for atmospheric optical scintillation. INTECH Open Access Publisher
37. Wang HX, Hu H, Zhang TY et al (2011) Modeling and simulation of error performance in optical-wireless communication systems in weak turbulence J Syst Simul 23(04):788–792
38. Tu YJ, Zhou XL, Zhang HJ et al (2012) Free space optical communication performance based on mixed noise model. Inf Electron Eng 10(06):649–653

39. Mohamed AENA, Rashed ANZ, El-Nabawy AEM (2012) The effects of the bad weather on the transmission and performance efficiency of optical-wireless communication systems. Int J Comput Sci Appl 1(3):4–7
40. Yao WM, Rao JH, Zhang XH et al (2013) Research on FDPIM performance in underwater optical-wireless communication. Laser Technol 37(05):605–609
41. Ansari IS, Yilmaz F, Alouini MS (2013) Impact of pointing errors on the performance of mixed RF/FSO dual-hop transmission systems. IEEE Wireless Commun Lett 2(3):351–354
42. Yao WM, Rao JH, Zhang XH et al (2013) Performance study of underwater optical-wireless communication FDPIM. Laser Technol 37(05):605–609
43. Nistazakis HE, Stassinakis AN, Muhammad SS et al (2014) BER estimation for multi-hop RoFSO QAM or PSK OFDM communication systems over gamma gamma or exponentially modeled turbulence channels. Opt Laser Technol 64(01):106–112
44. Varotsos GK, Stassinakis AN, Nistazakis HE et al (2014) Probability of fade estimation for FSO links with time dispersion and turbulence modeled with the gamma–gamma or the I-K distribution. Int J Light Electron Opt 125(24):7191–7197
45. Zhu D, Zhang F, Zhou P et al (2014) Wideband phase noise measurement using a multifunctional microwave photonic processor. IEEE Photonics Technol Lett 26(24):2434–2437
46. Ke XZ, Liu M (2015) Diversity reception technology in turbulent channel optical-wireless communication. J Opt 35(01):88–95
47. Yang BS (2015) Research on the symbol error rate performance of MPPM based optical-wireless communication systems in turbulent channels. Xi’an University of Electronic Science and Technology, Xi’an
48. Ke XZ, Tian XC (2015) Modeling and simulation of two-dimensional correlated K-distribution turbulent channels. J Opt (4):29–31
49. Tian HX (2015) Analytical BER performance in differential n-PSK coherent transmission system influenced by equalization enhanced phase noise. Opt Commun 334(01):222–227
50. Lu F, Zhao D, Liu CB (2016) The effect of non kolmogorov atmospheric turbulence on the intensity scintillation of Gaussian array beams. Infrared Laser Eng 45(07):105–110
51. Chen M, Ke XZ (2016) Research on the effect of atmospheric turbulence on the performance of laser communication systems. Infrared Laser Eng 45(08):115–121
52. Fan XK, Zhang L, Tong SF et al (2017) The influence of sky background light on space laser communication systems. Prog Laser Optoelectron 54(07):102–110
53. Zhang GF (2017) Optimal relay layout algorithm for free space optical communication system. Nanjing University of Posts and Telecommunications, Jiangsu Nanjing
54. Ke XZ, Wang CH, Chen D (2018) Phase noise analysis of subcarrier modulation system in Malaga atmospheric turbulence channel. J Commun 39(11):80–86
55. Wang CH (2019) Research on phase noise characteristics and compensation technology of optical-wireless subcarrier modulation. Xi’an University of Technology, Xi’an
56. Nafkha A, Demni N (2020) Closed-form expressions of ergodic capacity and MMSE achievable sum rate for MIMO Jacobi and Rayleigh fading channels. IEEE Access 8(1):149476–149486
57. Xing FY (2020) Research on some key technologies of cooperative communication in underwater optical-wireless network. Dalian University of Technology, Liaoning, Dalian
58. Liu WJ (2020) Study on non-ideal transmission characteristics of underwater LED optical-wireless communication system. Xi’an University of Posts and Telecommunications, Xi’an
59. Wang LJ (2020) Research on clustering algorithm of static node AD hoc network based on UWOC. University of Electronic Science and Technology of China, Sichuan Chengdu
60. Fu YQ, Duan Q, Zhou L (2020) Gamma Gamma Performance study of underwater optical-wireless communication system under strong ocean turbulence and aiming error. Infrared Laser Eng 49(02):110–117
61. Zhang L (2020) Fast self-learning modulation recognition method for smart underwater optical communication systems. Opt Express 28(25):38223–38240
62. Lee-Leon A, Yuen C, Herremans D (2021) Underwater acoustic communication receiver using deep belief network. IEEE Trans Commun 69(06):3698–3708

63. Majlesein B, Gholami A, Ghassemlooy Z (2021) Investigation of the scattering noise in underwater optical-wireless communications. Sci 3(2):27–29
64. Han ZD (2021) Research on visible light indoor positioning system based on neural networks. Xi'an University of Technology, Xi'an
65. Ke XZ, Qin HH, Yang SJ, et al (2021) Night background light noise model of Vehicle-to-everything visible light communication system. J Radio Sci, 36(06):986–990
66.]Ke XZ, Wu JL, Yang SJ (2021) Research progress and prospects of atmospheric turbulence for optical-wireless communication. J Radio Sci 36(03):323–339
67. Yang RK, Han JX, Wu FP et al (2021) Research on the simulation method of double generalized Gamma distribution light intensity scintillation sequence in atmospheric turbulence. Acta Photonica Sin 50(10):26–30
68. Shen H, Yu LK, Zhou YX et al (2021) The intensity scintillation of plane and spherical waves under aperture smoothing. Prog Laser Optoelectron 58(23):5–9
69. Miao M, Li X (2022) Performance analysis of FSO systems over a lognormal-Rician turbulence channel with generalized pointing errors. J Lightwave Technol 40(13):4206–4216
70. Li Z, Liao ZW, Liang JY et al (2023) Research and prospects of atmospheric turbulence models. Opt Commun Technol 47(03):9–17
71. Zhang JL, Gao GJ, Wang JJ (2022) Theoretical and experimental verification of underwater optical-wireless communication under noisy light. J Beijing Univ Posts Telecommun 45(03):102–106
72. Chen E, Buscaino B, Kahn JM (2022) Phase noise analysis of resonator-enhanced electro-optic comb-based analog coherent receivers. J Lightwave Technol 40(21):7117–7128
73. Li Z, Wang FG, Liang JY et al (2023) Research progress in LED visible light communication. J Light Eng 34(01):29–44
74. Jin CQ (2023) Wideband multichannel Nyquist-spaced long-Haul optical transmission influenced by enhanced equalization phase noise. Sensors 23(03):1493–1493
75. Boisrobert CY, Debeau J, Hartog AH (1976) Sinusoidal modulation of a CW GaAs laser from 9 MHz to 1.1 GHz. Opt Commun 19(2):305–307
76. Schimpe R, Bowers JE, Koch TL (1986) Characterisation of frequency response of 1.5 μm InGaAsP DFB laser diode and InGaAs PIN photodiode by heterodyne measurement technique. Electron Lett 22(9):453–454
77. Wieczorek H (1995) 1/f noise in amorphous silicon nip and pin diodes. J Appl Phys 77(7):3300–3307
78. Fang ZH (1997) Equivalent noise models in fiber optical communication systems. Wuhan Univ J Nat Sci 2(2):180–184
79. Haas SM, Shapiro JH (2003) Capacity of optical-wireless communications. IEEE J Sel Areas Commun 21(8):1346–1357
80. Guo CL, Zeng D, Li J et al (2004) Noise model of Photomultiplier tube Nucl Electron Detect Technol 24(2):117–120
81. Bao JL, Zhuang YQ, Du L et al (2005) Flicker noise model of optocoupler. Acta Photonica Sinica 34(09):1359–1362
82. Gaoxin J, Zhang XC, Chen Y (2007) Device model and numerical simulation of InGaAs/InP SAGCM-APD. Semicond Optoelectron 28(05):617–622
83. Wang XY (2008) Design of optical receiver for image optical-wireless communication system. Changchun University of Science and Technology, Changchun, Jilin
84. Hu T, Si HY (2010) Design and research of preamplifier circuit for photoelectric detectors. Appl Optoelectron Technol 25(01):52–55
85. Wang K (2012) Impact of background light induced shot noise in high-speed full-duplex indoor optical-wireless communication systems. Opt Express 19(22):21321–21332
86. Yan ZG, Lin YL, Yang J et al (2012) Statistical distribution function of random noise characteristic quantities in photodetectors. J Phys 61(20):88–96
87. Kharraz O, Forsyth D (2013) Performance comparisons between PIN and APD photodetectors for use in optical communication systems. Int J Light Electron Opt 124(13):1493–1498

88. Yan Q, Li Z, Hong Z et al (2019) Photon-counting underwater optical-wireless communication by recovering clock and data from discrete single photon pulses. IEEE Photonics J 11(05):1–15
89. Mostafapour E, Ghobadi C, Nourinia J et al (2022) Performance analysis of mobile adaptive networks in VLC multiplicative SPAD channel noise conditions. Opt Commun 5(24):128760–128763
90. Kolmogorov AN (1941) The local structure of turbulence in an incompressible viscous fluid for very large Reynolds numbers. Sov Phys Usp 30(42):301–305
91. Kolmogorov AN (1941) Dissipation of energy in locally isotropic turbulence. Akad Nauk SSSR Dokl 32(4):299–303
92. Karman VT (1948) Progress in the statistical theory of turbulence. Proc Natl Acad Sci U S A 34(11):530–539
93. Tatarski VI (1961) Wave propagation in a turbulent medium. Phys Today 14(12):69–78
94. Karman VT (1963) From low-speed aerodynamics to astronautics. Pergamon Press, New York
95. Greenwood DP, Tarazano DO (1974) A proposed form for the atmospheric microtemperature spatial spectrum in the input range. Rome Air Development Center
96. Hill RJ, Clifford SF (1978) Modified spectrum of atmospheric temperature fluctuations and its application to optical propagation. J Opt Soc Am 68(7):892–899
97. Andrews LC (1992) An analytical model for the refractive index power spectrum and its application to optical scintillations in the atmosphere. J Mod Opt 39(9):1849–1853
98. Lipatack PA (2004) Optical wave propagation through non-Kolmogorov atmospheric turbulence. University of Central Florida
99. Zilberman A, Golbraikh E, Kopeika NS (2008) Propagation of electromagnetic waves in Kolmogorov and non-Kolmogorov atmospheric turbulence: three-layer altitude model. Appl Opt 47(34):6385–6391
100. Toselli I, Andrews LC, Phillips RL et al (2008) Free space optical system performance for laser beam propagation through non Kolmogorov turbulence for uplink and downlink paths. Opt Eng 47(2):115–122
101. Whitman AM, Beran MJ (1970) Beam spread of laser light proagting in a random medium. J Opt Soc Am 60(12):1595–1602
102. Poirier JL, Korff D (1972) Beam spreading in a turbulent medium. J Opt Soc Am 62(7):893–898
103. Fante RL (1975) Electromagnetic beam propagation in turbulent media. Proc IEEE 63(12):1669–1692
104. Fan CY (1999) Atmospheric turbulence effect of Gaussian beam wave oblique propagation. J Quantum Electron 16(6):519–525
105. Andrews LC, Al-Habash MA, Hopen CY et al (2001) Theory of optical scintillation: Gaussian-beam wave model. Waves Random Media 11(3):271–291
106. Pan F, Ma J, Tan L et al (2005) Scintillation characterization of multiple transmitters for ground-to-satellite laser communication. Inferaed Compon Appl 56(40):448–454
107. Zhang YX, Wang GG (2007) Average intensity and short-term beam expansion of oblique atmospheric transmission laser beams. Infrared Laser Eng 36(2):167–170
108. Ma BK, Guo LX, Cui JQ (2007) Study of the oblique range scintillation problem of light waves based on the modified Rytov method. J Basic Sci Text Colle Univ 20(2):176–180
109. Qian XM. Zhu WY, Rao RZ (2008) Numerical simulation analysis of turbulent effects of oblique propagation of ground to air laser in the atmosphere. Infrared Laser Eng 37(5):787–792
110. Wei HY, Wu ZS (2008) The broadening and drift characteristics of laser beam oblique propagation in atmospheric turbulence. J Radio Wave Propag 23(4):611–615
111. Wei HY, Wu ZS, Peng H (2008) Scattering characteristics of diffuse targets in slant-range atmospheric turbulence. Phys Lett 57(10):7–11
112. Chu X, Liu Z, Wu Y (2010) Comparison between relay propagation and direct propagation of Gaussian-Schell-model beam in turbulent atmosphere along a slant path. Chin Phys B 19(9):73–86

113. Zhang XX, Dan YQ, Zhang B (2012) Beam extension of partially coherent light propagating diagonally in turbulent atmosphere. Acta Opt Sin 32(12):102–111
114. Duan ML, Li JH, Wei JL (2013) Influence of different propagation paths on the propagation of laser in atmospheric turbulence. Optoelectron Lett 9(6):0477–0480
115. Zhang Y, Liu L, Wang F et al (2014) Average intensity and spectral shifts of a partially coherent standard or elegant Laguerre-Gaussian beam beyond paraxial approximation. Opt Quant Electron 46(13):365–379
116. Fayed HA, El Aziz AA, Aly AM et al (2016) Irradiance scintillation index on slant atmospheric turbulence: simple approach. Opt Eng 55(5):56113–56113
117. Leng K, Wu WY, Gong YC et al (2018) Simulation study on laser transmission characteristics in the marine atmosphere. Laser Infrared 48(12):1480–1485
118. Wang L, Wang J, Yuan C et al (2020) Beam wander of partially coherent twisted elliptical vortex beam in turbulence. Optik 218(15):128–134
119. Yang RK, Li FJ, Wu FP et al (2022) Study on the influence of dust turbulent atmosphere on the performance of free space quantum communication. Acta Phys Sin 71(22):28–38
120. Sayan MF, Gerekciolu H, Baykal Y (2022) Multimode laser beam scintillations in weak atmospheric turbulence for vertical link laser communications. Waves Random Complex Media 32(4):14–20
121. Xu Y, Xu Y (2023) Scintillation index and bit error rate of partially coherent twisted Gaussian beams in turbulent atmosphere. Opt Quant Electron 55(6):75–79
122. Silbaugh EE, Welsh BM, Roggemann MC (1996) Characterization of atmospheric turbulence phase statistics using wave-front slope measurements. J Opt Soc Am A 13(12):2453–2460
123. Rao RZ, Wang SP, Liu XC et al (2000) Experimental study on laser beam drift in turbulent atmosphere. China Laser 10(11):1011–1015
124. Biswas A, Wright MW (2002) Mountain-top-to-mountain-top optical link demonstration. IPN Prog Rep 42(149):1–27
125. Wang LL, Ke XZ (2004) Research and simulation of turbulence effects on optical communication links. J Light Scattering 28(03):250–255
126. Shi XY, Wang YJ (2005) Numerical analysis of beam spread experiments with different obscuration ratios of emission systems. J Hangzhou Dianzi Univ 41(06):94–96
127. Phillips RL, Andrews LC, Stryjewski J et al (2006) Beam wander experiments: terrestrial path. Atmos Opt Model, Meas, Simulat II 63(03):151–162
128. Wu XJ, Wang HX, Song B et al (2015) Study on the measurement and transmission characteristics of light intensity fluctuations in different environments. Optoelectron Laser 26(6):1138–1145
129. Wang HX, Song B, Wu XJ et al (2016) Experimental study on beam drift and spot spread under different underlying surfaces. Prog Laser Optoelectron 53(08):70–76
130. Wang HQ, Li Y, Hu Q et al (2018) Experimental study on the fluctuation characteristics of nighttime light intensity in Lanzhou region. Acta Photonica Sin 47(04):194–201
131. Zhai C (2021) Anisotropic non-Kolmogorov turbulence spectrum with anisotropic tilt angle. Photonics 8(11):521–530

Chapter 2
Detectors and Their Noise Models

In a optical-wireless communication system, the photoelectric signal is converted by a photodetector at the receiving end, and the noise in the detector directly affects the photoelectric-conversion efficiency. In this chapter, the photodetector noise-generation mechanisms are described, and the corresponding mathematical models are presented. The mechanisms include scattered-particle noise, generation-compound noise, 1/f noise, and thermal-detector noise The noise models of the avalanche photodetector, original negative detector, photomultiplier tube, four-quadrant detector, quantum-dot infrared detector, and balanced detector are analyzed.

Photodetectors are key devices that convert optical signals into electrical signals, and the effectiveness and stability of their output signals are important indicators of their working performance. In this chapter, detectors commonly used in optical-wireless communication systems are summarized, including the detector's working principle, performance indicators, application scenarios, and noise models. Finally, the directions and challenges worthy of further study and discussion in this field are discussed.

2.1 Photodetectors

A optical-wireless communication system includes channel coding and decoding, signal modulation and demodulation, optical amplification and coupling, optical transceiver antennas, and acquisition, alignment, and tracking systems. A laser or light-emitting diode is used as the signal light source, and the signal is modulated after channel coding.

Modulation is divided into active and passive, and active modulation can be divided into direct and indirect. A laser is directly output or amplified by an optical amplifier, which includes semiconductor optical amplifiers and optical-fiber amplifiers. A single or multiple antennas are used for transmission in free space. The

X. Ke and C. Ke, *Noise Models in Optical-Wireless Communication Systems*,
Optical Wireless Communication Theory and Technology,
https://doi.org/10.1007/978-981-97-7550-7_2

transmission channels can be atmospheric, indoor-scattering, ultraviolet-scattering, and underwater.

The receiving end uses a single or multiple antennas for optical-signal reception. The optical-coupling module couples the signal into a single-mode fiber, after preamplification. It uses direct or coherent detection to convert the optical signal to an electrical signal, and then after demodulation and decoding, it delivers the information to the destination. The photoelectric detector is the core of the receiving end.

2.1.1 Photodetector Classification

Photodetectors are a class of devices that can convert optical-radiation signals into electrical signals. Photodetectors commonly used in the field of optical-wireless communication can be divided into two categories, based on their physical-mechanism characteristics: photon detectors, based on the photoelectric effect, and thermal detectors, based on the temperature-change effect [1]. Figure 2.1 shows the classifications of some of the photodetectors commonly used in the field of optical-wireless communication, based on their external and internal photoelectric effects.

When incident light illuminates the surface of a photoelectric-based detector, a photon of the incident light collides with the electrons in the material to produce new photoelectrons, which overflow from the detector surface, resulting in an external photoelectric effect. If the charge carriers generated inside the detector material do not overflow from the surface, but are bound to the inside of the detector material, it is called an internal photoelectric effect.

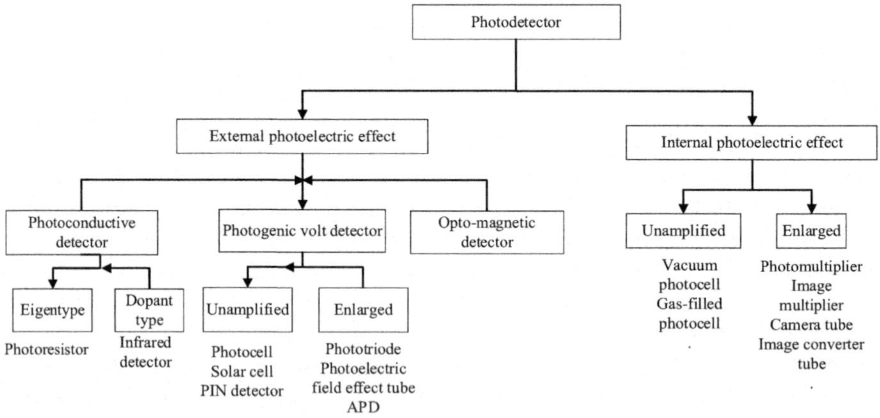

Fig. 2.1 Photodetector classification

The external photoelectric effect generally includes photoconductive, photovoltaic, and magneto-optical effects. The internal photoelectric effect includes the photon-emission effect. The quantum-well and the quantum-dot infrared detectors use the photoconductive effect. For photocells, the positive–intrinsic–negative detector (PIN photodetector) and avalanche photodetector are photovoltaic-effect detectors. Photoresistors belong to detectors using the photoconductive effect. Vacuum photocells, gas-filled photocells, and photomultiplier tubes take advantage of the photon-emission effect.

In a thermal detector, a detector material that absorbs incident light will produce a thermoelectric effect, resulting in a change in resistivity. Commonly used thermal detectors include thermistors and thermopiles.

2.1.1.1 Photoconductive Effect

When incident light irradiates the surface of a semiconductor material, the material absorbs energy. At this time, the carrier concentration gradually increases, the conductivity increases, and the resistance decreases; this is the photoconductive effect.

A photoconductive detector, which is a uniform intrinsic semiconductor with a photoconductive effect, can be represented by the model shown in Fig. 2.2. A voltage U is applied to both ends of the semiconductor to form electrodes. The current generated by the circuit can be detected by a galvanometer. When the optical metric is changed, the current of the circuit changes; that is, the resistance of the detector reverses.

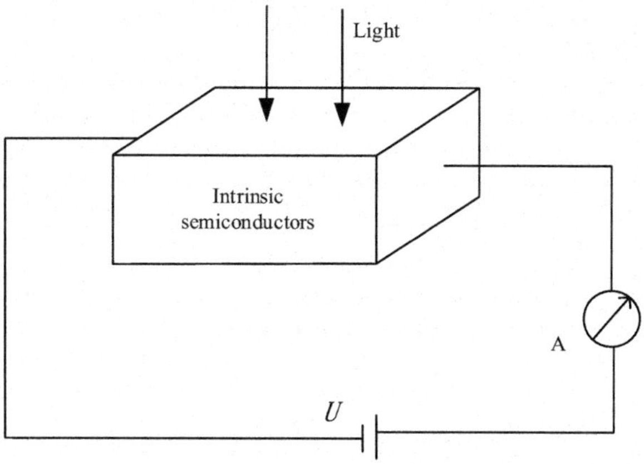

Fig. 2.2 Simple model of a photoconductive detector

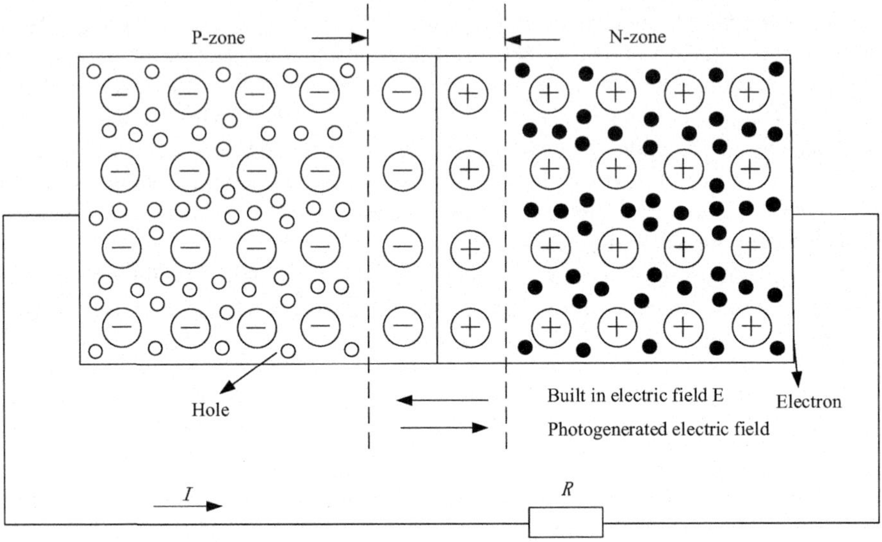

Fig. 2.3 Photogenerated voltage effect of a P–N junction

2.1.1.2 Photovoltaic Effect

Figure 2.3 shows a schematic diagram of the photovoltaic effect with the presence of charge carriers in the P–N junction, with two types of carriers, majority and minority. Most carriers will constantly drift in the interior of the P–N junction, forming a self-established electric field E inside the P–N junction. When the P–N junction and the area around it are irradiated with light, a few charge carriers will be present near the P–N junction.

Under the action of the self-built electric field E, the electrons drift to the N region and the holes drift to the P region. The N region, where electrons drift, carries a negative charge. The P region, to which the holes drift, has a positive charge. The electric field generated by the control of a few carriers is an additional electric field, that is, the photogenerated electric field.

When a reverse voltage is applied to the P–N junction, the resulting reverse current gradually increases as the reverse voltage increases. The resulting photoelectric signal can be observed using photocurrent I.

2.1.1.3 Photoelectron-Emission Effect

According to the quantum theory of light, when incident light strikes the surface of the device material, if the frequency of the incident light increases, the energy absorbed by the electron increases, and part of the energy is used for collision loss. If the remaining energy is sufficient to overcome the barrier on the surface of the

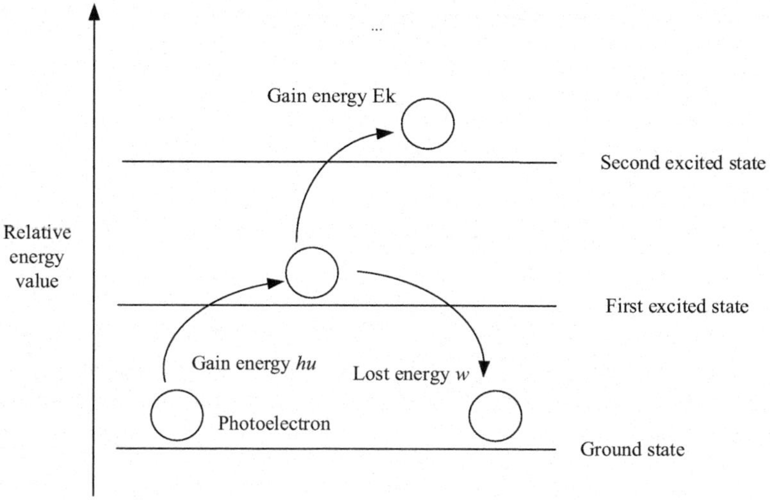

Fig. 2.4 Photoelectron energy-level transition diagram

material, the electrons overcoming the barrier will overflow from the surface of the material and enter the space environment.

The photoelectron energy-level transitions are shown in Fig. 2.4. The maximum kinetic energy of the photoelectrons that overflow the surface of the material is described by Einstein's equation:

$$E_k = hv - w, \tag{2.1}$$

where w is the work function of the material in joules, h is Planck's constant, and $E_k = 1/2^{mv^2}$ is the kinetic energy of a photoelectron spilling over the surface of the material. The mass of the photoelectron is represented by m and its velocity is represented by v.

When the incident light irradiates the surface of the material, the kinetic energy of the photoelectrons does not change with the change in light intensity, but is only related to the frequency of the incident light. When the incident-light frequency increases, the kinetic energy increases; when the incident-light frequency decreases, the kinetic energy decreases. When photoelectrons leak from the surface of the material, there is a critical threshold value; that is, when the kinetic energy is 0, the photoelectrons simply reach the surface of the material. In this case, the frequency of the incident light is the limiting frequency.

2.1.2 Photodetector Performance Indicators

2.1.2.1 Quantum Efficiency

The quantum efficiency is defined as the average number of electrons released by an incident photon, η, and is expressed as

$$\eta = \frac{Ih\nu}{Pe} \tag{2.2}$$

where η is the quantum efficiency, I is the average value of the photocurrent, e is the electron charge, P is the optical power, h is Planck's constant, and ν is the frequency of the incident light. Detectors with higher quantum efficiencies have better performance.

2.1.2.2 Noise-Equivalent Power

The noise-equivalent power (NEP) reflects the detection capability of the detector and is generally described under the premise of a unit signal-to-noise ratio, that is,

$$NEP = \frac{PV_s}{V_n}, \tag{2.3}$$

where NEP is the noise-equivalent power, V_s is the voltage intensity of the signal, V_n is the voltage intensity of the noise, and P is the incident light power. The larger the NEP, the weaker the detection ability; the smaller the NEP, the stronger the detection ability.

2.1.2.3 Responsivity

The ratio of the output optical power to the input optical power of the detector is the responsivity R, which can be subdivided into voltage and current responsivities. The voltage responsivity R_u is defined as the ratio of the output voltage to the incident light power, that is,

$$R_u = \frac{U_s}{P}, \tag{2.4}$$

where R_u is the voltage responsivity, U_s is the output voltage of the photodetector, and P is the incident-light power.

The current responsivity R_i is defined as the ratio of the output current of the photodetector to the incident-light power, as follows:

$$R_i = \frac{I_s}{P}, \tag{2.5}$$

where R_i is the current responsivity, I_s is the output current of the photodetector, and P is the incident-light power.

2.1.2.4 Spectral Responsivity

The spectral responsivity $R(\lambda)$ is defined as the ratio of the output voltage or current of the detector to the incident light power, when it is illuminated by monochromatic light of wavelength λ:

$$R_V(\lambda) = \frac{V_S(\lambda)}{P(\lambda)} \tag{2.6}$$

$$R_I(\lambda) = \frac{I_S(\lambda)}{P(\lambda)} \tag{2.7}$$

where $R_V(\lambda)$ and $R_I(\lambda)$ are the spectral responsivities, $V_S(\lambda)$ is the output voltage, $I_S(\lambda)$ is the output current, and $P(\lambda)$ is the power of the incident light.

2.1.2.5 Frequency Responsivity

The frequency responsivity $R(f)$ is affected by the incident-light frequency and is defined as

$$R(f_\tau) = \frac{R_0}{\left[1 + (2\pi f_\tau)^2\right]^{1/2}} \tag{2.8}$$

where $R(f_\tau)$ is the responsivity when the frequency is f_τ and R_0 is the responsivity when the frequency is 0. The time constant τ of the detector is a fixed value, usually determined by the material of the detector and the external circuit.

2.1.2.6 Detectivity and Normalized Geochemical-Exploration Measures

The detectivity D reflects the sensitivity of the detector and is the reciprocal of NEP:

$$D = \frac{1}{NEP} \tag{2.9}$$

where D is the detection degree and NEP is the noise-equivalent power of the detector. The normalized geochemical exploration measure D* is defined as follows:

$$D^* = D \cdot (A_d \Delta f)^{1/2} \tag{2.10}$$

where D^* is the normalized geochemical-exploration measure, D is the detectivity, A_d is the area of the detector, and Δf is the amplifier bandwidth.

2.2 Photodetector Types

2.2.1 *Source Noise*

According to the different emission and modulation mechanisms of the system, source noise can be divided into laser light source noise and noise in LED modulated signal light in visible light communication using LED as the light source. When the communication signal modulates the laser emission, the source noise can be roughly divided into two parts: the inherent thermal noise of the laser itself, and the distortion of the laser emission compared to the modulated signal. The study of the above noise/ distortion of lasers has started since the invention of lasers in the 1960's. In 1966, Pauwels derived the series expressions for the amplitude and phase noise of lasers from the Van del Pol equation of quantum mechanics, but their expressions were too complex [2]. In the same year, Leeson obtained the spectral approximation to 1/ f noise from a simpler spectral model [3]. In 1970, Ziel investigated the scattering grain noise model for GaAs lasers [4], and the results were widely quoted. In 1982, Harder et al. [5] solved the semiconductor rate equation to obtain the light intensity and spectral undulation of a semiconductor laser, and obtained the laser's general equivalent circuit of the lase. In 1989, Petermann gave an exhaustive summary of the various types of noise in lasers [6]. Due to the complex form of the solution of the rate equation, it is often necessary to describe it by the series method, and this signal nonlinear distortion is also a kind of noise affecting the communication system. In 1992, Kuo conducted theoretical analysis and experimental research on the influence of gain saturation, relaxation oscillation, spatial hole burning, and other factors on the nonlinear noise of LD [7]. In 1994, Bihan et al. [8] used perturbation method and Bessel function method to solve the rate equation, derived the expression of LD nonlinear intermodulation distortion, and the results show that the FM signal distortion is more obvious than the AM signal. In 1997, Chai et al. [9] analyzed the effect of phase noise of semiconductor laser on the characteristics of high-speed optical fiber communication system, and the results show that the dispersion characteristics of optical fiber make the phase noise of laser will be transformed into intensity noise at the receiving end, the sensitivity of the optical receiver deteriorates and shows the 'saturation' phenomenon on the BER curve. In 2002, Fukuda et al. [10] elucidated experimentally and theoretically the reason for the increase of the residual spectral linewidth during the degradation of the device based on the multi-quantum-well distributed feedback lasers, and proposed a correlation between the 1/f noise and the degradation of the semiconductor lasers. model for the correlation between degradation of semiconductor lasers. In the same year, Lawrence conducted a detailed study on nonlinear distortion in LDs driven by FM signals, and showed that factors such as the modulation regime, the modulation frequency, the magnitude of the injected current, and the length of the external cavity affect the nonlinear dynamic distortion of the lasers [11].

In 2006, Yu et al. [12] designed a new type of low noise fiber laser using a composite cavity structure, which effectively suppressed the intensity noise of the light source. In 2010, Domenico et al. [13] proposed an approximate formula to handle the relationship between laser frequency noise spectrum and laser linear, which can be applied to any noise spectral density. In 2012, Liu et al. [14] studied the effects of laser phase noise on modulation format selection, parameter optimization, and carrier phase estimation performance in coherent fiber optic communication systems. Simulation results showed that when the carrier phase estimation method is used in coherent fiber optic communication systems, the optimal average length of the symbol sequence in the algorithm is inversely proportional to the laser phase noise. In 2015, Mahmoud conducted a detailed study on the impact of LD noise and distortion in Radio Over Fiber on the system, pointing out that intermodulation distortion is more severe than harmonic distortion during laser modulation [15]. In 2016, Chen et al. [16] derived the expression of nonlinear intermodulation distortion in lasers using the Bessel function, studied the impact of nonlinear intermodulation distortion on the carrier intermodulation noise ratio and bit error rate performance of the system, and provided the selection of the optimal modulation index of the system and the corresponding relationship with the nonlinear coefficient. In 2016, Chen proposed a novel non-optical heterodyne detection method based on subcarrier modulation, and conducted system simulation experiments under additive noise, multiplicative noise, and mixed noise, providing experimental basis for the application of this non-optical heterodyne detection method [17]. In 2017, Song et al. [18] qualitatively analyzed the phase noise of fiber lasers with different pump powers, seed source powers, and wavelengths using power spectra and integral spectra. The results showed that the higher the pump power, the greater the proportion of low-frequency phase noise; For different seed source powers, the proportion of phase noise in the test frequency band is basically at the same level; The proportion of phase noise in the high-frequency range varies for different seed wavelengths tested, with the highest proportion occurring when the seed wavelength is 1560.48 nm. In 2019, Bai derived the relationship between laser linewidth, frequency noise power spectral density, and phase noise power spectral density. A method for calculating laser linewidth by measuring laser phase noise was obtained, and numerical simulation and characteristic analysis were conducted on the output laser spectral line of the laser [19]. In 2022, Du et al. [20] proposed and conducted a theoretical study aimed at exploring the noise co driving chaotic synchronization phenomenon based on a three-stage distributed Bragg reflection semiconductor laser.

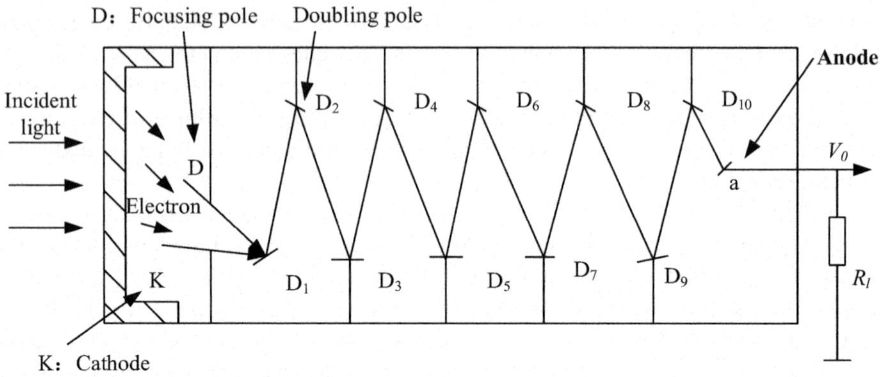

Fig. 2.5 Schematic diagram of a photomultiplier tube

2.2.2 *Photomultiplier Tubes*

2.2.2.1 Working Principle

The principle of the photomultiplier tube is illustrated in Fig. 2.5. The electron-focusing system includes a focusing pole *D*, photocathode *K*, and anode *a*, from which the incident light enters into the interior of the photomultiplier tube. Under the action of the photocathode, the incident light will converge to form a beam of light that strikes the first multiplier pole. Under the excitation of the first multiplier pole, the reflected beam produces a large number of secondary electrons, which are driven by an electric field to the second multiplier pole, producing more secondary electrons; this continues to D_{10}. The last incident beam passes through D_{10}. The pole multiplication reaches the anode, and the output photocurrent at load R_LV is formed on V_0.

2.2.2.2 Research Progress

In 1972, Shun et al. [21] developed several new types of photocathode materials and analyzed the corresponding processing technology. A new secondary-emission technology was used in the photomultiplier, which improved its gain characteristics.

In 1989, Li [22] discovered how the polarization effects of the grating and photomultiplier tubes varied with the wavelength. The experimental results showed that the polarization effect of the grating was greatly affected by the change in wavelength, in the range of 420–740 nm. Therefore, the polarization effect of the grating must be considered when measuring the polarization.

In 2000, Chow [23] analyzed basic photomultiplier-control technology and its improvements. Through the continuous improvement of basic devices, a photomultiplier tube can be used as a weak light-detection device.

In 2004, Guo et al. [24] discussed the noise sources of photomultiplier tubes, including loose-particle, thermal, generation-compound, and 1/f noise, and established the corresponding noise model.

In 2014, Chowdhuri et al. [25] analyzed the spectral detection based on a photomultiplier array; the detection method had a fast time response of 10 μs and a spatial resolution of 3 cm.

In 2015, Maltseva et al. [26] used photomultiplier tubes to detect Cherenkov light; the detector's spatial resolution reached 3 m for 20-m-long fibers. In addition, the resolution could be improved by optimizing the parameters of the optical fiber and photomultiplier, and the spatial resolution of the detector reached 0.5 m.

In 2016, Chen et al. [27] analyzed the performance of a silicon photomultiplier tube, designed a nuclear-radiation detector, developed a radiation-measurement probe, and tested its radiation performance. This detector has some reference value for the design of nuclear-radiation detectors.

In 2017, Zhao et al. [28] developed a one-dimensional single-photon position-sensitive silicon photomultiplier tube, with a single-photon position resolution of 393.4 μm. It improved from 393.4 to 56.2 μm when the number of photoelectrons was increased from one to seven.

In 2018, Takahashi et al. [29] analyzed the detection principle of low-intensity light and estimated the average signal generated by a single photoelectron, using the fact that photoelectrons conform to the Poisson distribution.

In 2019, Elorrieta et al. [30] described the detailed characteristics of a three-inch photomultiplier, including the dark count rate, probability of associated pulses, timing characteristics, gain, and temperature dependence of the peaks and valleys. The effect of conductive objects near the photocathode on photomultiplier noise was also shown.

In 2020, Sun et al. [31] designed a system consisting of ultraviolet (UV) light and highly sensitive photomultiplier tubes to simulate very cloudy harbor water. It transmitted data at a rate of 85 Mbit/s using on–off control modulation, and established a non-direct-looking link with a transmission distance of 30 cm.

In 2021, Ning et al. [32] investigated the performance of devices under various light intensities and photomultiplier-tube gains, and proposed a small-volume receiver-control system and related adaptive-control strategy to improve stability and reliability.

Table 2.1 summarizes the relevant properties of photomultiplier tubes. Photomultiplier tubes have applications in the ultraviolet communication field, as well as weak optical-signal detection.

2.2.2.3 Typical Application Scenarios

Figure 2.6 shows the structure of a UV communication system that consists of transmitting, receiving, and atmospheric channels [33]. The transmitting component modulates the original electrical signal and transforms it into a signal suitable

Table 2.1 Research progress of photomultiplier tubes

Researcher	Year	Development
Shun	1972	Improved the gain characteristics of photomultiplier tubes
Li	1989	Grating polarization effect must be considered in polarization measurement
Zhou	2000	Photomultiplier tubes can be used as weak light-detection devices
Guo	2004	Discussed the sources of noise and built a noise model
Maltseva	2015	Detector spatial resolution reached 3 m
Chen	2016	Developed nuclear-radiation detector
Zhao	2017	Single-photon position resolution is 393.4 μm
Takahashi	2018	Conversion of photons to photoelectrons
Elorrieta	2019	Effect of conductive objects near the photocathode
Sun	2020	Non-direct-looking link with a data rate of 85 Mbit/s
Ning	2021	Small-volume receiver-control system

for transmission in the channel. The signal is disrupted by interference in the atmospheric channel, and the receiving part must recover the original signal from the received disrupted signal.

The receiving part is composed of an ultraviolet photomultiplier tube, pre-processing circuit, demodulation module, and beacon module. The ultraviolet detector collects the ultraviolet signal and converts the corresponding optical signal into an electrical signal. The pre-processing circuit amplifies and filters the electrical signal, and the demodulation module demodulates it and converts it into a baseband digital signal. The UV photomultiplier is currently the most ideal UV detector. It has been widely used for non-direct UV communication because of its large detection area, high gain, and low dark current.

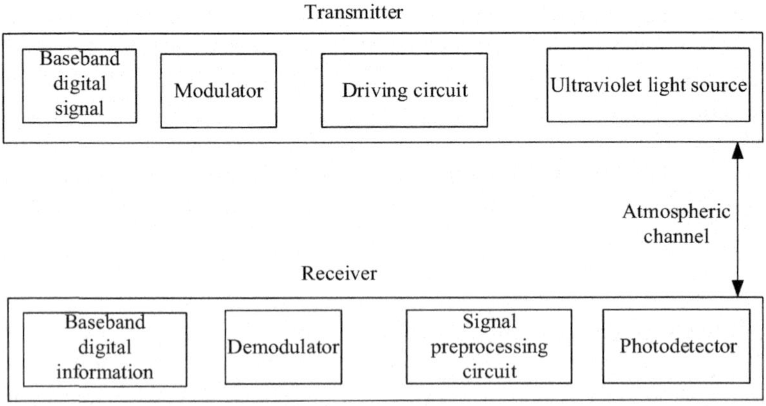

Fig. 2.6 Structure of a UV communication system [33]

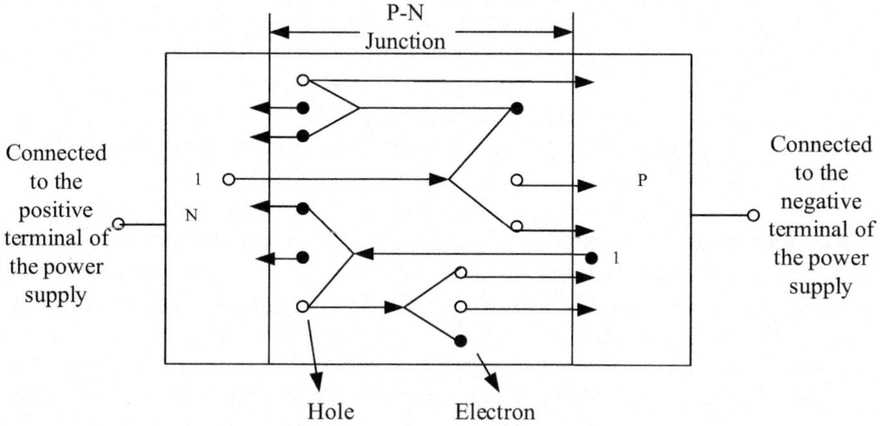

Fig. 2.7 Schematic diagram of the avalanche breakdown

2.2.3 Avalanche Photodetector

2.2.3.1 Working Principle

An avalanche photodetector is based on an avalanche photodiode (APD). Figure 2.7 shows a schematic diagram of the avalanche breakdown. When incident light illuminates the diode surface, the light is absorbed by the P–N junction to form a photocurrent. The photocurrent is affected by the reverse bias and increases with the increase in reverse bias. When the reverse bias is continuously increased, the photocurrent exponentially increases. At this time, the so-called "avalanche" phenomenon occurs.

2.2.3.2 Research Progress

In 1980, Yang et al. [34] studied avalanche-photodiode technology and proposed a new avalanche photodiode that had high gain and low noise and could be used for long-wavelength detection.

In 1990, Watkins et al. [35] observed the characteristics of laser-induced electrical failures in biased silicon avalanche photodiodes. When monitoring current–voltage characteristic changes, two types of changes were observed: the first was a large increase in leakage current, which can be modeled by introducing defects into the depletion region through deep melting transients. The second type was a catastrophic failure, in which an electrical short circuit occurred in the device after irradiation; it can be modeled by excessive current density in the photodiode junction.

In 1996, Li et al. [36] designed a superlattice avalanche photodetector. The optimal structural parameters were obtained as follows: the thickness of the i-Si avalanche zone was 1.8–2 μm. The carrier concentration in the p-Si zone was 1.018 cm^{-3}, and the thickness was 17 nm. The total thickness of the superlattice was 340 nm. The detector could operate in the range of 1.3–1.6 μm.

In 2005, Ng et al. [37] concluded that collision ionization in the InGaAs absorption layer increased the excess noise in the InP avalanche photodiode. Moreover, strict limits needed to be imposed on the doping in the charge-control layer if the avalanche noise was to be minimized.

In 2009, Guo et al. [38] used passive-suppression technology to determine that dark current does not reflect Pass-through characteristics, which is an obvious difference between photocurrent and dark current. Through the above research, it was concluded that to improve the signal-to-noise ratio of a single-photon detector, the through-through characteristics could be used to properly select the reverse-bias voltage of an avalanche photodiode in Geiger mode.

In 2010, Chitnis et al. [39] investigated a compact readout circuit that could be integrated within a single single-photon avalanche detector. The avalanche process in photodetectors can be suppressed by using the smallest transistor size.

In 2011, Zhang et al. [40] analyzed the scattered-particle noise and 1/f noise of avalanche photodiodes, and established their signal-to-noise ratio (SNR) model.

In 2012, Guo et al. [41] studied the relationship between dark noise and the detector temperature. The results showed that when the temperature was reduced from 60 to − 40 °C, the noise-equivalent power decreased by one order of magnitude. If the cooling, spatial filtering, and spectral filtering methods were combined, the sensitivity of the detector increased by 7.7 times.

In 2013, Adamo et al. [42] analyzed novel silicon photomultiplier tubes manufactured on silicon p-type substrates; they could be used as sensitive power meters for low-photon fluxes.

In 2014, Youn et al. [43] studied an 850-nm photoelectric integrated-circuit receiver and found that the signal-to-noise characteristics of avalanche photodiodes depended on the reverse-bias voltage.

In 2015, Chen et al. [44] studied a low-voltage germanium-waveguide avalanche photodetector with a gain × bandwidth product higher than 100 GHz, and developed a photon receiver based on this photodetector. Using the bit-error rate measurement, the avalanche gain was 6 and the sensitivity was improved by 5.8 dB at a − 5.9-V bias.

In 2016, Goykhman et al. [45] studied a silicon plasma Schottky photodetector and achieved a responsivity of 0.37 A/W with a reverse bias of 3 V.

In 2017, Yin et al. [46] simulated an avalanche photodetector using benzocyclobutene as a bonding-layer material, with a detection efficiency of 96.7%.

In 2018, Jukic et al. [47] discovered that the excess noise of avalanche photodiodes was related to the avalanche gain. By analyzing the structure of the avalanche photodiode, it was found that this correlation could be suppressed by modulation doping.

In 2019, Huffaker [48] proposed a new avalanche photodiode consisting of a vertical InGaAs array that could be used for single-photon detection, greatly reducing the avalanche volume and number of filling traps.

In 2020, Zhou et al. [49] studied a linear array of 1×64 highly uniform 4H-SiC avalanche photodiodes for UV detection; it achieved a highly uniform breakdown voltage and a fluctuation of less than 0.5 V.

In 2021, Eid et al. [50] studied the pulse broadening and signal-fiber bandwidth of a silicon-doped germanium-fiber link in terms of the temperature variation, and found that the bit-error rate of the silicon avalanche photodetector at the receiving point was related to the temperature.

In 2022, Li et al. [51] studied the excess noise of avalanche photodiodes, analyzed two noise-measuring methods, including the direct-power measurement method and phase-sensitive detection method, and analyzed and discussed their advantages and disadvantages. On this basis, three methods of reducing the excess noise were obtained: selecting materials with a low impact-ionization coefficient ratio, reducing the thickness of the multiplier layer, and using APD impact-ionization engineering.

Table 2.2 shows the development of avalanche photodiodes. The gain and sensitivity of the detector are discussed in the above literature. A noise model of the avalanche photodetector was established by analyzing the source of the noise, noise-suppression techniques were studied, and the InGaAs/InAlAs detector as a mature material was further analyzed.

Table 2.2 Research progress on avalanche photodiodes

Researcher	Year	Development
Yang	1980	APDs can be used for long-wavelength detection
Watkins	1990	Observed the characteristics of laser-induced electrical failures
Li	1996	Superlattice avalanche photodetector is studied and optimized
Ng	2005	Strict restrictions on the doping of charge-control layers
Guo	2009	Proper selection of the reverse bias of an avalanche photodiode
Chitnis	2010	Avalanche process in the photodetector is suppressed
Zhang	2011	SNR models of granular noise and 1/f noise are established
Guo	2012	Sensitivity of the detector is improved by 7.7 times
Adamo	2013	Used as a sensitive power meter for low photon flux
Youn	2014	SNR depends on the reverse bias voltage
Chen	2015	Sensitivity is improved by 5.8 dB
Goykhman	2016	Responsivity of 0.37 A/W was achieved
Yin D	2017	Detection efficiency reached 96.7%
Jukic	2018	Excess noise can be suppressed by modulation doping
Huffaker	2019	Vertical InGaAs arrays can be used for single-photon detection
Zhou	2010	High uniform breakdown voltage is achieved
Eid	2021	Bit-error rate is temperature-dependent
Li	2022	Excess noise of avalanche photodiodes is studied

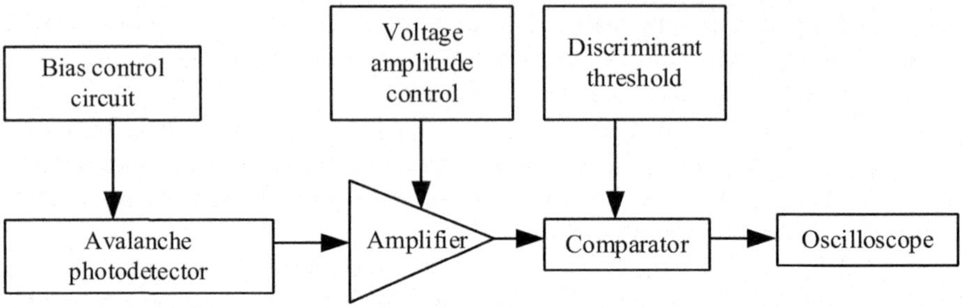

Fig. 2.8 Block diagram of the receiving system

2.2.3.3 Typical Application Scenarios

Figure 2.8 shows the principle block diagram of a detection and receiving system. A weak optical signal that is diffusely reflected back is converted into an electrical signal by an avalanche photodetector and amplified by an amplifier. Then, it enters a comparator for comparison. The bias control circuit provides the avalanche photodetector with the corresponding bias voltage, giving the detector an optimal multiplier factor under the optimal bias voltage. The voltage amplitude control adjusts the output-signal amplitude. Adjusting to the optimal amplitude can reduce the blind area of the ranging system. The threshold-circuit comparison voltage is discriminated to output a useful signal level. The comparison voltage must be simultaneously lower than the signal voltage and higher than the noise voltage. Finally, further processing is conducted using the oscilloscope.

2.2.4 PIN Photodetectors

2.2.4.1 Working Principle

The PIN detector separates the p-type region from the n-type region by adding a lightly doped thin layer in the middle of the P–N junction. The structural diagram of the PIN diode is shown in Fig. 2.9. The junction capacitance of the PIN detector is very small, and the depletion-layer width of the P–N junction is directly related to the reverse bias. When the reverse bias increases or decreases, the depletion-layer width increases or decreases correspondingly. The PIN detector structure is simple.

2.2.4.2 Research Progress

In 1986, Shu B researchers [52] successfully manufactured a silicon PIN photodetector based on high-resistance monocrystalline silicon using a mature semiconductor planar process.

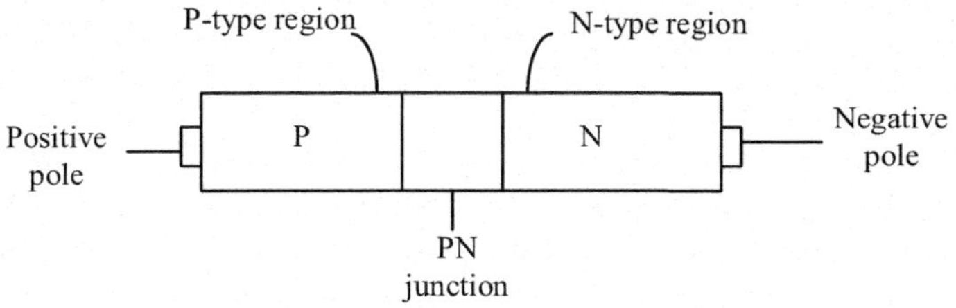

Fig. 2.9 Structural diagram of a PIN diode

In 1992, Zhang et al. [53] developed a state-space model for PIN photodetectors. This model produced the detector's time-domain and frequency-domain responses to any form of optical excitation, including terminal, uniform, and exponential illumination. Transmission time limits and circuit RC time constants were included in the model.

In 1997, Zhang et al. [54] used a coaxial package for the InGaAs PIN photodetector chip to measure the C–V characteristics and transient response of the coaxial package device. The results showed that the transient response of the device was greatly improved; the rise time was reduced from 85 ps to less than 25 ps, the half-height width was reduced from 210 to 85 ps, the capacitance was reduced by approximately 0.4 pF, and the bandwidth was increased to more than 6 GHz.

In 2004, Jutzi et al. [55] investigated a SiGe PIN photodetector and achieved an optical responsiveness of 160 mA/W at 1300 nm.

In 2010, Giannopolous et al. [56] designed a vertical-cavity surface-emitting laser that was integrated with a PIN photodetector monolith for optical-position sensing. The design was integrated into a silicon platform to form a microsystem. By using a metal grating as a position meter, the sensor microsystem could measure the difference in the reflected power of the grating as it traveled parallel to the sensor.

In 2011, Shao et al. [57] studied the growth and preparation of a back-lit AlGaN/GaN PIN solar blind UV photoelectron detector. The peak responsivity of the photodetector was about 0.055 A/W at 270 nm at a 2.5-V reverse bias, and the dark-current measurement of a 1500-μm-diameter device was approximately 25 pA at a 0.3-V bias.

In 2012, Guan et al. [58] analyzed various photoelectric-conversion circuit noises where the PIN detector was located, and proposed a method to reduce the noise in the photoelectric conversion circuit to improve the signal-to-noise ratio and measurement accuracy.

In 2014, Wu et al. [59] studied the threshold characteristics of a PIN detector in optical-wireless communication. The experiment showed that the detector reached the depth saturation at a 1064-nm wavelength, resulting in serious intercode crosstalk; thus, it could not meet the communication indicators.

In 2015, Chaudhari et al. [60] proposed a low-cost, low-temperature process with precise-thickness films, which was suitable for the low-cost industrial manufacturing of PIN-based detectors.

In 2016, Rouse et al. [61] manufactured a Ge-based PIN detection device on 300-mm silicon wafers using high-throughput, large-area complementary metal-oxide semiconductor technology.

In 2018, Xu et al. [62] used a laser light source with a 10-ns pulse width and a PIN detector for photoelectric conversion. When the applied bias voltage was 89.449 V, the output current signal needed to be amplified by a trans-resistance amplifier, and the amplified voltage amplitude reached − 3.7 V.

In 2019, Shu et al. [63] studied the influence of the main noise sources of a 10 Gbit/s PIN photodetector circuit on the output current signal using an adaptive-filtering algorithm and a peak-pulse median removal algorithm. The results showed that the above denoising method could effectively reduce the measurement error of the output current, improve the measurement accuracy, and increase the stability.

In 2021, Wang et al. [64] designed a highly responsive silicon-based PIN detector with a 4.1 mm × 13.8 mm surface area and a 420-μm thickness. The photocurrent signal was extracted through a ring aluminum electrode, and the detector had a response of 0.6 A/W at a constant laser light source with an 860-nm wavelength.

In 2022, Xiao et al. [65] designed a high-speed monolithic PIN photodetector based on InP and integrated it with a multi-quantum-well semiconductor optical amplifier. The chip allowed for the separate design of the multi-quantum-well semiconductor optical amplifier and PIN photodetector. The PIN photodetector was manufactured with a 300-pA dark current at 3 V.

Table 2.3 illustrates the development of the PIN detector. The above literature mainly discusses the industrial structure and preparation of PIN detectors and their corresponding noise models, as well as the related applications of PIN detectors in optical-wireless communication.

Table 2.3 Research progress of PIN photodetectors (PD)

Researcher	Year	Development
Zhang	1992	Established a state-space model of the PIN photodetector
Zhang	1997	Measured the transient response of a PIN PD with a coaxial package
Jutzi	2004	Optical response of 160 mA/W is achieved at 1300 nm
Giannopolous	2010	PIN photodetector monolithic set created
Shao	2011	Peak responsivity is about 0.055 A/W at 270 nm
Guan	2012	Noise of photoelectric conversion circuit is reduced
Wu	2014	Observed the threshold characteristics of a PIN detector
Chaudhari	2015	Low-cost industrial manufacturing of PIN detectors
Rouse	2016	Ge-based PIN detection devices were manufactured
Xu	2018	Voltage amplitude reached − 3.7 V
Shu	2019	Reduced the measurement error of the output current
Wang	2021	Responsivity of the detector at a constant laser source
Xiao	2022	Dark current is 300 pA at 3 V

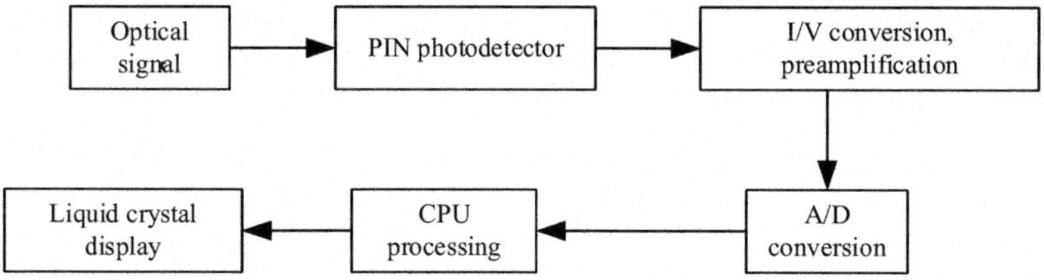

Fig. 2.10 Schematic diagram of an optical power meter

2.2.4.3 Typical Application Scenarios

Owing to the small dynamic range of traditional optical-power detection and the errors in amplification and switching, a weak optical signal is amplified according to the response characteristics of a PIN photodiode, and a single-chip microcomputer is used for data-correction processing. When it is finally combined with liquid–crystal display (LCD) technology, an optical power-detection meter can be produced.

The basic working-principle block diagram of the optical power meter is shown in Fig. 2.10. The optical signal first passes through the PIN photodetector, which converts the optical signal into an electrical signal; then, the electrical signal goes through an I/V transformation and pre-amplification processing. The analog signal obtained after processing is converted to digital; the digital signal represents the power size. Then, the CPU processes and analyzes the data. Finally, the power size is displayed on the LCD.

2.2.5 Four-Quadrant Detectors

2.2.5.1 Working Principle

Following Cartesian-coordinate requirements, diodes with the same performance are placed in the four quadrants, and the detector formed by such a combination is called a four-quadrant detector. A very narrow interval exists between the four quadrants, which is called the dead zone. The detector's detection ability is affected by the dead zone; therefore, the dead-zone area should be reduced as much as possible when making the detector.

Figure 2.11 shows the surface structure of the four-quadrant photodetector. The incident light irradiates the surface of the circular area in the figure, which is called the sensitive surface. The sensitive surface is divided into four parts, namely, A, B, C, and D, and the corresponding areas of these four parts are S_A, S_B, S_C, and S_D, respectively. The corresponding electrodes inside each of these four parts will generate impedance currents I_A, I_B, I_C, and I_D, respectively.

As the incident beam moves, the spot on the detector's surface moves with it, so the true area of light in each quadrant changes, as does the intensity of the light

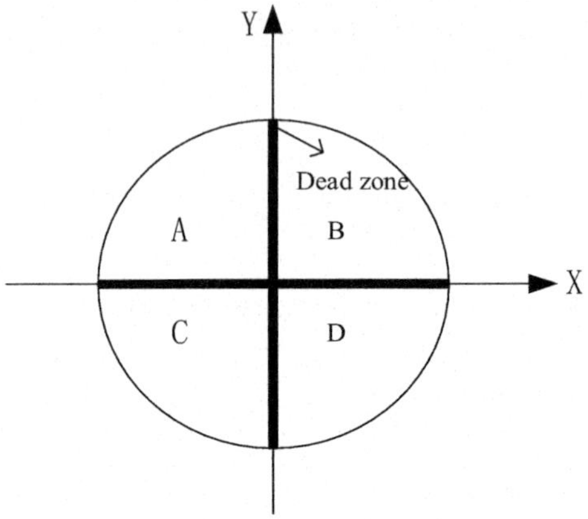

Fig. 2.11 Four-quadrant detector surface-structure diagram

received. The lateral and longitudinal offsets of the light can be determined by the amount of current generated. The formulas for determining the spot position by the area of light received are as follows:

$$\sigma_{\mathrm{x}} = \frac{(S_A + S_D) - (S_B + S_C)}{S_A + S_B + S_C + S_D} = \frac{(I_A + I_D) - (I_B + I_C)}{I_A + I_B + I_C + I_D} \tag{2.11}$$

$$\sigma_y = \frac{(S_A + S_B) - (S_C + S_D)}{S_A + S_B + S_C + S_D} = \frac{(I_A + I_B) - (I_V + I_D)}{I_A + I_B + I_C + I_D} \tag{2.12}$$

where S_A, S_B, S_C, and S_D are respectively the corresponding spot areas of the four quadrants; I_A, I_B, I_C, and I_D are respectively the impedance currents generated by the corresponding four quadrant electrodes. σ_{x} and σ_y are the calculated values for the X and Y axes, respectively, which can be used to analyze the position information of the light spot on the target plane.

2.2.5.2 Research Progress

In 1979, researchers [66] designed a quadrant detector with a new structure. A P–N junction was used for isolation on the same silicon chip. In addition, a quadrant detector was successfully developed using a four-quadrant patterned oxidation-mask diffusion process.

In 1996, Feng et al. [67] discussed some of the key factors involved in optical roaming using a four-quadrant photodetector, such as the error-data extraction and target-spot size on the active quadrant of the photodetector, in search of more flexible software and hardware design methods.

In 2002, Wang et al. [68] proposed a method to improve the detection accuracy of the high-point position. A Gaussian spot was used as the incident-light model to analyze the solution relationship between the actual position of the spot and the output signal of the four-quadrant detector. Then, the output current signal of the detector was amplified through an analog circuit, and the output voltage signal with noise was digitally filtered. The results showed that the maximum spot-position detection error of this method was only 0.0277 mm, and the root mean square error was 0.0065 mm, when the four-quadrant detector was in a high-background noise environment.

In 2004, Kuang et al. [69] determined the offset of the two directions of the light spot by measuring the change in the position of its centroid from the laser beam.

In 2005, Chi et al. [70] analyzed the noise of a four-quadrant infrared detector, including loose-particle noise, thermal noise, generation-compound noise, 1/f noise, and the cause of the noise formation.

In 2010, Xin et al. [71] analyzed the factors affecting the tracking accuracy of a four-quadrant detector, including spot size, spot location, and detector noise. The analysis results showed that when using the commonly used four-quadrant centroid algorithm, it was necessary to adopt an appropriate optical system and signal-detection circuit to achieve high-precision tracking-error detection.

In 2013, Yang et al. [72] made a detailed analysis of the dead zone of a four-quadrant detector and showed the influence of the dead zone on the detection sensitivity; the larger the dead zone area, the greater the impact.

In 2014, Schalkwyk et al. [73] made a forward-looking intrinsic day-blind UV four-quadrant Schottky detector based on AlGaN with a Schottky barrier height of 1.22 ± 0.07 eV.

In 2016, Tian et al. [74] proposed an optical-system design method that controlled the spot-size design on a four-quadrant detector by adjusting the entry/exit pupil ratio of the non-focal system, according to the characteristics of inter-satellite laser communication. The optical distortion, spot pattern, and circumferential energy were analyzed and evaluated. The focal-free system was proved to be applicable to a beacon system for tracking and detection.

In 2018, Grajower et al. [75] used a new silicon-based plasma four-quadrant photodetector to operate in the short-wave infrared band; it could accurately track various waist beams at telecommunication wavelengths.

In 2019, Wang et al. [76] proposed a new polynomial-fitting algorithm, using a circular spot with Gaussian energy as the incident-light model. When the radius of the incident spot was 0.5 mm, within a [− 0.5 and 0.5 mm] detection range, the new fifth-order polynomial-fitting algorithm was adopted, and the maximum error was 1.353 μm. The root-mean-square error was 0.4596 μm, which improved the accuracy of the four-quadrant detector.

In 2020, Liu et al. [77] designed a four-quadrant detector with a five-level amplifier circuit that could be used for low-light detection. The magnification could be adjusted according to the power of the incident light, and the centroid position of the light spot could be intuitively judged. The results showed that the four-quadrant detector had a high resolution, low static noise, and a simple design.

In 2021, Safi et al. [78] studied a surface-to-air free-space optical link for hovering unmanned aerial vehicles (UAVs), and a four-quadrant photodetector array was used at the optical receiver to mitigate the adverse effects of hovering fluctuations by expanding the receiver's field of view.

Table 2.4 Research progress of four-quadrant detectors

Researcher	Year	Development
Feng	1996	Discussed some of the key factors involved in optical roaming
Kuang	2004	Maximum spot-position detection error was only 0.0277 mm
Wu	2005	Offset of two spot directions was determined
Zhao	2010	Noise of a four-quadrant infrared detector was analyzed
Yang	2013	A suitable optical system was needed to detect the tracking error
Schalkwyk	2014	Influence of the dead zone on detection sensitivity was studied
Tian	2016	Proposed a forward-looking intrinsic day-blind UV four-quadrant detector
Grajower	2018	Proved that a focus-free system can be applied to a beacon system
Wang	2019	Can accurately track all types of beam waists
Liu	2020	A four-quadrant detector with a five-stage amplifier circuit was designed

Table 2.4 shows the development of the four-quadrant photodetector. The above literature introduces the application of the four-quadrant detector to position detection and laser collimation, further analyzes the factors that affect the detection accuracy and the causes of related noise, and expounds the improvement in sensitivity of the four-quadrant detector.

2.2.5.3 Typical Application Scenarios

Figure 2.12 shows a block diagram of a spot-tracking system, based on a four-quadrant detector and fast deflection mirror. The light signal first passes through the parallel optical tube, and the fast deflection mirror reflects the beacon light from there onto the lens. The lens, with an entry pupil diameter of H, projects the light spot onto the four-quadrant detector used for defocusing.

The data-acquisition system processes the electrical signal and sends it to the multi-point control unit, which controls the deflection angle of the fast deflecting mirror to adjust the incident direction of the optical path, forming a closed-loop tracking system. The mobile electric platform under the lens can control the size of the light spots, and then detect the root mean square error of the infinite integral algorithm under different light spots.

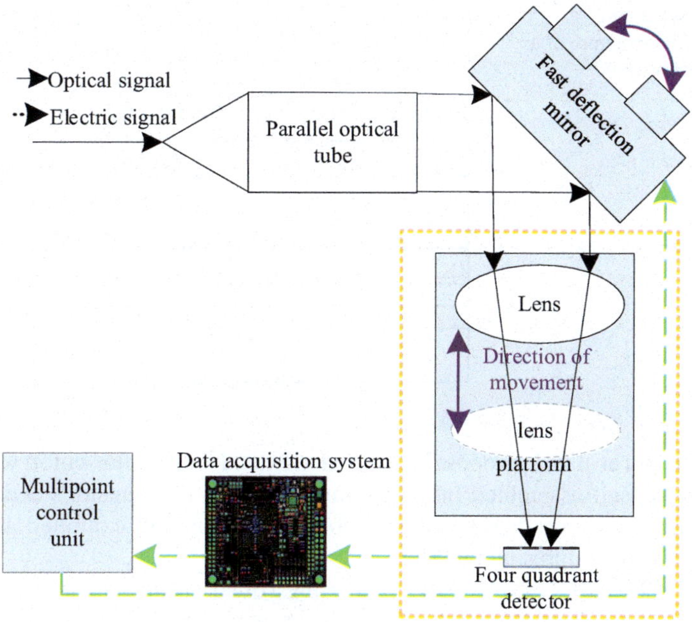

Fig. 2.12 Block diagram of a spot-tracking system

2.2.6 Quantum-Dot Infrared Detectors

2.2.6.1 Working Principle

The quantum-dot infrared detector is composed of periodic composite layers of quantum dots. Figure 2.13 shows a schematic diagram of the structure of the quantum-dot infrared detector. Each quantum dot in the composite layer has a large number of boundary states, which can receive more electrons. When infrared light is incident on the photosensitive region of the detector, electrons transition from the ground state to the excited or continuous states, which changes the conductivity of the detector, which finally detects the infrared light.

2.2.6.2 Research Progress

In 1999, Sheng et al. [79] analyzed the noise of a quantum-well infrared detector, including loose-particle noise, thermal noise, generation-compound noise, and 1/f noise, and explored the negative impact of each noise on the detector.

In 2006, Zhao et al. [80] experimentally verified an AlGaAs/GaAs multi-quantum-well infrared detector and showed that the dark current did not appear to change with the temperature when the temperature was less than 40 K. When the temperature was higher than 40 K, the dark current increased rapidly with an increase in temperature.

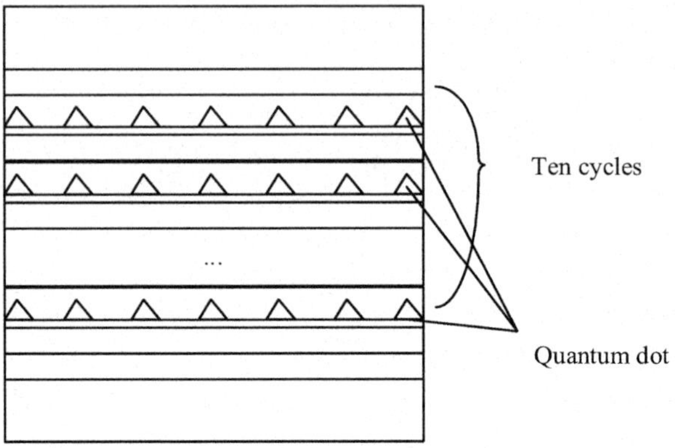

Fig. 2.13 Schematic diagram of the structure of a quantum-dot infrared detector

In 2010, Hill et al. [81] proposed a method to extend the detector-cutoff wavelength by combining self-assembled InSb quantum dots with the detector's active region. This extended the cutoff wavelength from 4.2 to 6 m, and exhibited an infrared response at temperatures up to 225 K.

In 2011, Chien [82] developed a model using a quadrant detector at room temperature to optimize a photodetector based on its structure, and the photoelectron capture and transit processes were studied using Monte Carlo simulations.

In 2012, Gustafsson et al. [83] studied quantum dots based on InSb. These quantum dots were used as active materials for a photon detector, which reduced the detector noise.

In 2013, Liu et al. [84] proposed a noise model for a quantum-dot infrared detector. They studied the dark current of the detector considering the influence of micron-scale and nano-scale electron transports on the excitation energy, and then deduced the noise model.

In 2014, Kim et al. [85] proposed and analyzed a novel detector structure for pixel-level multispectral infrared imaging, and then studied the device performance of a raster integrated quantum-dot well photodetector under back-illumination. By adjusting the grating parameters, a wide range of spectral responses was obtained.

In 2015, Srinivasan et al. [86] used molecular-beam epitaxy to investigate the fabrication of 20-layer silica-doped InAs dots in the structure of quantum-dot infrared photodetector devices.

In 2016, Karow et al. [87] realized a monolithic integrated circuit of a laser and detector, pushed the research on micro-scale sensor technology to the low-photon quantum limit, and paved the way for on-chip photoelectric-feedback experiments.

In 2017, Kim [88] studied the effects of micron and nanoscale external electrons and electron transport on the noise of quantum-dot infrared detectors by computational simulation. The results showed that under a 2545 kV/cm applied electric field, the noise increased with an increase in the applied electric field and temperature. When the control temperature was less than 80 K, the noise increased very quickly. When the control temperature was greater than 80 K, the noise increased

Table 2.5 Research progress of quantum-dot infrared detectors

Researcher	Year	Development
Sheng	1999	Loose-particle noise and thermal noise were analyzed
Zhao	2006	Dark current increases rapidly with the increase of temperature
Hill	2010	Extended the detector cutoff wavelength
Chien	2011	Investigated photoelectron capture and transit processes
Lau	2013	Noise model of a quantum-dot infrared detector is proposed
Kim	2014	Wide range of spectral responses was obtained
Karow	2016	Realized a monolithic integrated circuit of laser and detector
Jin	2017	Noise is affected by the applied electric field
Wang	2018	Temperature affects the detector's dark current
Baira	2019	Larger quantum dots with higher aspect ratios
Gansen	2020	Noise model of 1/f noise is derived

more slowly. Thus, it could be concluded that the lower the temperature, the more obvious the impact of noise on an applied electric field.

In 2018, Wang [89] established a model and algorithm that could represent the dark current of quantum-dot infrared detectors. The model was simplified by combining the actual detector situation. It was found that the temperature, quantum-dot density, and side dimension affected the dark current of the detector.

In 2019, Baira et al. [90] evaluated the size and shape of self-organizing GeSn/Ge quantum dots and found that larger quantum dots with higher aspect ratios might have higher oscillator strengths and longer radiation lifetimes.

In 2020, Gansen et al. [91] investigated a 1/f noise source in a quantum-dot field-effect transistor and produced a 1/f noise model.

Table 2.5 shows some of the progress in quantum-dot infrared detectors. The above studies investigated the influence of the dark current and temperature on noise generation and the influence of noise on detector performance by analyzing the quantum-dot junctions of quantum-dot infrared detectors.

2.2.6.3 Bottlenecks and Breakthroughs

Although gallium nitride (GaN), indium antimonide (InSb), mercury cadmium telluride (HgCdTe), and other bulk semiconductor materials can broaden the spectral detection range of CMOS detectors, these bulk semiconductor materials are not compatible with silicon-based CMOS detectors and require flip bonding and other complex processes. This results in a high cost and low pass rate in the detector-manufacturing process. In addition, ultraviolet and infrared photon energies are very different, and cross-band coupling is extremely difficult.

The solution properties of colloidal quantum-dot materials can be used to pattern a quantum-dot thin film after the quantum-dot infrared material is coupled to a

CMOS readout circuit. This allows direct lithography to be used in the fabrication of multi-band wide-spectrum detectors from ultraviolet to infrared by plane coupling. Using lead sulfide (PbS) quantum dots, the team of Professor Jiang Tang from Huazhong University of Science and Technology has fabricated near-infrared quantum-dot chips, and demonstrated good imaging performance under 970-nm infrared irradiation.

To overcome the band bottleneck, Professor Xin Tang from the Beijing Institute of Technology proposed the use of mercury telluride (HgTe) and perovskite colloidal quantum-dot materials to broaden the CMOS detection range, improve the detector-manufacturing scale, realize wafer-level processing, greatly reduce the detector production cost, and improve the process pass rate. Because the bulk material band gap is zero, mercury telluride (HgTe) colloidal quantum dots can be used in short-wave infrared, medium-wave infrared, and long-wave infrared detectors.

This breakthrough greatly simplified the complexity of the optical hardware and software of multiband cameras, reduced the cameras' size and weight, and helped realize the micro-platform design of multiband cameras. The team used a direct-photolithography process to form colloid quantum dots into a two-color or multicolor focal plane array, and broke through the traditional block semiconductor-material flip-bonding process to complete multicolor detection and detector coupling. In addition, they realized multiband detection and multicolor imaging capabilities in one detector, greatly reduced the production cost of multicolor detectors, and improved the production-process scale and product pass rate.

2.2.6.4 Typical Application Scenarios

To better use space resources, countries constantly launch satellites and space stations to establish detection sites, all competing to occupy the optimal positions. Space photoelectric systems play a very important role in this respect, because the new quantum-dot infrared detectors have high detection sensitivity, high working temperature, and anti-radiation ability. It would be extremely beneficial to apply quantum-dot infrared detectors to space photoelectric systems.

During laser communications between satellite and ground or satellite and satellite, the space environment is complicated because of the long transmission distance, which mainly causes two problems: The first is that the intensity attenuation intensifies as the communication distance increases. The second is the harsh space environment, which requires photoelectric devices to have better temperature characteristics and radiation-resistance performance. A quantum-dot infrared detector not only has a high working temperature and radiation resistance, but also has advantages in structure and volume. Therefore, it has broad application prospects in space environments.

2.2.7 Balanced Detectors

2.2.7.1 Working Principle

Figure 2.14 shows the principle diagram of a balanced detector receiving coherent light [92]. When the local light and signal light fully meet the coherence conditions, the light field interferes and the photodetector converts the light to an electrical signal. The signal light is represented by E_S, and the local oscillator light is represented by E_L. Assuming that the power divider is an ideal device, that is, linear, symmetrical, and lossless:

$$\begin{bmatrix} E_1 \\ E_2 \end{bmatrix} = \exp(j\psi_r)\begin{bmatrix} 1 & \exp\left(j\frac{\pi}{2}\right) \\ \exp\left(j\frac{\pi}{2}\right) & 1 \end{bmatrix}\begin{bmatrix} E_S \\ E_L \end{bmatrix} \tag{2.13}$$

where w_{IF} denotes the reflected phase shift of the power divider, E_S denotes the signal light, and E_L denotes the local oscillator light.

$$I_1(\mathrm{t}) - I_2(\mathrm{t}) = 2\frac{\eta e}{h\nu}\sqrt{P_S P_L}\cos(w_{IF}t + \phi_S - \phi_L) \tag{2.14}$$

where w_{IF} is the frequency of the intermediate frequency signal; ϕ_S and ϕ_L are the phases of the signal light and local oscillator light, respectively; P_S and P_L are the optical powers of the signal light and local oscillator light, respectively; η is the quantum efficiency of the photodetector; e is the amount of charge; h is Planck's constant; ν is the optical frequency; and I_0 is amplified by the amplifier to generate the current. Because the balanced detector can suppress the common-mode signal, the noise is reduced, and the detector sensitivity is improved.

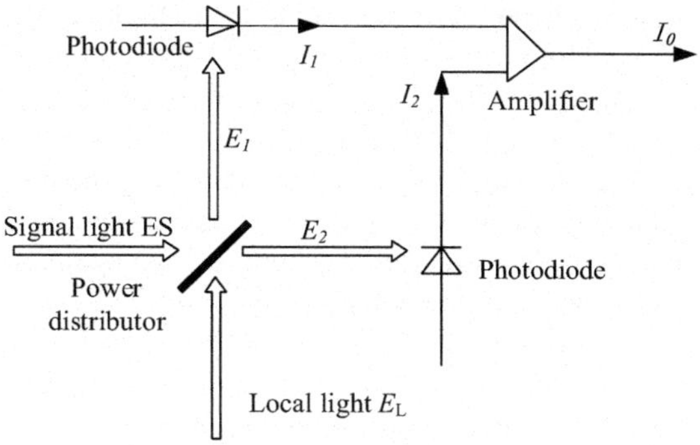

Fig. 2.14 Block diagram of a coherent-light-receiving balanced detector [92]

2.2.7.2 Research Progress

In 2014, Dai et al. [93] designed a set of high-speed balanced-detection systems that achieved coherent detection at different rates. For a 2.5-Gbit/s signal rate, the direct-detection sensitivity reached − 26 dBm and the dynamic range was − 26 to 3 dBm. When the local oscillator light was 3 dBm, the sensitivity reached − 46.5 dBm and the dynamic range was − 46.5 to 3 dBm.

In 2015, Guo et al. [94] analyzed the output noise of balanced heterodyne detection and established a relevant noise model. The simulation results showed that, compared with the signal-to-noise ratio of common coherent detection, the signal-to-noise ratio of a balanced heterodyne-detection method could be improved by reasonably setting the structural parameters of the detector.

In 2016, Shi et al. [95] designed a coherent-detection test scheme based on the composition of a balanced detector. When the communication rate was 5 Gbit/s, the common-mode rejection ratio was higher than 27 dB. Balanced detection was adopted to achieve coherent demodulation, and the minimum signal optical power was improved by 8 dB compared with single-tube detection.

In 2017, Munshid [96] designed and constructed a heterodyne detector based on photonic crystal fiber, and found that the relationship between the wavelength shift and temperature change was linear.

In 2018, Yong et al. [97] estimated the dispersion of coherent optical communication systems based on heterodyne detection, and investigated the nonlinear limitations of 10-Gbit/s non-return to zero electron predistortion systems by simulation.

In 2019, Zhang et al. [98] designed a balanced detector in a high-speed terahertz time-domain spectroscopy system. Two low-noise photodiodes in series were used to directly detect the difference in the photocurrents of two laser beams and reduce the detector noise.

In 2021, Hu et al. [99] studied a novel heterodyne detection scheme, and compared it to the heterodyne detection of a single-ended photodetector. The new scheme allowed for the use of a simple transmitter structure, reducing the receiver bandwidth to a value far below the sign rate, while preserving the ability to compensate for digital dispersion.

As shown in Table 2.6, the aforementioned literature focuses on the application of balanced detectors in coherent optical communication systems based on heterodyne detection. The structure can suppress the noise, and the noise model of the detectors is obtained simultaneously.

2.2.7.3 Typical Application Scenarios

The principle of coherent detection is illustrated in Fig. 2.15. The light emitted from the laser acts as a carrier and is modulated by the signal either directly or externally. At the receiving end, the spatial optical signal W_s is coupled into the fiber via a fiber coupler, and the lightwave generated by the local oscillator W_L is superimposed on the received signal in the mixer, producing an intermediate frequency (IF)

Table 2.6 Research progress of balanced detectors

Researcher	Year	Development
Bae	1993	Excess intensity noise of local vibration light is suppressed
Dai	2014	Sensitivity can reach – 46.5 dBm
Guo	2015	Improved SNR ratio of the balanced heterodyne detection method
Shi	2016	Common-mode rejection ratio is higher than 27 dB
Munshid	2017	Relationship between wavelength shift and temperature is linear
Yong	2018	Ten Gbit/s non-return-zero electron predistortion systems
Zhang	2019	Detector noise is reduced
Hu	2021	Receiver bandwidth is reduced to a value well below the sign rate

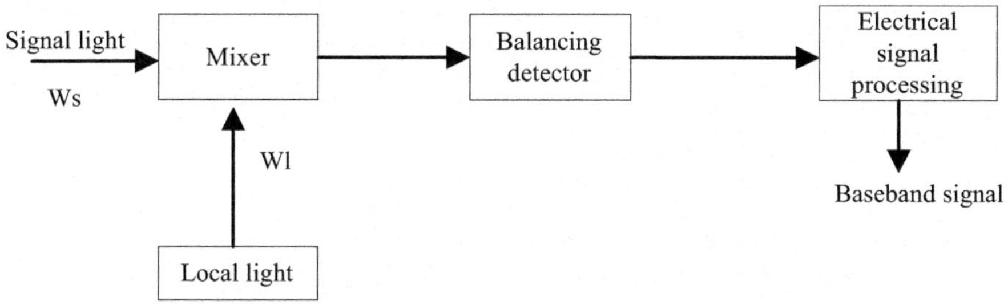

Fig. 2.15 Schematic diagram of coherent detection

signal at the output end of the balanced detector. Zero-difference detection can be directly obtained from the baseband signal and balanced-detector output signal after IF amplification, filtering, demodulation, and restoration of the digital signal to the sending end.

2.2.8 *Double Balanced Probes*

2.2.8.1 Working Principle

Double-balanced detection is based on the structural composition of a balanced heterodyne detection system. The double-channel balanced receiving technology is shown in Fig. 2.16, which uses four photodetectors that are almost identical. The two detectors at the receiving end are a group, forming two groups of balanced heterodyne detectors [100].

The signal light and local light enter the port at the input end of a 90° optical mixer. After a quarter of the waveplate, the local light can be converted into circularly polarized light. At this time, the two present a 45° angle. Through the beam-splitting prism, the signal and local lights can be shot into half of the wave plate after the

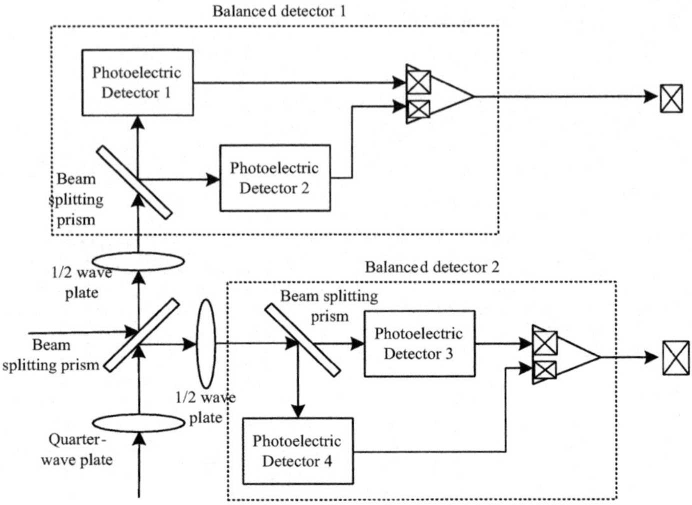

Fig. 2.16 Double-balanced probe

mixed-beam splitting; at this time, the two will present a 22.5° angle. Finally, four light beams are obtained, based on the polarization beam splitter. Two orthogonal IF current signals can be obtained by combining two beams of light with opposite phases and inputting them into a balanced detector.

$$\begin{cases} I : I_0 - I_{180} = 2k_1k_4 \cos\left[\omega_{IF} - \left(\varphi + \frac{\pi}{4}\right)\right] \\ Q : I_{90} - I_{270} = 2k_2k_3 \sin\left[\omega_{IF} - \left(\varphi + \frac{\pi}{4}\right)\right] \end{cases} \tag{2.15}$$

where I and Q are the 90° phase difference between the two channels, that is, the in-phase component and the orthogonal component required by the demodulation information. I_0, I_{180}, I_{90}, and I_{270} are the four optical-mixer outputs obtained through the polarization splitter prism; k_1, k_2, k_3, and k_4 represent the local light and signal light in the directions of the component, φ is the phase difference, and ω_{IF} represents the intermediate frequency. When the two subtracted photocurrents after the DC component are completely offset, and the intensity noise related to the DC component is also eliminated, the AC term is proportional to the square root of the local light power. Therefore, the noise intensity has a much smaller influence.

2.2.8.2 Research Progress

In 2003, Li et al. [101] verified the physical mechanism of the balanced detector and built a 1550-nm balanced heterodyne detection system. The results showed that the signal-to-noise ratio performance of the double-balanced detector was much better than that of the single-ended detector.

In 2010, Gao et al. [102] applied dual-balanced detection technology to a weak signal-detection field; for example, in Lidar, the polarization factor of incident light will cause a polarization mixing error, and the error of the dual-balanced detection system was discussed. The results showed that the vertical orthogonal factor of the ellipse formed by polarized light on the polarization splitter was 0.523, when the polarization degree was in the range of (0, 0.02).

In 2013, Zhu [103] established a dual optical-path balanced detection system using 3×3 couplers to solve the corrosion and electromagnetic-interference problems of traditional acoustic microphones. This enabled the differential processing of two equal-amplitude signals. Experiments showed that the system was stable and the waveform distortion was small. It extracted and recovered speech signals in the 300–3400 Hz range.

In 2013, Huan [104] studied the weak laser-signal detection of a double-balanced heterodyne detection system, starting from the aspects of the system structure and a mathematical model. Through simulation analysis, they deduced a conclusion that conformed to the mathematical model, which laid the foundation for research on weak laser-detection technology.

In 2015, Zheng et al. [105] proposed a new dual-balanced differential configuration method with a photon-detection efficiency of 10% for a 1550-nm laser at 10 MHz. The results showed that this configuration effectively resolved ultra-weak avalanche signals.

In 2015, Peng et al. [106] introduced a new photoelectric heterodyne receiving technology based on double-balanced coherent detection, explained the working principle of double-balanced coherent detection, and further analyzed double-balanced and balanced coherent-detection systems.

In 2016, Li et al. [107] designed a double-balanced sound receiving system based on the Mach–Zehnder ring structure to increase the measuring distance of a laser Doppler velocimeter. The SNR expression of the double-balanced sounding system was derived. Simulation results showed that the signal-to-noise ratio of the double-balanced detection system was higher than that of common coherent detection.

In 2017, Huang et al. [108] considered the shortcomings of common ground-fault detection methods and proposed a new two-bridge detection method. This method can be used for fault detection, which not only solves the problem caused by the dead zone, but also does not affect the distributed capacitance. The feasibility of this method was experimentally verified.

Table 2.7 shows the development of the dual-balanced detector. The above literature describes the working principle of double-balanced detection technology and analyzes whether it can be applied to the detection of weak signals and other fields. A comparison with common coherent detection technology shows that double-balanced detection technology can improve the system signal-to-noise ratio.

Table 2.7 Research progress on double-balanced detectors

Researcher	Year	Development
Li	2003	Improved the SNR performance of double-balanced detectors
Gao	2010	Applied dual-balanced technology to weak signal detection
Zhu	2013	Used a 3 × 3 coupler to establish a balanced detection system
Huan	2013	Weak laser-signal detection in a double-balanced detection system
Zheng	2015	Proposed a double-balanced differential configuration method
Peng	2015	Studied double-balanced coherent detection
Li	2016	Created a DB detection system based on the Mach–Zehnder ring structure
Huang	2017	Proposed a new two-bridge detection method

2.2.8.3 Typical Application Scenarios

Figure 2.17 presents a simulation block diagram of a coherent-optical communication system. In the simulation, a pseudo-random bit sequence forms the modulated signal, and the signal light is modulated by quaternary phase-shift keying (QPSK). In the experiment, the oscillator light and signal light are introduced into the mixer for mixing. The mixer signal is output to the double-balanced detection system, and its signal is converted to analog after digital signal processing. The receiver decision circuit measures the signal-to-noise ratio to determine whether the system performance is good or bad by observing the bit-error rate.

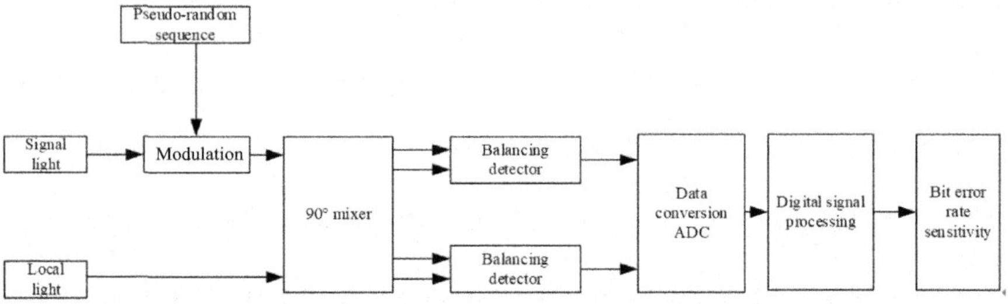

Fig. 2.17 Schematic diagram of a coherent-optical communication-system simulation

2.3 Photodetector Noise

2.3.1 Sources of Photodetector Noise

2.3.1.1 Loose-Particle Noise

When incident light irradiates a detector surface, the photoelectrons oscillate randomly, and scattered noise is generated. The electrons emitted by the electron tube in any short time τ will fluctuate near the mean value. From the perspective of the mean-square deviation of the fluctuation, the power of the scattered particle noise is

$$i_n^2 = 2eI\Delta f \tag{2.16}$$

where i_n^2 represents the power of the scattered particle noise, *e* represents the electron charge, Δf represents the noise bandwidth, and I is the average photocurrent generated by the detector.

For the cases of incident light and no incident light, the dark-current noise power and light-radiated scattered-particle noise power are, respectively,

$$i_{nd}^2 = 2eI_d\Delta f \tag{2.17}$$

$$i_{np}^2 = 2eI_p\Delta f \tag{2.18}$$

where i_{nd}^2 denotes the dark-current noise power, i_{np}^2 denotes the light-radiated scattered-particle noise power, *e* denotes the amount of charge in an electron, and Δf represents the noise bandwidth of the detector. The average photocurrent without incident light is represented by I_d and the average photocurrent with incident light is represented by I_p.

2.3.1.2 Generation-Compound Noise

The constant movement of charge carriers in a semiconductor leads to the random generation and recombination of electrons and holes; hence, the average carrier concentration will fluctuate continuously. The noise generated at this time is called generation-compound noise, and the expression is

$$i_{ng-r}^2 = 4eIM\Delta f \tag{2.19}$$

where the power generating the generation-compound noise is represented by i_{ng-r}^2, I represents the average photocurrent, Δf represents the noise bandwidth of the detector, and M represents the gain.

2.3.1.3 Thermal Noise

The random motion of the charge carriers in the dissipative element produces thermal noise. For resistances, these electrons produce thermal noise as long as the resistance is in thermal equilibrium, even if no voltage is applied.

An AB-electrode equivalent diagram is shown in Fig. 2.18. The resistance between the poles of AB is denoted by R. At an absolute temperature T, the electrons between the poles of AB are in constant random thermal motion.

Because there are voltage fluctuations at both ends of AB, the average power of the thermal noise obtained is

$$U_{nJ}^2 = 4KT\Delta fR \tag{2.20}$$

where U_{nJ}^2 denotes the average power of thermal noise, the Boltzmann constant is denoted by k, the measurement bandwidth is denoted by Δf, the absolute temperature is denoted by T, and R is the resistance between the AB electrodes. If the noise is caused by the current,

$$i_{nJ}^2 = \frac{4KT\Delta f}{R} \tag{2.21}$$

where i_{nJ}^2 represents the average power of the thermal noise.

The root-mean-square values of the thermal-noise current and voltage are typically calculated as follows:

$$\sqrt{i_{nJ}^2} = \left(\frac{4KT\Delta f}{R}\right)^{1/2} \tag{2.22}$$

$$\sqrt{U_{nJ}^2} = (4KT\Delta fR)^{1/2} \tag{2.23}$$

where $\sqrt{i_{nJ}^2}$ is the root mean square value of the thermal noise current and $\sqrt{U_{nJ}^2}$ is the root mean square value of the thermal noise voltage.

Fig. 2.18 Equivalent diagram of an AB electrode

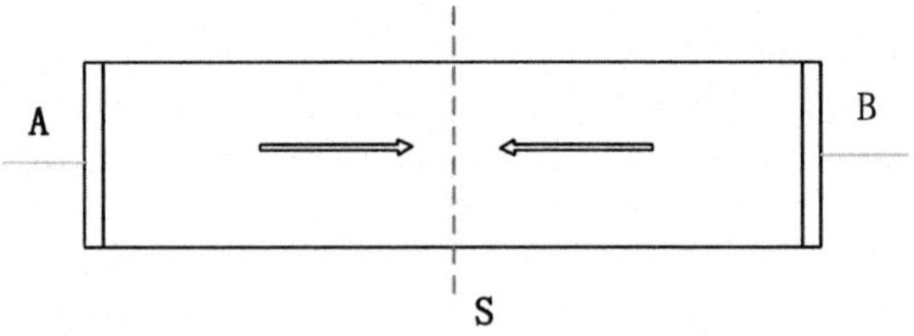

2.3.1.4 1/f Noise

The detector-manufacturing process introduces unnecessary trace impurities or small nonuniform particles into the sensitive layer. Under the action of current, microburst pulses are generated between these particles, resulting in 1/f noise.

The current mean-square value of 1/f noise can be expressed as

$$i_{nf}^2 = k_1 \frac{I^b \Delta f}{f^a} \tag{2.24}$$

where i_{nf}^2 represents the current mean square value of 1/f noise, and the proportionality coefficient is k_1. k_1 is affected by various factors, including the detector manufacturing process, electrode contact, and size of the detector. Δf is the working bandwidth, a is a constant related to the material of the detector, usually between 0.8 and 1.3. Current I is in the device; b is related to I, b usually taking the value 2.

2.3.2 Typical Photodetector-Noise Models

2.3.2.1 Noise Model of an Avalanche Photodetector

Thermal Noise

For an avalanche photodiode, the thermal noise current satisfies Eq. (2.18), if the resistance inside the diode is ignored, and only the effect of the load of the current in which the diode is located is considered.

Loose-Particle Noise Generated by the Dark Current

The continuous random movement of carriers in the P–N junction generates loose particle noise. Then [109],

$$i = i_p + i_s \tag{2.25}$$

where i is the total current flowing through the P–N junction. The scattered-particle noise current is represented by i_s and the average value of the current is i_p. If the bandwidth of the receiving system is B, the effective value of the scattered noise can be expressed as

$$i_s = \sqrt{2ei_pB} \tag{2.26}$$

where i_s is the effective value of the scattered particle noise, e is the electronic charge, the average current value is i_p. In a dark environment, for an avalanche photodiode with a multiplication factor M, the dark current is

$$i_d = i_b + Mi_t \tag{2.27}$$

where i_d is the dark current, i_b is the surface leakage current, i_t is the volume leakage current, and M is the multiplication factor. The dark current of the detector causes loose-particle noise, and the effective value of the noise current is expressed as [109]

$$i'_s = \sqrt{2ei_tBM^2 + 2ei_bB} \tag{2.28}$$

where i'_s is the effective value of the noise current.

Excess Noise

The gain of an avalanche photodiode represents the ratio of the output photocurrent to the incoming photocurrent [109]:

$$M = \frac{I_p - I_d}{I_{p0} - I_{d0}} \tag{2.29}$$

where the multiplication factor is M, I_p is the photocurrent of the avalanche photodiode, I_d is the dark current of the avalanche photodiode, I_{p0} represents the photocurrent at unit gain, and I_{d0} represents the dark current at unit gain.

Assuming that the avalanche multiplication region is uniform, the overnoise factor is related to the gain and ionization coefficient, that is,

$$F(M) = kM + (1 - k)\left(1 - \frac{1}{M}\right) \tag{2.30}$$

where $F(M)$ is the overnoise factor, k is the ionization coefficient, and M is the multiplication factor. In device design, materials with smaller k values should be selected for the multiplication region as much as possible.

2.3.2.2 PIN Photodetectors

Thermal Noise

Thermal noise is the current or voltage fluctuation caused by the random movement of charge carriers in the material at a certain temperature. The thermal noise of a PIN detector satisfies Eq. (2.20).

Loose-Particle Noise

The expression for the scatter-noise current satisfies Eq. (2.14).

Generation-Compound Noise

The mean-square current of PIN-detector generation-compound noise is

$$i_{nf}^2 = \frac{cI^\alpha}{f^\beta}\Delta f \tag{2.31}$$

where i_{nf}^2 is the mean square value of the current of generation-compound noise, the average value of the output current is expressed by I, the operating frequency is expressed by f, α is approximately 2, Δf is the operating bandwidth, and c is the proportionality constant. The values of α, β, and c can be measured empirically and are related to variations in the device surface.

1/f Noise

The mean-square current of the 1/f noise of a PIN detector is expressed as

$$i_{nf}^2 = \frac{cI^\alpha}{f^\beta}\Delta f \tag{2.32}$$

where i_{nf}^2 is the mean square value of the current of 1/f noise; the working frequency is expressed by f, α is approximately 2, between 0.8 and 1.5, Δf is the operating bandwidth, c is a constant of proportionality, and the values of a α, b β, and c can be measured experimentally and are related to the surface undulation state of the device.

2.3.2.3 Photomultiplier Tubes

Signal-to-Noise Ratio

Suppose the photomultiplier is order K, and the average number of electrons collected by its anode is n_a. If each order of multipliers is the same and conforms to the Poisson distribution [110],

$$SNR_a \approx \left|\frac{\eta n_p(\delta - 1)}{\delta}\right|^{1/2} \tag{2.33}$$

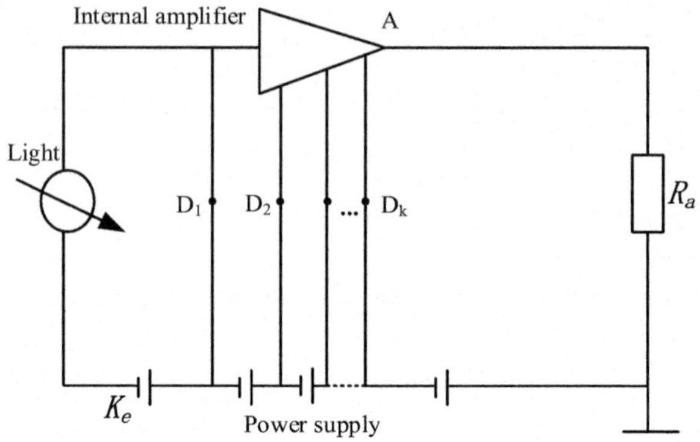

Fig. 2.19 Equivalent-circuit diagram of a photomultiplier

where SNR_a is the signal-to-noise ratio of the photomultiplier tube, ηn_p is the average number of photoelectrons, and δ is the gain. If $\delta \gg 1$, then

$$SNR_a \approx \sqrt{\eta n_p} \tag{2.34}$$

Scattered-Particle Noise Current

The mean square current value of the scattered noise is [110]

$$I_{Npe}^2 = 2q^2 \eta I_p \Delta f \tag{2.35}$$

where the charge is q, and forms a series of current pulses. The spectrum of the current pulse consists of many harmonics; Δf is the harmonic interval, I_p is the rate reached by the average photon, and η is the quantum efficiency. The mean-square current of the scattered-particle noise of the photomultiplier tube is

$$I_{N_a} = 2qI_{pe}\Delta f \left(1 + \frac{1}{\delta 1} + \frac{1}{\delta 2} + \cdots + \frac{1}{\delta_1 \delta_2 \cdots \delta_k}\right) m_k \tag{2.36}$$

where $m_k = \prod_{i=1}^{k} \delta_i$ is the total gain of the photomultiplier and I_{pe} is the photocathode current.

Simplified Photomultiplier Model

The simplified noise model of a photomultiplier consists of a photocathode, several secondary emission multipliers, and an anode. An equivalent circuit diagram of the photomultiplier is shown in Fig. 2.19.

In Fig. 2.19, D1, D2, … Dk represent the first k multipliers, K_e is the photocathode, A is the output anode of the inner amplifier, and R_a is the anode load resistor.

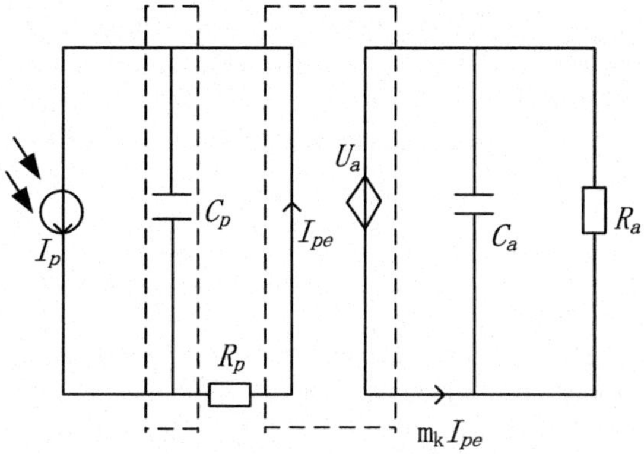

Fig. 2.20 Simplified photomultiplier noise model

The photomultiplier tube can also be viewed as a current source whose anode current is an m_k multiple of the photocathode current:

$$I_a = m_k I_{pe} = \left(\prod_{i=1}^{k} \delta_i\right) I_{pe} \tag{2.37}$$

where I_{pe} is the cathode current and $\prod_{i=1}^{k} \delta_i$ is the total gain of the photomultiplier tube.

In Fig. 2.20 [110], R_a is the anode resistance, I_{pe} is the cathode current, $m_k = \prod_{i=1}^{k} \delta_i$ is the total gain of the photomultiplier, I_p is the average photon-arrival rate, C_a is the anode capacitance, R_p is the cathode resistance, C_p is the cathode capacitance, and U_a is the anode voltage source.

2.3.2.4 Four-Quadrant Detector

Thermal Noise

The thermal noise of the four-quadrant detector satisfies Eq. (2.20).

1/f Noise

The 1/f noise-voltage power-spectrum density in the four-quadrant detector is expressed as follows [111]:

$$S_V(f) = AV^{\beta}/f^{\gamma} \tag{2.38}$$

where $S_V(f)$ is the power spectrum density of the 1/f noise voltage, A is related to the detector material and is a constant, β is the exponential factor, V is the noise voltage, f is the frequency, and γ is the frequency exponent.

In 1969, Hooge proposed his famous empirical formula and mobility fluctuation model:

$$\frac{S_I(f)}{I^2} = \frac{S_R(f)}{R^2} = \frac{S_v(f)}{V^2} = \frac{\alpha_H}{fN} \tag{2.39}$$

where the power spectral density of the noise current is represented by $S_I(f)$, the power spectral density of the resistance is represented by $S_R(f)$, and the power spectral density of the voltage is represented by $S_v(f)$. The Hooge coefficient is α_H; it can be used to measure the 1/f device noise level, where f is the frequency and N is the total number of carriers.

Generation-Compound Noise

When generating four-quadrant detector generation-compound noise, the power spectral density is satisfied as follows [111]:

$$S_I(f) = \sum_{i=1}^{m} \frac{C_i \tau_i}{1 + \left(2\pi f_{\tau_i}\right)^2} \tag{2.40}$$

where $S_I(f)$ is the power spectral density of generation-compound noise, τ_i is the characteristic time constant of generation-compound noise, m is the number of deep energy levels, C_i represents the amplitude of the generation-compound noise component of time constant τ_i, and f_{τ_i} is the frequency of the characteristic time constant.

The characteristic time constant τ_i can be expressed as

$$\frac{1}{\tau_i} = CT^2 \exp(-Ea/kT) \tag{2.41}$$

where k is Boltzmann's constant, C is a constant, Ea is the activation energy of the deep-level impurity, and T is the temperature.

2.3.2.5 Quantum-Dot Detector

Dark Current

The barrier has moving carriers, and the dark current can be calculated from the density of the carriers [84]:

$$I_{dark} = 2evuEA\left(1 + \left(\frac{uE}{v_s}\right)^2\right)^{-1/2}\left(\frac{m_b kT}{2\pi h^2}\right)^{3/2}\exp\left(-\frac{E_{0,micro}\exp(-E/E_0) + E_{0,nano-}\beta E}{KT}\right) \tag{2.42}$$

where e is the electronic charge, v is the electron drift velocity, A is the area of the detector, E is the bias electric-field intensity and can be approximated by dividing the applied voltage by the thickness of the intrinsic region of the detector. μ is the electron mobility, v_s is the maximum moving velocity of an electron, m_b is the effective mass of an electron, k is Boltzmann's constant, T is the temperature, and h is a normalized Planck's constant. $E_{0,micro}$ is the excitation energy of a micron-scale electron transport under zero pressure bias, $E_{0,nano}$ is the excitation energy of a nano-scale electron transport under zero pressure bias, E_0 describes how fast or slow the excitation energy of the micron-scale electron transport varies with the electric field, and β describes how fast or slow the excitation energy of the nano-scale electron transport varies with the electric field.

Noise Gain

The noise gain can be calculated by the ratio of the electron recombination time τ_r to the electron transit time τ_d [84]:

$$g_n = \frac{\tau_r}{\tau_d} \tag{2.43}$$

where g_n is the noise gain.

In general, the electron capture by quantum dots is considered to be symmetric, and the specific calculation form is as follows:

$$\frac{1}{\tau_T} = N_t(4\pi DR_t)\left\{1 - \left[\frac{D}{V_t R_t^2}\tan h\left(\frac{V_t R_t^2}{D}\right)\right]\right\}^{1/2} \tag{2.44}$$

where R_t is the effective radius of the quantum dot, V_t is the capture speed of a quantum dot capturing an electron, τ_T is the capture time of a quantum dot capturing an electron, N_t is the quantum number of points, and D is the band-diffusion coefficient, which is related to electron mobility μ:

$$\mu = \frac{eD}{kT} \tag{2.45}$$

Equations (2.40) and (2.41) are substituted into Eq. (2.39) and sorted to obtain the noise gain, which can be written as

$$g_n = \frac{uE\left(1+\left(\frac{uE}{v_s}\right)^2\right)^{-1/2}}{N_t(4\pi R_t ukT/e)\left\{1-\left[\frac{ukT}{ev_tR_t^2}\tan h\left(\frac{ev_tR_t^2}{ukT}\right)\right]^{1/2}\right\}} \tag{2.46}$$

where E is the bias electric-field intensity of the quantum-dot infrared detector, μ is the mobility of the electron, and v_s is the saturation movement speed of the electron.

Noise

The noise of quantum-dot detectors mainly comes from the electron production–recombination process and can be written as

$$i_n = \sqrt{4eg_nI_d} \tag{2.47}$$

where e is the basic electron charge and g_n is the noise gain, which is approximately equal to the photoconductance gain, and I_d is the dark current.

According to the dark-current and gain-calculation methods, Eqs. (2.39) and (2.43) are substituted into Eq. (2.44) to obtain the formula for calculating quantum-dot infrared detector noise [84]:

$$i_n = \sqrt{\frac{8e^2u^2E^2\left(1+\left(\frac{uE}{v_s}\right)^2\right)^{-1}\left(\frac{m_bkT}{2\pi h^2}\right)^{3/2}exp\left(-\frac{E_{0,micro}\exp(-E/E_0)+E_{0,nano}-\beta E}{KT}\right)}{N_t(4\pi R_t ukT/e)\left\{1-\left[\frac{ukT}{ev_tR_t^2}\tan h\left(\frac{ev_tR_t^2}{\mu KT}\right)\right]^{1/2}\right\}}} \tag{2.48}$$

where the bias electric-field intensity is expressed by E, m_b is the effective mass of the electron, μ is the electron mobility under a zero voltage bias, $E_{0,\ micro}$ is the excitation energy generated with a micron-magnitude electron transmission, and $E_{0,\ nano}$ is the excitation energy generated with a nanometer-magnitude electron transmission. The size will be affected by the change of the electric field, β describes the change of the speed, and v_s represents the maximum speed that can be reached when the electron is moving.

2.3.2.6 Balanced Detector

Thermal Noise

The thermal noise of the balanced detector satisfies Eq. (2.20).

Loose-Particle Noise

The mean-square value of the scattered noise current is expressed as

$$i_{sn}^2 = 2eP_LB \tag{2.49}$$

where e denotes the electron charge and B denotes the noise bandwidth.

2.3.2.7 Double-Balanced Detector

Loose-Particle Noise

The power spectral density of a single sideband of scattered noise is expressed as [112]

$$G_{shot}^2 = 2 \cdot q \cdot (\eta \cdot P + I_d) \cdot \Delta f \tag{2.50}$$

where G_{shot}^2 is the single-sideband noise power spectral density, q is the electron charge, η is the device-response factor, I_d corresponds to the dark current, Δf is the single-sideband effective noise bandwidth, and P is the total received power.

Thermal Noise

The main source of thermal noise is cross-resistance amplification; the noise can be represented as follows [112]:

$$\sigma_{thermal}^2 = (i_{TIA})^2 \cdot \Delta f \tag{2.51}$$

where $\sigma_{thermal}^2$ represents thermal noise, i_{TIA} is the single-side differential input noise current either, and Δf is the single-sideband effective noise bandwidth.

Relative-Intensity Noise

The relative-intensity noise caused by local vibration light can be expressed as [112]

$$\sigma_{RIN} = 2(\eta P_L)^2 \cdot RIN \cdot \Delta f \tag{2.52}$$

where σ_{RIN} represents the relative intensity noise, P_L is the local vibration power, RIN is the relative-intensity noise coefficient, and η is the equipment-response coefficient.

2.4 Summary

In this chapter, specific detectors were analyzed, including the avalanche photodetector, PIN photodetector, photomultiplier, four-quadrant detector, quantum-dot infrared detector, and balanced detector. Their thermal noise, loose-particle noise, generation-compound noise, and 1/f noise were analyzed from the noise-generation mechanism, and the noise model of related detectors was established. This provided a basis for studying the signal quality at the receiving end of a optical-wireless communication system.

In the future, the following work is worthy of further study and discussion.

1. Develop new composite materials with a high carrier mobility, longer lifetime, and diffusion distance, as well as a large photoconductance gain, to improve the response speed and quantum efficiency of the detector.
2. Explore new integration methods to increase the photodetector integration and reduce the number of detectors and search time. In addition, the resolution of the detector-channel wavelength should be enhanced and the cost reduced.
3. Further reduce the carrier concentration, noise, and impact of the junction area and pressure-welding size on the dark current of the detector, and thereby improve the sensitivity of the photodetector.
4. Improve the detector-surface process state, avoid the existence of uneven particles or unnecessary trace impurities in the photosensitive layer, and reduce the 1/f noise of the detector.
5. Improve the detector structure, reduce the composite loss of photogenerated carriers in the dead zone of the device surface, and improve the detector detection efficiency.
6. Ultra-high speed, low-voltage, low-loss lithium-niobate thin-film electro-optic modulators have introduced a new era in the sustainable and rapid development of information technology. Whether future optical-interconnection technology can transmit the massive amounts of information in sufficient time is a fundamental issue concerning the sustainability of information-technology development.

References

1. Ke XZ, Dong K (2022) Optical-wireless communication. Springer Press, Beijing, pp 10–13
2. Pauwels H (1966) Phase and amplitude fluctuations of the laser oscillator. IEEE J Quantum Electron 2(3):54–62
3. Leeson DB (1966) A simple model of feedback oscillator noise. Proc IEEE 54(2):329–330
4. Van DZA (1970) Noise in solid-state devices and lasers. Proc IEEE 58(8):1178–1206
5. Harder C, Katz J, Margalit S et al (1982) Noise equivalent circuit of a semiconductor laser diode. IEEE J Quantum Electron 18(3):333–337
6. Petermann K (1991) Laser diode modulation and noise. Sold and distributed in the U.S.A. and Canada by Kluwer Academic Publishers, pp 56–60

7. Kuo CY (1992) Fundamental second-order nonlinear distortions in analog AM CATV transport systems based on single frequency semiconductor lasers. J Lightwave Technol 10(2):235–243
8. Bihan JL, Yabre G (1994) FM and IM intermodulation distortions in directly modulated single-mode semiconductor lasers. IEEE J Quantum Electron 30(4):899–904
9. Chai YJ, Yang ZX (1997) Research on influence of laser phase noise on characteristics of high-speed IM/DD fiber communication system. J Commun 18(7):1–5
10. Fukuda M, Hirono T, Kurosaki T et al (2002) 1/f noise behavior in semiconductor laser degradation. IEEE Photonics Technol Lett 5(10):1165–1167
11. Lawrence JS, Kane DM (2002) Nonlinear dynamics of a laser diode with optical feedback systems subject to modulation. IEEE J Quantum Electron 38(2):185–192
12. Yu BL, Zhen SL, Zhu J et al (2006) Experimental study on low noise fiber laser. Acta Opt Sin 26(2):217–220
13. Domenico GD, Schilt S, Thomann P (2010) Simple approach to the relation between laser frequency noise and laser line shape. Appl Opt 49(25):4801–4807
14. Liu JH (2012) Influence of laser phase noise on performance of coherent fiber communication system. J Xi'an Univ Posts Telecommun 17(5):25–28
15. Mahmoud SWZ, Mahmoud A, Ahmed M (2015) Noise performance and nonlinear distortion of semiconductor laser under two-tone modulation for use in analog CATV systems. Int J Nume. Model Electron Networks Devices Fields 29(2):280–290
16. Chen D, Ke XZ, Zhang L (2016) Intermodulation distortion characteristics of laser in turbulent channel. Acta Photonica Sin 45(2):99–103
17. Chen JN (2016) Experimental study on key technologies of subcarrier modulation non-optical domain extrinsic error detection in optical-wireless communication. Xi'an University of Technology
18. Song ZY, Yao GB, Zhang LL et al (2017) Influence factors of phase noise in single frequency fiber laser. Infrared Laser Eng 46(3):90–93
19. Bai Y (2019) Characterization method of laser linewidth at 2m band and development and application of single longitudinal mode thulium doped fiber laser. Beijing Jiaotong University
20. Du YC, Zhang R, Wang LS et al (2022) Chaotic synchronization of distributed Bragg reflection semiconductor laser driven by common noise. Acta Opt Sin 42(23):2314003
21. Shun Y (1972) Photomultiplier tubes for laser ranging. Laser Infrared 2(8):35–37
22. Li G (1989) Polarization effects of grating monochromator and photomultiplier tubes. J Qufu Normal Univ 2(1):34–40
23. Zhou RF (2000) Outlook of photomultiplier tubes. Optoelectron Technol 4(2):84–89
24. Guo CL, Zeng D, Li J et al (2004) Noise modeling of photomultiplier tubes. Nucl Electron Detect Technol 2(1):117–120
25. Chowdhuri MB, Ghosh J, Manchanda R et al (2014) Measurement of spatial and temporal behavior of H(α) emission from Aditya tokamak using a diagnostic based on a photomultiplier tube array. Rev Sci Instrum 85(11):101–108
26. Maltseva Y, Emanov FA, Petrenko AV et al (2015) Distributed beam loss monitor based on the Cherenkov effect in an optical fiber. Phys Usp 58(5):516–519
27. Chen XQ, Tang QY, Wang XH (2016) Development of small radiation measuring probe based on silicon photomultiplier tube. Mech Eng Autom 5(2):179–180
28. Zhao T, Yu P, Miao Q et al (2017) One-dimensional single-photon position-sensitive silicon photomultiplier and its application in Raman spectroscopy. Opt Express 25(19):22820–22828
29. Takahashi M, Inome Y, Yoshii S et al (2018) A technique for estimating the absolute gain of a photomultiplier tube. Nucl Instrum Methods Phys Res 894(10):1–7
30. Elorrieta M, Classen L, Reubelt J et al (2019) Characterisation of the Hamamatsu R12199-01 HA MOD photomultiplier tube for low temperature applications. J Instrum 14(3):3015–3021
31. Sun X, Kong M, Alkhazragi O et al (2020) Non-line-of-sight methodology for high-speed optical-wireless communication in highly turbid water. Opt Commun 461(1):125264–125271
32. Ning J, Gao G, Liang ZJ et al (2021) Adaptive receiver control for reliable high-speed underwater optical-wireless communication with photomultiplier tube receiver. IEEE Photonics J 13(4):1–7

33. Wang HD (2022) Analysis of current status and development trend of vacuum ultraviolet photomultiplier tubes. Vac Electron Technol 4(2):23–28
34. Yang WZ (1980) Current status and future of long-wavelength detectors for optical communications. Semicond Photonics 2(3):45–61
35. Watkins SE, Zhang CZ, Walser RM et al (1990) Laser-induced failure in biased silicon avalanche photodiodes. NIST Spec Publ 12(1):95–104
36. Li GZ (1996) Analysis and optimum design on GexSi1-x/Si strained-layer superlattices avalanche photodetector. Acta Opt Sin 16(6):17–24
37. Ng JS, Tan CH, David JPR et al (2005) Effect of impact ionization in the InGaAs absorber on excess noise of avalanche photodiodes. IEEE J Quantum Electron 41(8):1092–1096
38. Guo JP, Liao CG, Wei ZG (2009) Effect of reverse bias voltage on the penetration characteristics of avalanche photodiodes in Geiger mode. Opt Photonics 7(5):9–11
39. Chitnis D, Collins S (2010) A flexible compact readout circuit for SPAD arrays. Proc Spie 7780(1):101–108
40. Zhang Y, Liu BQ, Zhou B et al (2011) Signal to noise ratio analysis of "cat's eye" target active detection system based on avalanche photodiode. Laser Infrared 41(11):1240–1243
41. Guo S, Ding QX, Yang Y (2012) Research on noise suppression technology of avalanche photodetectors. Electro-Optics Control 19(3):69–73
42. Adamo G, Agro D, Stivala S et al (2013) Measurements of silicon photomultipliers responsivity in continuous wave regime. IEEE Trans Electron Devices 60(11):3718–3725
43. Youn JS, Lee MJ, Park KY et al (2014) SNR characteristics of 850-nm OEIC receiver with a silicon avalanche photodetector. Opt Express 22(1):900–907
44. Chen HT, Verbist J, Verheyen P et al (2015) High sensitivity 10Gb/s Si photonic receiver based on a low-voltage waveguide-coupled Ge avalanche photodetector. Opt Express 23(2):815–822
45. Goykhman I, Sassi U, Desiatov B et al (2016) On-chip integrated, silicon-graphene plasmonic Schottky photodetector with high responsivity and avalanche photogain. Nano Lett 16(5):3005–3013
46. Yin D, Yang X, He T et al (2017) InGaAs/InAlAs avalanche photodetectors integrated on silicon-on-insulator waveguide circuits. J Opt Technol 84(5):350–357
47. Juki T, Brandl P, Zimmermann H (2018) Determination of the excess noise of avalanche photodiodes integrated in 0.35-μm CMOS technologies. Opt Eng 57(4):44101–44112
48. Farrell AC, Meng X, Ren D et al (2019) InGaAs-GaAs nanowire avalanche photodiodes toward single-photon detection in free-running mode. Nano Lett 19(1):582–590
49. Zhou X, Tan X, Lv Y et al (2020) High-uniformity 1×64 linear arrays of silicon carbide avalanche photodiode. Electron Lett 56(17):895–897
50. Eid M, Urooj S, Alwadai NM et al (2021) AlGaInP optical source integrated with fiber links and silicon avalanche photo detectors in fiber optic systems. Indonesian J Electric Eng Comput Sci 23(2):847–854
51. Li ZB, Li YX, Ma X et al (2022) Measurement and suppression method of excess noise in avalanche photodiodes. Infrared Technol 44(4):343–350
52. Lin YF, Sun YR, Lu ZL (1986) Design of double layer anti-reflection film for silicon PIN photodetector. J Hangzhou Univ (04):435–443
53. Zhang JM, Conn DR (1992) State-space modeling of the PIN photodetector. J Lightwave Technol 1(5):10–18
54. Zhang YG, Cheng Z (1995) SMA coaxially packaged high speed photodetectors. Semicond Optoelectron 2(1):23–30
55. Jutzi M, Berroth M, Woehl G et al (2004) SiGe PIN photodetector for infrared optical fiber links operating at 1.25 Gbit/s. Appl Surf Sci 224(1):170–174.
56. Giannopolous AV, Kasten AM, Hardy N et al (2010) Position sensing using an integrated VCSEL and PIN photodetector microsystem. Proc SPIE—Int Soc Opt Eng 76(3):15–25
57. Shao HM, Zhang SL, Xie S et al (2011) Research of AlGaN/GaN PIN solar-blind ultraviolet photodetector with back-illumination. J Optoelectron Laser 22(7):984–986
58. Guan MJ, Zhao Dong'e (2012) Noise study based on PIN-type photoelectric conversion circuit. Electron Test 2(5):35–38

59. Wu F, Li HZ, Du Ch M et al (2014) Characterization of optical threshold of PIN detector in free optical communication. Mod Electron Technol 37(5):12–15
60. Chaudhari P, Singh A, Topkar A et al (2015) Fabrication and characterization of silicon based thermal neutron detector with hot wire chemical vapor deposited boron carbide converter. Nucl Instrum Methods Phys Res 779(11):33–38
61. Rouse C, Zeller JW, Efstathiadis H et al (2016) Development of low dark current SiGe near-infrared PIN photodetectors on 300 mm silicon wafers. Opt Photonics J 6(5):61–68
62. Xu ZP, Jin CG, Yu Q et al (2018) Lidar reference light detection with PIN detectors. Infrared Laser Eng 47(10):267–273
63. Shu B, Yan K, Zhong S (2019) RLS denoising for photocurrent detection circuit of PIN-TIA photodetector. Mod Electron Tech 24(1):1–4
64. Wang N, Zhao BQ, Wang S et al (2021) Optoelectronic monolithic integration of PIN detector and transimpedance amplifier. Infrared Laser Eng 50(9):307–312
65. Xiao F, Han Q, Ye H et al (2022) InP-based high-speed monolithic PIN photodetector integrated with an MQW semiconductor optical amplifier. Jpn J Appl Phys 61(1):12005–12013
66. Shun Y (1979) Precision active tracking technology using four-quadrant detector. Laser Infrared S1:25–38
67. Feng LG, Deng RL (1996) Discussion on some problems in four-quadrant optoelectronic tracking technology. Infrared Laser Eng 25(1):12005–12013
68. Wang X, Xiu QS, Liu G et al (2020) A method for improving the detection accuracy of the spot position of the four-quadrant detector in a free space optical communication system. Sensors 20(24):7164–7173
69. Kuang ZF, Feng QB, Feng JY et al (2004) Characterization of a four-quadrant detector for laser collimation. Opt Technol 12(4):387–389
70. Wu QB, Chen HQ, Liu Y (2005) Research on noise detection technology of four-quadrant infrared photodetector. Opt Photonics 28(5):28–31
71. Zhao X, Tong SF, Jiang HL (2010) Characterization of four-quadrant detectors. Opt Precision Eng 18(10):2164–2170
72. Yang GB, Zhang ZF, Zhai YS et al (2013) Study on the effect of dead zone on the detection range and sensitivity of four-quadrant detectors. Adv Lasers Optoelectron 50(6):156–161
73. Schalkwyk SV, Meyer WE, Nel JM et al (2014) Implementation of an AlGaN-based solar-blind UV four-quadrant detector. Phys B 439(1):93–96
74. Tian YM, Wang YZ, Jiang YJ et al (2016) The design of a focal optical system of beacon tracking system in inter-satellite laser communication. Guidance Fuze 11(4):54–57
75. Grajower M, Desiatov B, Mazurski N et al (2018) Integrated on-chip silicon plasmonic four quadrant detector for near infrared light. Appl Phys Lett 113(14):143103–143110
76. Wang X, Su X, Liu G et al (2019) Investigation of high-precision algorithm for the spot position detection for four-quadrant detector. Optik 203(6):163941–163950
77. Liu PF, Song CL, Ma YC (2020) A four-quadrant detector design for dim light detection. Foreign Electron Meas Technol 39(5):99–104
78. Safi H, Dargahi A, Cheng J (2021) Beam tracking for UAV-assisted FSO links with a four-quadrant detector. IEEE Commun Lett 25(12):3908–3912
79. Sheng HM, Wang JK, Gao Q et al (1999) Noise characterization of GaAs/AlGaAs quantum well infrared detectors. Semicond Technol 5(1):26–29
80. Zhao YL, Li XJ, Cai DM et al (2006) Dark current characterization of AlGaAs/GaAs multi-quantum well infrared detectors. Infrared Laser Eng 15(5):68–71
81. Hill CJ, Soibel A, Keo SA et al (2010) Mid-infrared quantum dot barrier photodetectors with extended cutoff wavelengths. Electron Lett 46(18):1286–1271
82. Chien LH (2011) Quantum dot infrared photodetectors based on structures with potential barriers: modeling and optimization. Phys E Low Dimensional Syst Nanostruct 14(5):11–16
83. Gustafsson O, Karim A, Berggren J et al (2012) Photoluminescence and photoresponse from InSb/InAs-based quantum dot structures. Opt Express 20(19):21264–21271
84. Liu HM, Yang CH, Liu X et al (2013) Noise characterization of quantum dot infrared detectors. Phys Lett 62(21):454–459

85. Kim JO, Ku Z, Krishna S et al (2014) Simulation and analysis of grating-integrated quantum dot infrared detectors for spectral response control and performance enhancement. J Appl Phys 115(16):101–112
86. Srinivasan T, Mishra P, Jangir SK et al (2015) Molecular beam epitaxy growth and characterization of silicon—doped InAs dot in a well quantum dot infrared photo detector (DWELL-QDIP). Infrared Phys Technol 70(1):6–11
87. Karow MM, Munnelly P, Heindel T et al (2016) On-chip light detection using monolithically integrated quantum dot micropillars. Appl Phys Lett 108(8):110–121
88. Kim Y-H (2017) Effects of imposed conditions and two types of electron transport on the noise of quantum dot infrared detectors. J Photonics 46(4):81–86
89. Wang Z (2018) Research on dark current modeling of quantum dot-based infrared detectors. Electron Des Eng 26(23):89–93
90. Baira M, Aljaghwani M, Salem B et al (2019) Investigation of GeSn/Ge quantum dots' optical transitions for integrated optics on Si substrate. Results Phys 12(1):1732–1736
91. Gansen EJ, Nickel TB, Venner JM et al (2020) Sources of 1/f noise in QDOGFET single-photon detectors. Phys E 118(1):61–70
92. Gao XM, Xing T, Gao WQ et al (2022) Optical communication technology 04):37–45
93. Dai YH, Ai Y, Xiao W et al (2015) Research on balanced detectors for high-speed coherent optical communication. J Photonics 44(1):179–185
94. Guo LR, Hu YH, Li Z et al (2015) Characterization of micro-Doppler effect laser balanced outlier detection. Laser Infrared 45(3):247–251
95. Shi QY, Ai Y, Liang HS et al (2016) Research and testing of balanced detectors in coherent optical communication. Sci, Technol Eng 16(16):207–211
96. Munshid MA (2017) Design and implementation of heterodyne detection based on photonic crystal fiber sensor. Adv Nat Appl Sci 11(8):639–644
97. Yong L, Yang A, Peng G et al (2018) Chromatic dispersion estimation based on heterodyne detection for coherent optical communication systems. Opt Eng 57(1):11–16
98. Zhang HF, Su B, Jones David R et al (2019) Design of a balanced detector in a high-speed terahertz time-domain spectroscopy system. J Terahertz Sci Electron Inf 17(5):755–759
99. Hu Q, Borkowski R, Schuh K et al (2021) Optical field reconstruction of real-valued modulation using a single-ended photoreceiver with half-symbol-rate bandwidth. J Lightwave Technol 39(4):1194–1203
100. Huai YC (2013) Simulation research of double balanced heterodyne laser detection system. Software 34(04):132–134
101. Gao L, Wang CH, Li YC et al (2010) Balanced detection and reception experiment of 1.55 um coherent laser wind radar. J Photonics 39(6):1064–1069
102. Gao L, Wang CH, Li YC et al (2010) Theoretical analysis of polarization mixing error in double-balanced outlier detection. Infrared Laser Eng 39(3):422–426
103. Zhu J (2013) Double balanced detection of optical fiber acoustic sensor based on Sagnac interferometer. Electron Sci Technol 2(1):22–31
104. Huai YC (2013) Simulation and study for dual-balanced heterodyne laser detection system based on the Opti System. Software 34(4):132–134
105. Zheng F, Zhu G, Liu X et al (2015) Double balanced differential configuration for high speed InGaAs/InP single photon detector at telecommunication wavelengths. Optoelectron Lett 11(2):21–30
106. Peng C, Zhao CM, Zhang HY et al (2015) Simulation study on double-balanced coherent detection technology of laser wind radar. Aerosp Return Remote Sens 36(6):55–63
107. Li Y, Zhang S, Zhou J et al (2016) Research on the application of double-balanced detection in laser velocimetry. Intense Laser Part Beam 28(10):22–26
108. Huang J, Zhu W (2017) A new method for detecting ground fault detection in DC system based on double balanced bridge. Electric Meas Instrum 54(8):75–79
109. Liu HX, Zhou B, Gao YC (2018) Research on noise characteristics and influencing factors of APD detection system. Laser Technol 42(06):862–867

110. Guo CL, Zeng D, Li J et al (2004) Noise model of photomultiplier tube. Nucl Electron Detect Technol (02):117–120
111. Zhou HW (2013) Design and Application research of four-quadrant detector noise test system. Xidian University, Xi'an
112. Chen M (2018) Research on channel noise in optical-wireless coherent communication and its influence on coherent detection performance. Xi'an University of Technology, Xi'an

Chapter 3
Atmospheric Turbulence

When laser signals are transmitted through atmospheric channels, they are subject to interference from atmospheric attenuation and turbulence, which affects the reliability and stability of optical-wireless communication. This chapter first introduces the generation of atmospheric turbulence and then introduces the impact of atmospheric attenuation on light transmissions, including atmospheric absorption and atmospheric scattering. Finally, the impact of atmospheric turbulence on laser propagation is studied from four aspects: light-intensity fluctuation, beam drift, beam expansion, and arrival-angle fluctuation.

3.1 Turbulence Formation

As shown in Fig. 3.1, turbulence is initially defined as an irregular motion in a fluid [1]. The formation of clouds in the atmosphere and the movement of water in rivers reflect this phenomenon. There are many turbulent flows in daily life, such as the blood flow in our own veins and arteries, as well as the flow of natural gas, oil, and water from the exploration end to the interior of cities.

Although turbulence is irregular, it has the following three characteristics [2]:

1. Irregularity and unpredictability: Turbulence is irregular in both space and time and exhibits unpredictable random patterns.
2. Statistical characteristics: A certain statistical order appears in the irregularity of the turbulent motion. The averages and correlations are consistent and predictable.
3. Mixing and enhanced diffusion: The fluid undergoes complex and circuitous paths, resulting in the extensive mixing of different parts of the fluid. This mixing significantly enhances the diffusion and increases the transport of momentum, energy, heat, and other advections.

X. Ke and C. Ke, *Noise Models in Optical-Wireless Communication Systems*,
Optical Wireless Communication Theory and Technology,
https://doi.org/10.1007/978-981-97-7550-7_3

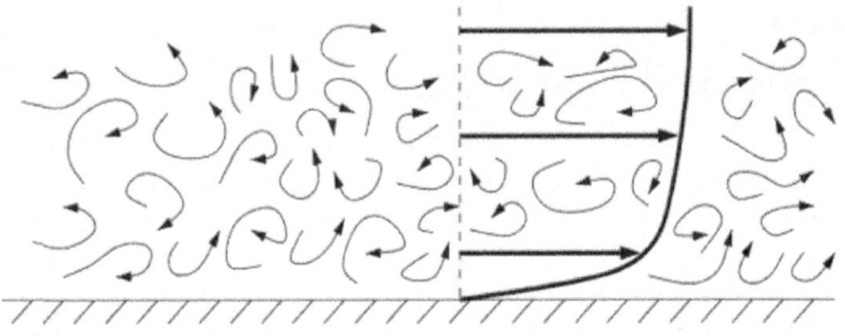

Fig. 3.1 Schematic diagram of irregular turbulent motion [2]

advections.

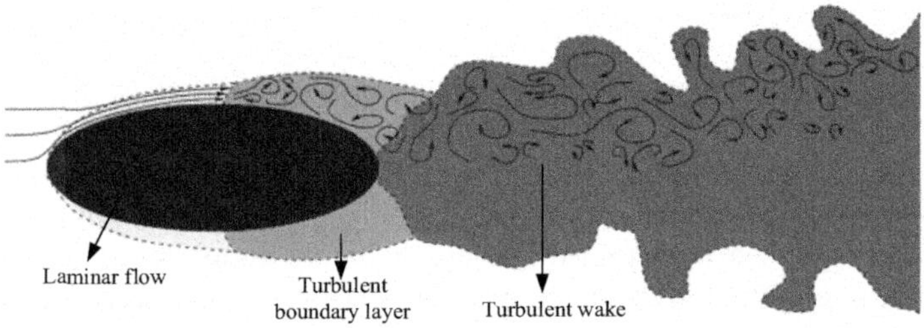

Fig. 3.2 Schematic diagram of a fluid motion state [3]

Figure 3.2 shows the different motion states during the fluid motion. The fluid exhibits a laminar flow at its upstream edge, a turbulent boundary layer downstream, and leaves a turbulent wake.

At the end of the nineteenth century, Reynolds carefully observed the transition from laminar to turbulent flows (see Fig. 3.3) in a series of experiments. He slowly dripped static water from a water tank into glass tubes of different diameters and then changed the flow rate of the fluid inside the tubes. He then observed the flow inside the different tubes. Reynolds believed that [1] a dimensionless critical value might exist. Once the critical value was exceeded, the laminar flow would produce a "bending" motion. The above experiments confirmed the existence of a critical value.

Reynolds used similarity theory to define a dimensionless quantity called the Reynolds number [1]:

$$\mathrm{Re} = \frac{\rho \upsilon L}{\mu} = \frac{\upsilon L}{\nu}, \tag{3.1}$$

where L represents the feature length (m), υ represents the characteristic velocity (m/s), ρ indicates the density of the fluid, μ represents the viscosity coefficient of a

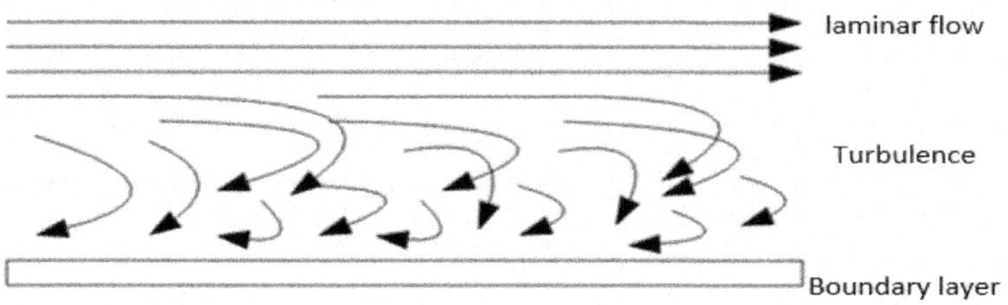

Fig. 3.3 Schematic diagram of laminar and turbulent conversions [1]

fluid, with dimensions of $MT^{-1}L^{-1}$, and $\nu = \frac{\mu}{\rho}$ represents the coefficient of motion viscosity, with dimensions M^2T^{-1}.

For smaller Reynolds values, the viscosity dominates and the flow is laminar, whereas for larger Reynolds values, the inertial terms dominate and the flow becomes more complex and turbulent. The larger the Reynolds number, the more complex the flow.

According to Eq. (3.1), when the speed of atmospheric flow υ exceeds a certain value, turbulence may occur. The first turbulence that appears is a huge vortex with a scale equivalent to the overall characteristic scale of the fluid. Its scale is L_0, defined as the outer turbulence scale, which is equivalent to the height of the airflow above the ground, usually in the range of several meters to hundreds of meters. When the turbulence-motion speed becomes increasingly large, the large turbulence vortices become unstable and split into smaller vortices that remove energy from them, and the corresponding internal Reynolds number decreases. This process continues. Richardson calls it the cascade property of turbulence [4].

Richardson's turbulent-energy cascade theory suggests that turbulence is composed of turbulence vortices of different scales, which gradually transport energy from the large-scale vortices to the small-scale vortices. The energy of the initial large-turbulence vortex comes from solar radiation and various meteorological factors. The large-turbulence vortex cannot maintain a stable state; therefore, it splits into secondary small-turbulence vortices, which split into lower-level small-turbulence vortices for the same reason [5].

Although such a cascade relationship exists, the splitting of turbulent vortices does not stop, owing to factors such as fluid viscosity. When the energy of a turbulent vortex is equal to its kinetic energy, it no longer splits. The smallest turbulent vortex has an internal scale l_0, which is generally several millimeters in magnitude.

From the overall turbulent-motion process, owing to the internal-friction force converting the flow energy into thermal-energy loss, energy must be used to compensate for the dissipation loss to maintain this turbulent motion. T^* represents the energy converted from the average kinetic energy to turbulent kinetic energy per unit time and ε represents the average dissipation rate. Therefore, the energy held by the turbulence is E'; the change over time can be expressed as [4]

$$\frac{dE'}{dt} = T^* - \varepsilon. \tag{3.2}$$

Equation (3.2) is the turbulent-energy balance equation. If $T^* = \varepsilon$, then E' is a constant, and the turbulence is in a stable state. When the fluctuation energy increases over time, i.e., $T^* > \varepsilon$, turbulence will continue to develop. When it decreases, the turbulence will gradually tend to disappear.

3.2 Atmospheric Refractive-Index Fluctuations

The atmospheric refractive index is a random field that changes slowly over time [2]. The random field of the atmospheric refractive index can be represented by $n(R, t)$, where t represents the time position and $R = (x, y, z)$ represents the space position. The change in the atmospheric refractive index with time t is relatively slow, and the function $n(R)$ of the spatial position R is typically used to represent the sample function of the random field.

$\langle n(R)\rangle$ is the mean of the sample function $n(R, t)$. When $\langle n(R)\rangle$ is independent of the spatial position R, the statistics are uniform. $B_n(R)$ is the covariance function of the stochastic process, which is independent of the specific spatial location and is related only to the location interval $R = R_1 - R_2$. The expression for $B_n(R)$ is

$$B_n(R) = \big\langle [n(R_1) - \langle n(R_1)\rangle][n^*(R_2) - \langle n^*(R_2)\rangle]\big\rangle, \tag{3.3}$$

where $\langle\ \rangle$ represents the statistical average, R_1 and R_2 represent the positions of any two points in space, $n(R)$ is a function of the spatial position R, and $n^*(R)$ represents a conjugate-complex function of $n(R)$.

$n(R)$ is locally statistically uniform, and the structural function $D_n(R)$ is generally used to describe the statistical characteristics of locally statistically uniform random fields [2]. The expression is as follows:

$$D_n(R) = \big\langle n(R_1 + R) - n(R_1)^2\big\rangle, \tag{3.4}$$

where $D_n(R)$ is a structural function whose relationship with the covariance function $B_n(R)$ is as follows:

$$D_n(R) = 2[B_n(0) - B_n(R)]. \tag{3.5}$$

Because $R = R_1 - R_2$, when $R_1 = R_2$, $B_n(R)$ becomes $B_n(0)$.

The power-spectral density $\Phi_n(K)$ and covariance function $B_n(R)$ form a Fourier-transform pair; thus, they contain the same information, but in different forms. The relationship between the two is as follows:

$$B_n(R) = \int e^{-iKR}\Phi_n(K)dK. \tag{3.6}$$

$$\Phi_n(K) = \left(\frac{1}{2\pi}\right)^3 \iiint_{-\infty}^{\infty} e^{-iKR}B_n(R)d^3R, \tag{3.7}$$

where $K = (\kappa_x, \kappa_y, \kappa_z)$ represents the vector wavenumber (rad/m) and $\kappa = |K|$ modulus also represents the vector wavenumber (rad/m). κ is also known as the spatial wavenumber; it is a reciprocal with the spatial scale l; that is $\kappa \sim 1/l$. e^{-iKR} is a complex function. Because the vector distances of $B_n(R)$ and $D_n(R)$ can be simplified as scalar distances $R = |R_1 - R_2|$, the obtained values $\Phi_n(\kappa)$ can be expressed as

$$\Phi_n(\kappa) = \frac{1}{2\pi^2\kappa}\int_0^{\infty} B_n(R)\sin(\kappa R)RdR. \tag{3.8}$$

The relationship $\Phi_n(\kappa)$ with $D_n(R)$ is derived from this:

$$\Phi_n(\kappa) = \frac{1}{4\pi^2\kappa^2}\int_0^{\infty} \frac{\sin(\kappa R)}{\kappa R}\frac{d}{dR}\left[R^2\frac{d}{dR}D_n(R)\right]dR. \tag{3.9}$$

3.3 Atmospheric Refractive-Index Structure Constant

Owing to the turbulent motion in the atmosphere, all properties related to flowing gases, such as temperature, refractive index, and aerosol-particle distribution, undergo turbulent mixing. Because the atmospheric refractive index is a function of temperature, the characteristics of the refractive-index field are closely related to the characteristics of the temperature field. The turbulence caused by temperature fluctuations is called optical turbulence, and the metric used to measure the intensity of this refractive-index fluctuation is called the atmospheric refractive-index structure constant C_n^2. In the case of uniform isotropic turbulence, it is defined as [6]

$$C_n^2 = \left\langle \left[n(\vec{x}) - n(\vec{x}+\vec{r})\right]^2 \right\rangle r^{-2/3},\ l_0 \ll r \ll L_0, \tag{3.10}$$

where n is the atmospheric refractive index, $\vec{x}$ and $\vec{r}$ are position vectors, r is the size of vector $\vec{r}$, and l_0 and L_0 are the inner and outer turbulence scales, respectively. C_n^2 is not a true constant, but is related to the atmospheric conditions and altitude.

The boundary-layer turbulence is mainly affected by surface parameters, such as temperature, humidity, pressure, and water vapor. A change in weather causes

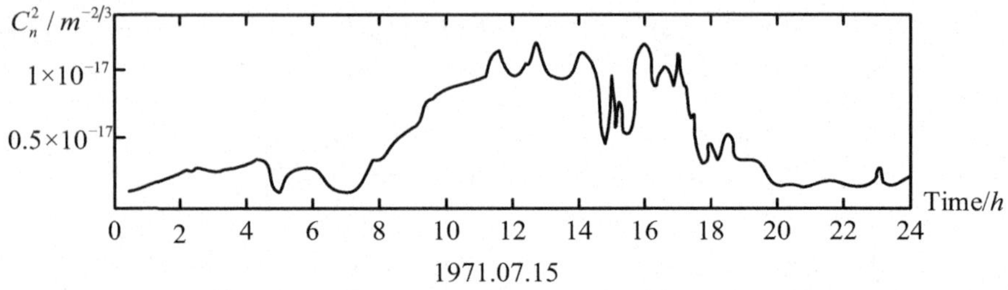

Fig. 3.4 Characteristics of the atmospheric structure constant over a 24-h period [6]

a change in the boundary-layer turbulence. Therefore, according to data from the Rocky Mountains near Boulder, Colorado, 2 m above the ground, a day–night change diagram was drawn, as shown in Fig. 3.4.

From Fig. 3.4, it can be observed that the turbulence at night and in the morning is very weak. As the sun rises, C_n^2 rapidly increases. When the sun sets, it significantly decreases again. The typical measurement value is $10^{-17}\,\text{m}^{-2/3} \prec C_n^2 \prec 10^{-12}\,\text{m}^{-2/3}$.

Hufnagel summarized a $C_n^2(h)$ empirical formula suitable for the 3–24-km range based on measured data [7]:

$$C_n^2(h) = 0.0059(\frac{v}{27})^2(10^{-5}h)^{10}e^{-h/100} + 2.7\times10^{-16}e^{-h/1500} + Ae^{-h/100}, \tag{3.11}$$

where h is the height (m); v is the wind speed at height h (m/s); A is the standard $C_n^2(h)$ value when on the ground, that is $C_n^2(0)$; the unit is $\text{m}^{-2/3}$.

In Eq. (3.11), v is often expressed as

$$v = [\tfrac{1}{15}\text{km}\int_{5\,\text{km}}^{20\,\text{km}} v^2(h)dh]^{1/2}, \tag{3.12}$$

where v refers to the root-mean-square wind speed between 5 and 20 km in height. Hufnagel believed that v followed a normal distribution. He presented typical v data indicating that the mean value was 27 m/s and the standard deviation was 9 m/s. The commonly used Burton wind-speed model is represented as [7]

$$v(h) = \omega_s h + v_g + 30\exp[-(\frac{h-9400}{4800})^2], \tag{3.13}$$

where v_g is the wind speed near the Earth's surface and ω_s is the rotational speed of a satellite, relative to an observer on the Earth's surface.

In practical applications, the v values of A in the International Telecommunication Union Radiocommunication Sector (ITU-R) model are often considered to have three

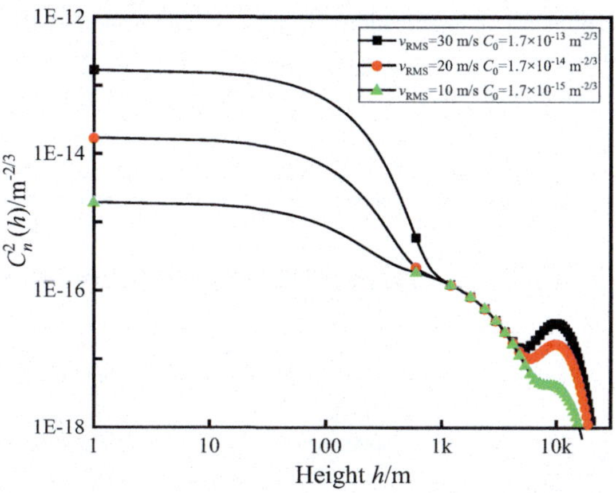

Fig. 3.5 Distribution of C_n^2 atmospheric structure constants with height h under different wind speeds v_{RMS} and near-ground atmospheric structure constants [8]

typical values [7]:

$$\begin{cases} A = 1.7 \times 10^{-15}\,\mathrm{m}^{-2/3}, v = 10\,\mathrm{m/s} \\ A = 1.7 \times 10^{-14}\,\mathrm{m}^{-2/3}, v = 20\,\mathrm{m/s} \\ A = 1.7 \times 10^{-13}\,\mathrm{m}^{-2/3}, v = 30\,\mathrm{m/s} \end{cases}. \tag{3.14}$$

Based on the typical values mentioned above and the ITU-R model [8], Fig. 3.5 summarizes the analysis of the variation of the atmospheric structure constant with height.

From Fig. 3.5, it can be seen that the atmospheric structure constant $C_n^2(h)$ is the largest at the ground surface, and is mainly $C_n^2(0)$ influenced by the ground surface [8]. As the altitude increases, the atmospheric refractive-index structure constant gradually decreases; however, there is a rebound followed by a rapid decrease at an altitude of 10 km. This is the result of the combined influence of atmospheric absorption, the atmospheric structure constant at the surface, and the wind speed at high altitudes. At altitudes of 1–4 km, the atmospheric structure constant is basically unaffected by wind speed and surface conditions $C_n^2(0)$; elevations above 10 km are mainly influenced by wind speed.

3.4 Atmospheric-Turbulence Power-Spectrum Model

The power-spectrum model of atmospheric refractive-index fluctuations varies in different turbulent environments. To further study the light-transmission characteristics of atmospheric turbulence, scholars have proposed various atmospheric refractive-index power-spectrum models suitable for different conditions, as listed in Table 3.1.

Table. 3.1 Classification diagram of atmospheric-turbulence refractive-index power-spectrum models

Turbulence type	Power-spectrum type	Power spectrum	Influence of internal scale	Influence of external scale
Kolmogorov turbulence (power law value is 11/3)	Basic spectrum	Kolmogorov spectrum	Not considered	Not considered
	Mathematical-derivation spectrum	Tatarskii spectrum	Considered	Not considered
		Von Karman spectrum	Considered	Considered
	Physical-analysis spectrum	Greenwood–Tarazano spectrum	Considered	Considered
		Hill spectrum	Considered	Considered
		Andrews spectrum	Considered	Considered
Non-Kolmogorov turbulence (power law value is between 3 and 5)	Basic spectrum	Non-Kolmogorov spectrum	Not considered	Not considered
	Mathematical-derivation spectrum	Generalized modified non-Kolmogorov spectrum	Considered	Considered

The atmospheric refractive-index power-spectrum model has evolved from the initially assumed Kolmogorov spectrum, with an outer scale of $L_0 \rightarrow \infty$ and an inner scale of $l_0 = 0$, to the von Karman spectrum, which reflects the inner and outer scale characteristics, and the more accurate universal-Hill spectrum model. The development of these models, as well as the non-Kolmogorov spectrum [9], which is currently the most in line with actual atmospheric characteristics, also marks the continuous improvement of turbulence-theory research.

3.4.1 Kolmogorov Spectrum

The turbulence power spectrum is a function that describes the energy–density distribution at each wavenumber scale in a turbulent system. It can be used to characterize the contribution of turbulent energy at different length scales and energy-cascade processes [10]. From a physical perspective, the turbulence power spectrum can be used to represent the energy contribution and distribution at various wavenumber scales. In addition, it can help researchers understand the information transfer in the energy-cascade process; that is, how much energy is transferred during the transformation of large-scale vortex motion into small-scale vortices, thus playing a core role in the study of turbulence phenomena [11].

In 1941, Kolmogorov proposed three assumptions about the equilibrium state in uniform turbulence at a sufficiently high Reynolds number and sufficiently small scale, and thus established the basis of classic turbulence theory [10]. The first hypothesis is that small-scale turbulent motion is statistically isotropic; the second hypothesis is that the statistical properties of small-scale turbulent motion are determined by the energy-dissipation rate ε. The universal form uniquely determines the kinematic viscosity v, and this scale range is also known as the universal-equilibrium range. The third hypothesis is that large-scale turbulent-motion statistics have a unique ε-determined general form, independent of v. Turbulence that simultaneously satisfies these three assumptions is called Kolmogorov turbulence [10].

The Kolmogorov theory of locally isotropic turbulence allows for nonuniformity and anisotropy at large scales. It is assumed that as energy cascades towards smaller scales, the directional effects generated at larger scales become weaker, making the motion of sufficiently small eddies statistically uniform, isotropic, and independent of specific energy-generation mechanisms.

As shown in Fig. 3.6, according to the Kolmogorov theory [10], turbulence can be divided into input, inertia, and dissipation regions, based on the different sizes of spatial wavenumbers κ. In the figure, l_0 and L_0 represent the inner and outer turbulence scales, respectively. The region of $\kappa < 2\pi/L_0$ is the input region, where $\Phi_n(\kappa)$ is usually anisotropic. The area of $2\pi/L_0 < \kappa < 2\pi/l_0$ is the inertial zone, and the shape of $\Phi_n(\kappa)$ is determined by the physical laws that constrain the breaking of large turbulent vortices into smaller ones. When the turbulence κ continues to increase to $\kappa > 2\pi/l_0$, the turbulent region enters the dissipation zone, where the dissipation of turbulent energy far exceeds the generated kinetic energy. Therefore, when entering the dissipation zone, the speed of the decrease in $\Phi_n(\kappa)$ increases significantly.

Kolmogorov proposed that when r is less than L_0, the turbulence is locally statistically uniform and isotropic [10]. In the inertial subregion, the refractive-index structure function $D_n(R)$ is directly proportional to the two-thirds power of the scalar distance R [10]:

$$D_n(R) = C_n^2 R^{2/3}, \tag{3.15}$$

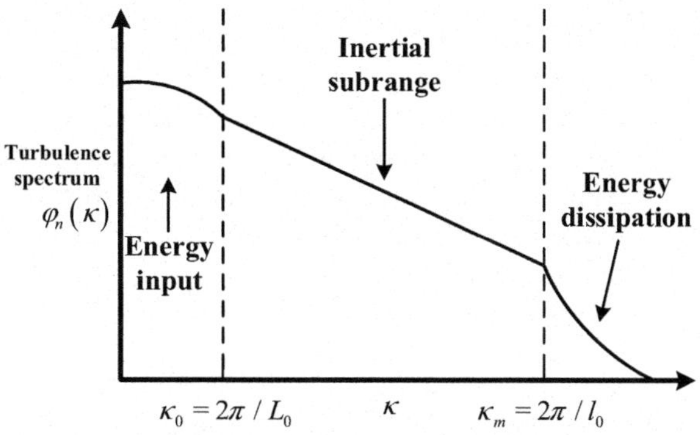

Fig. 3.6 Kolmogorov turbulence-spectral model [10]

where C_n^2 is the atmospheric refractive-index structure constant ($m^{-2/3}$). By introducing Eq. (3.7) into Eq. (3.15), the famous Kolmogorov spectrum can be obtained [10]:

$$\begin{aligned}\Phi_n(\kappa) &= \frac{\Gamma(a+1)}{4\pi^2}\sin[(a-1)\pi/2]C_n^2\kappa^{-a-2} \\ &= 0.033C_n^2\kappa^{-11/3},\ 1/L_0 \le \kappa \le 1/l_0,\end{aligned} \tag{3.16}$$

where $a = 5/3$ and $\Gamma(a)$ represents the gamma function.

Regarding the calculation of the 0.033 parameter, Kolmogorov believed that the energy-transfer rate within the inertial-scale range (usually between the energy-injection scale and the vortex-dissipation scale) was relatively independent of the scale and had a specific numerical value. In addition, he analyzed the experimental data between the turbulent-energy density and energy-transfer rate, and found that this value was approximately 0.033 [5].

It should be noted that the 0.033 parameter value in the Kolmogorov spectrum is not an exact constant, but an estimated value. In practical applications, it may deviate slightly, owing to interference from other factors. However, this parameter value is still widely used in turbulence research to quantitatively describe the turbulent-energy transfer process [10].

3.4.2 Tatarskii Spectrum

In the atmospheric boundary layer, the characteristic-length scale of turbulent motion is the turbulence scale, defined as follows [11]:

$$L = \frac{1}{\sigma^2}\int_0^\infty R(r)dr, \tag{3.17}$$

where σ^2 is the variance of the velocity component or temperature, r is the spatial displacement, and $R(r)$ is the sum of the autocorrelation functions of the velocity component or temperature. L is the turbulence scale.

Figure 3.7a shows the turbulent kinetic-energy (TKE) spectrum, where E(κ) is the turbulent kinetic energy per unit mass. Figure 3.7b shows the TKE spectral conversion, where T(κ) is the amount of energy transferred from wavenumber κ to all other wavenumbers.

The inertial subrange scale of the Kolmogorov spectrum in Eq. (3.16) is applicable only to smaller turbulent scales ($1/L_0 < \kappa < 1/l_0$). This spectrum ignores the effects of internal and external scales, which often results in inconsistent theoretical simulations and actual observations. Owing to the influence of atmospheric conditions, height

Fig. 3.7 Spectral map of corresponding turbulence scales [11]

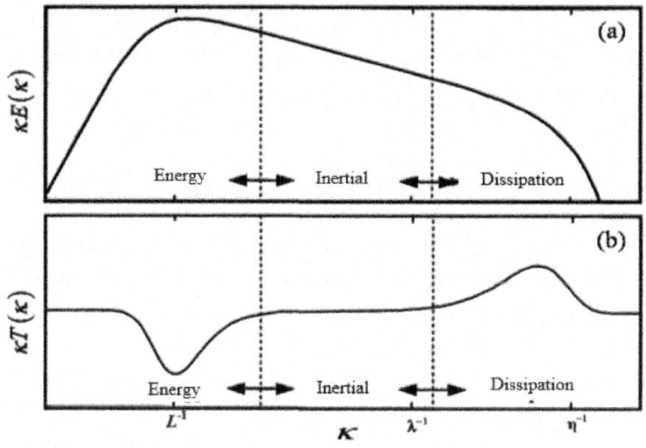

above the ground, and terrain structure on the turbulence scale in actual atmospheric-turbulence motion, there is no universal value for the turbulence scale.

To more accurately describe the power spectrum of atmospheric turbulence, Tatarskii obtained the spectrum function $\exp\left(\kappa^2/\kappa_m^2\right)$ by introducing an internal turbulence-scale parameter ($\kappa_m = 5.92/l_0$), which expanded the range of spatial wavenumbers to the dissipation area (i.e., $\kappa > 1/L_0$). The spectrum function has the form of a Gaussian function. Moreover, for convenience of mathematical description, Tatarskii improved the power spectrum used by Novikov in the study of turbulent wind-speed fluctuations and applied it to the study of atmospheric refractive-index fluctuations, establishing the Tatarskii spectrum [12]:

$$\Phi_n(\kappa) = 0.033C_n^2\kappa^{-11/3}\exp\left(\frac{\kappa^2}{\kappa_m^2}\right), \kappa > 1/L_0, \kappa_m = 5.92/l_0, \tag{3.18}$$

where C_n^2 is the structure constant of the atmospheric refractive index, k is the number of space waves, L_0 is the outer turbulence scale, l_0 is the inner turbulence scale, and exp is the exponential function.

The Kolmogorov power spectrum in Eq. (3.16) is widely used in theoretical calculations, owing to its concise expression. However, this power-spectrum model was only established for an inertial area of $1/L_0 < k < 1/l_0$. The Tatarskii spectrum in Eq. (3.18) increases the range of space waves to the scattering area, based on the Kolmogorov power spectrum (i.e., $k > 1/L_0$), but does not include the input area (i.e., $k < 1/L_0$). Therefore, the descriptions of the Kolmogorov and Tatarskii spectra are not sufficiently accurate in the input area (i.e., $k < 1/L_0$).

In addition, the definition domain of the Kolmogorov and Tatarskii spectra exists at the magic point $k = 0$. When the number of space waves k tends to 0, $\kappa \to 0$, $\kappa^{-\frac{11}{3}} \to \infty$, Eq. (3.16) becomes $\lim_{k\to 0}\Phi_n(\kappa) = 0.033C_n^2\kappa^{-\frac{11}{3}} \to \infty$, and Eq. (3.18) becomes $\lim_{k\to 0}\Phi_n(\kappa) = 0.033C_n^2\kappa^{-11/3}\exp\left(\frac{\kappa^2}{\kappa_m^2}\right) \to \infty$, which leads to an unreasonable result where $\Phi_n(\kappa)$ tends towards ∞.

This is because the total energy contained in a material system should be limited to any scale in space. Turbulent motion is composed of various vortex structures at different scales. Although the scale of the vortices can become increasingly smaller, when a certain microscale is reached, the size of the vortices tends to infinity and the corresponding energy tends to be limited.

Current research on turbulent energy spectra has led to the development of various models and theories that can be applied to different flow states and turbulent structure scales. In general, it is difficult to cover the turbulence characteristics of all flow situations using a single model. Therefore, it is necessary to choose a suitable model based on the actual situation for processing.

3.4.3 Von Karman Spectrum

When $k < 1/L_0$, the turbulence spectrum is within the input area. Because of wind shear and temperature gradients, energy is input into turbulence. The shape of the spectrum in this region depends on how specific turbulence occurs, and the characteristic physical quantities of random media are generally non-uniform and anisotropic [12]. When $k > 1/L_0$, the turbulence spectrum is in the inertial or dissipative region, and energy is converted from the turbulent flow field into other forms, such as heat. At this time, the velocity field becomes extremely complex, and a large number of small-scale vortex structures are generated and interact with each other. The statistical characteristics of these small-scale structures are uniform and isotropic [6].

The Kolmogorov spectrum is in the range of $1/L_0 < \kappa < 1/l_0$ and that of Tatarskii is in the range of $\kappa > 1/L_0$. Therefore, the Kolmogorov spectrum and Tatarskii spectrum are only uniform and isotropic in the inertial or dissipative region (i.e., $\kappa > 1/L_0$). The von Karman spectrum improves the Kolmogorov and Tatarskii spectrum models,$\lim_{k\to 0}\Phi_n(\kappa) = 0.33C_n^2\frac{\exp(0)}{(0+\kappa_0^2)^{11/6}} = 0.33C_n^2\kappa_0^{-\frac{11}{3}}$, expanding the κ range to 0–∞.

When the spatial wavenumber $\kappa \to 0$, the von Karman spectrum becomes $D_n(R) = 2[B_n(0) - B_n(R)]$, and at this point, $\Phi n(\kappa)$ is a finite value that exists simultaneously. Therefore, when the spatial wavenumber is 0, the $D_n(R)$ and $B_n(R)$ of the von Karman spectrum also exist simultaneously, which can overcome the inability of the Kolmogorov and Tatarskii spectra to integrate singular points. The von Karman spectral expression is [11]

$$\Phi_n(\kappa) = 0.033C_n^2\left(\kappa^2 + \kappa_0^2\right)^{-11/6}, 0 \le \kappa \le \infty, \kappa_0 = 2\pi/L_0. \tag{3.19}$$

To fully consider the influence of the internal and external turbulence scales, the internal turbulence-scale parameter κ_m was introduced into the von Karman model, and a modified von Karman spectrum was proposed. The spectral expression is [13]

$$\Phi_n(\kappa) = 0.033C_n^2 \frac{\exp\left(-\kappa^2/\kappa_m^2\right)}{\left(\kappa^2+\kappa_0^2\right)^{11/6}}, 0 \le \kappa \le \infty; \kappa_m = 5.92/l_0, \kappa_0 = 2\pi/L_0, \tag{3.20}$$

where $0 \le \kappa \le \infty$, $\kappa_0 = 2\pi/L_0$, and $\kappa_m = 5.92/l_0$. Because the κ range in the Kolmogorov spectrum is between $2\pi/L_0 < \kappa < 2\pi/l_0$, Eq. (3.20) can be transformed into the Kolmogorov spectrum in the $\kappa_0 < \kappa < \kappa_m$ inertial region, where $0 \le \kappa \le \infty$. $\kappa_0 = 2\pi/L_0$ when $\kappa \to 0$, $\lim_{k\to 0} \Phi_n(\kappa) = 0.33C_n^2 \frac{\exp(0)}{(0+\kappa_0^2)^{11/6}} = 0.33C_n^2\kappa_0^{-\frac{11}{3}}$. It can be seen that the von Karman turbulence spectrum artificially corrects the singularity problem of the Kolmogorov and Tatarskii spectra when the spatial wavenumber κ approaches zero.

Researchers have experimentally found a significant difference between the theoretical prediction results of the traditional von Karman spectrum and the observed values; therefore, it should be corrected to make it more consistent with the actual situation [13]. Equation (3.20) was modified based on the von Karman spectrum with a correction coefficient of 5.92, and its form was modified to be closer to the actual situation.

Researchers have inverted energy spectra based on meteorological measurement data. A large amount of field-measured data was collected, including wind-speed time series, power-spectral density, and relevant statistical information of fluid movement. After data processing and analysis, the modified energy-spectrum form was inverted, and the correction coefficient 5.92 was verified by a large amount of field-measured data and numerical simulation results [13]. The existence of this coefficient renders the revised energy spectrum more consistent with the actual situation and provides a better foundation for turbulence research [8].

3.4.4 Greenwood–Tarazano Spectrum

The von Karman spectrum only describes the fluid inside the pipeline, which has a unique characteristic scale: the diameter of the pipeline. However, actual atmospheric turbulence has many characteristic scales. Because of the height, boundary-layer thickness, and non-uniform ground scales in the atmospheric boundary layer, the von Karman spectrum cannot fully describe the actual atmospheric turbulence [14].

Because of the correlation between the spatial temperature structure function and the spatial power spectrum, and not relying on the Taylor freezing assumption, Greenwood et al. reconstructed the refractive-index spectrum using an experimentally measured spatial structure function and compared it with the von Karman spectrum. The results showed that the von Karman spectrum was not the optimal model for matching the data. Therefore, Greenwood et al. proposed the Greenwood–Tarazano spectrum based on the experimental data [14]:

Fig. 3.8 Schematic diagram of turbulent energy conversion [14]

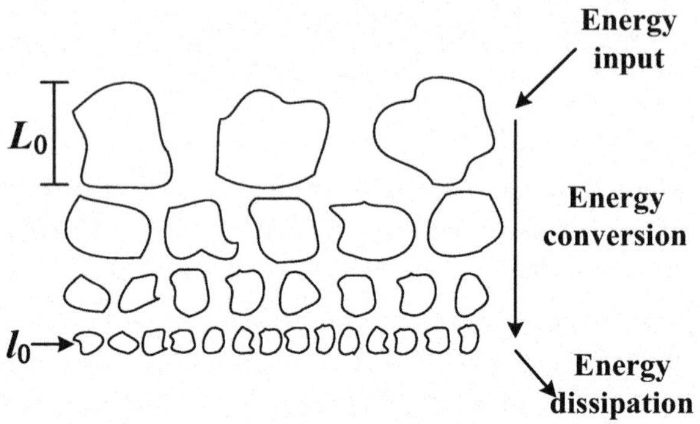

$$\Phi_n(\kappa) = 0.033C_n^2\left(\kappa^2 + \kappa\kappa_0\right)^{-11/6}, 0 \le \kappa \le \infty, \kappa_0 = 2\pi/L_0. \tag{3.21}$$

Based on Eq. (3.19) and experimental measurement data, Eq. (3.21) replaces one of the inner scale parameters κ in Eq. (3.19) with the outer scale parameter κ_0. Considering Figs. 3.6 and 3.8, it can be observed that when $\kappa > 2\pi/l_0$, the turbulent region enters the dissipation zone, $\Phi_n(\kappa)$. The dissipation of the turbulent energy in the dissipation zone far exceeds the generated kinetic energy. Therefore, when entering the dissipation zone, the speed of the decrease is significantly faster. Therefore, compared with the von Karman spectrum, the Greenwood–Tarazano spectrum decreases faster within the inertial subrange [14].

3.4.5 Hill Spectrum

Strictly speaking, the spectral models represented by Eqs. (3.16)–(3.21) are only correct within the inertial subrange, because their applicability in turbulent non-inertial subranges is only mathematical and cannot be reasonably explained from a physical perspective. Champagne et al. demonstrated through experimental measurements [29] that, because of the transition of the power spectrum from the inertial convection range to the viscous convection range occurring in the relatively-high wavenumber region, there is a $\kappa^{-11/3}$ power law, with a κ^{-3} power law. Hence, when the power-spectral density of the high wavenumber is close to $1/l_0$, small protrusions appear (see Fig. 3.9), while the power spectrum described in Eqs. (3.16)–(3.21) does not have this feature.

Figure 3.9 shows the Tatarskii, Andrews, and Hill spectral density curves [15]. There is a protrusion feature at $\kappa = 1/l_0$, followed by a rapid descent.

Based on experiments, Hill considered the influence of sudden changes in the high-wavenumber region of the power spectrum [15]; that is, the conversion of the power spectrum from the inertial convection range to the viscous convection range occurs in the relatively high wavenumber region, which has a $\kappa^{-11/3}$ power law. A

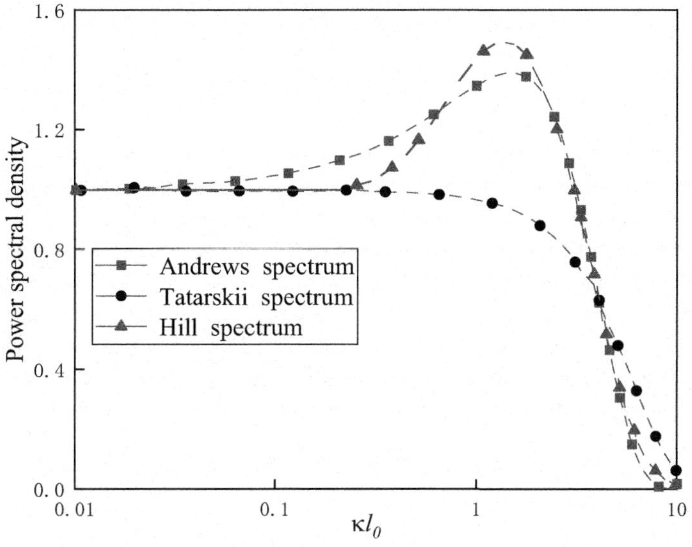

Fig. 3.9 Power-spectral density curves [15]

κ^{-3} power law called the Hill spectrum was proposed to accurately describe the inertial and dissipative regions of a numerical model [15]:

$$\Phi_n(\kappa) = 0.033C_n^2\kappa^{-11/3} \times \left\{\exp(-1.2\kappa^2 l_0^2) + 1.45\exp\left[-0.97(\ln \kappa l_0 - 0.452)^2\right]\right\}, 1/L_0 \leq \kappa \leq 1/l_0, \tag{3.22}$$

where exp represents the exponential function with a constant *e* as the base, and ln represents the logarithm with a constant *e* as the base. Because of the effect of the internal turbulent scale l_0, this spectral form is different from that of the Kolmogorov spectrum. Before entering a rapid exponential decline, there is no rapid decrease, but there is a small increase peak.

3.4.6 Andrews Spectrum

Because of the complex form of the Hill spectral model, it is not convenient for theoretical calculations [11]. Andrews used an interpolation method to modify the Hill spectral model, which provided the same theoretical model as the von Karman turbulence spectrum. This power spectrum describes the protrusions at high wavenumbers and has an easy-to-handle mathematical form. As shown in Eq. (3.23) [16],

$$\Phi_n(\kappa) = 0.033C_n^2\left[1 + a_1(\kappa/\kappa_l) - a_2(\kappa/\kappa_l)^{7/6}\right]\frac{\exp(-\kappa^2/\kappa_l^2)}{(\kappa_0^2 + \kappa^2)^{11/6}}, \tag{3.23}$$

where $a_1 = 1.802$, $a_2 = 0.254$. In addition, when $a_1 = a_2 = 0$, Eq. (3.23) becomes $\Phi_n(\kappa) = 0.033C_n^2\frac{\exp(-\kappa^2/\kappa_l^2)}{(\kappa_0^2+\kappa^2)^{11/6}}$, and at this point, the expression of the Andrews spectral model is consistent with that of the von Karman spectral model. When $\kappa_0 = l_0 = 0$, Eq. (3.23) becomes $\Phi_n(\kappa) = 0.033C_n^2\kappa^{-11/6}$, and at this point, the expression of the Andrews spectral model is consistent with that of the Kolmogorov spectral model.

The von Karman spectrum is essentially the same as the Kolmogorov spectrum in the inertial subrange. As the wavenumber increases, the von Karman spectrum sharply decreases monotonically [16]. The Hill and Andrews spectra differ significantly from the Kolmogorov spectrum in the high-wavenumber region of the inertial sub-range, reflecting the impact of high-wavenumber mutations on optical transmissions.

Figure 3.10 shows the variation in the atmospheric refractive-index power spectrum with spatial wavenumbers. In Fig. 3.10, with an outer scale of $L_0 = 10$ m and an inner scale of $l_0 = 1$ cm, the Kolmogorov, von Karman, and Andrews spectra continuously decrease with an increase in the spatial wavenumber. In the inertial-turbulence subregion, the three types of turbulence spectra have the same trend, following the "− 11/3" law.

When the spatial wavenumber continues to increase and reaches the dissipative region, the fluctuation of the Andrews spectrum in the dissipative region is suppressed, and the Kolmogorov and Von Karman spectra exhibit unstable states in the high-frequency turbulent dissipation region.

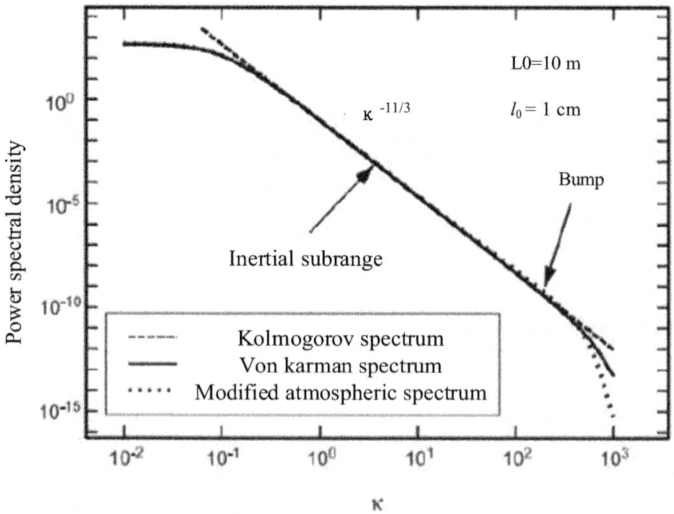

Fig. 3.10 Changes in atmospheric refractive-index power spectra with the spatial wavenumber [16]

3.4.7 Non-Kolmogorov Spectrum

The classic Kolmogorov theory, based on Richardson's energy cascade, indicates that all turbulent vortices are isotropic within the inertial-subscale range. However, in a free atmosphere, particularly in a stable stratified stratosphere, turbulence may become anisotropic on a large scale, and most of the anisotropy occurs in the boundary layer, that is, the ground, buildings, aircraft, etc. Turbulence does not always follow the Kolmogorov power-spectrum model [9], even within the inertial range. This is due to the extraction of turbulent energy by buoyancy, which results in a sharp decrease in the energy-transfer rate. Therefore, the kinetic energy is transmitted at a non-uniform rate throughout the entire spectrum, and it decreases as the wavenumber increases.

In 1962, Obukhov et al. discovered through experimental observations that large-scale turbulence exhibits significant anisotropy [17], but it is limited to the range of large-scale turbulence. Subsequently, Toselli et al. found that energy transfers could occur directly between large- and small-scale turbulence [17]. Nowadays, an increasing amount of experimental data indicates that in actual atmospheric turbulence, most turbulence is in an anisotropic state, namely, non-Kolmogorov turbulence [18].

To study the influence of atmospheric turbulence on non-Kolmogorov light-wave propagation, Andrews et al. established a non-Kolmogorov spectral model [18]:

$$\Phi_n(\kappa, \alpha) = A(\alpha)\tilde{C}_n^2\kappa^{-\infty}, 3 < \alpha < 4, \tag{3.24}$$

where κ is the spatial wavenumber, α is the power index of the power-spectral density of the refractive-index fluctuation, and $\tilde{C}_n^2$ is the refractive-index structure constant of non-Kolmogorov turbulence ($m^{3-\alpha}$). $A(\alpha)$ is a constant that maintains consistency between the refractive-index structure function and its power spectrum, $A(\alpha) = \Gamma(\alpha - 1)\cos(\alpha\pi/2)/\left(4\pi^2\right)$.

When $\alpha = 11/3$, $\tilde{C}_n^2 = C_n^2$, and $A(\alpha) = 0.033$, Eq. (3.25) becomes the refractive-index fluctuation power-spectral density model of Kolmogorov turbulence.

To understand the impact of internal and external scales on non-Kolmogorov atmospheric turbulence effects, Andrews et al. developed a generalized modified non-Kolmogorov spectral model [19]:

$$\Phi_n(\kappa, \alpha) = \frac{A(\alpha)\tilde{C}_n^2}{(\kappa^2 + \kappa_0^2)^{\alpha/2}}\exp(-\kappa^2/\kappa_m^2), 0 < \kappa < \infty, 3 < \alpha < 4, \tag{3.25}$$

where $\kappa_0 = 2\pi/L_0, \kappa_m = c(\alpha)/l_0$ and $c(\alpha)$ are the power-index α functions of the refractive-index fluctuation power-spectral density, which have the following form:

$$A(\alpha) = \frac{1}{4\pi^2}\Gamma(\alpha - 1)\cos\left(\frac{\pi}{2}\alpha\right) \tag{3.26}$$

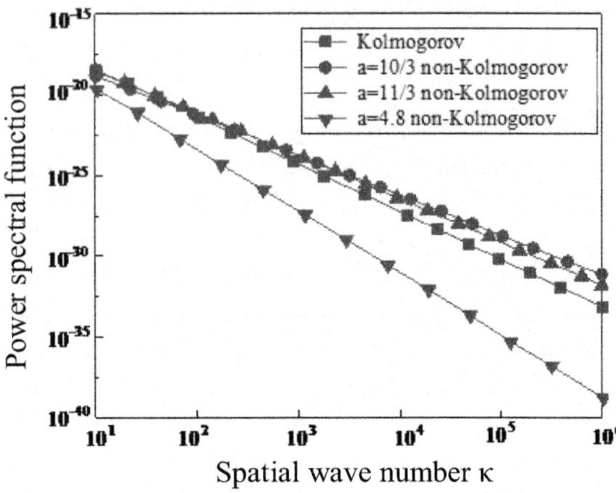

Fig. 3.11 Comparison of non-Kolmogorov and Kolmogorov spectra under different α values [1]

$$c(\alpha) = \left[\frac{2\pi}{3}\Gamma\left(5 - \frac{\alpha}{2}\right)A(\alpha)\right]^{1/(\alpha-5)}. \tag{3.27}$$

From Eq. (3.27), it can be seen that when $\alpha = 11/3$ and $A(\alpha) = 0.033$, the non-Kolmogorov spectrum could be equivalent to the modified von Karman spectrum. When $A(\alpha) = 0.033$, $\alpha = 11/3$, $L_0 \to \infty$ and $l_0 \to 0$, the non-Kolmogorov spectrum can be equivalent to the Kolmogorov spectrum.

As shown in Fig. 3.11, when $\alpha = 11/3$, $\tilde{C}_n^2 = C_n^2$, and $A(\alpha) = 0.033$, $\Phi_n(\kappa) = 0.033C_n^2\kappa^{-11/3}$ and the non-Kolmogorov spectrum becomes a Kolmogorov spectrum. When $\alpha = 10/3$, $\alpha = 11/3$ and $\alpha = 4.8$, the non-Kolmogorov spectrum significantly deviates from the Kolmogorov spectrum, and the larger the α in $\Phi_n(\kappa) = 0.033C_n^2\kappa^{-11/3}$, the greater the degree of deviation. However, in the inertial region (i.e., $\kappa < 1.1 \times 10^3$), the non-Kolmogorov spectrum at different α values has smaller deviations from the Kolmogorov spectrum. The above analysis shows that the Kolmogorov spectrum is a special case of the non-Kolmogorov spectrum.

Figure 3.12 shows the curve of parameter M varying with the generalized exponential factor α when the beam propagates horizontally in non-Kolmogorov atmospheric turbulence at different propagation distances z.

Figure 3.12 shows the influence of the generalized exponential factor α on parameter M at different transmission distances and shows that the size of α directly affects the transmission characteristics of the beam in atmospheric turbulence. That is, parameter M directly reflects the intensity of atmospheric turbulence, and the larger parameter M, the stronger the atmospheric turbulence [1]. From Fig. 3.12, it can be seen that when z is 1000, 1200, and 1500 m, the α corresponding to the maximum M value also increases correspondingly, but remains approximately $\alpha = 3.1$. When $\alpha > 3.8$, parameter M tends to flatten with the change of α.

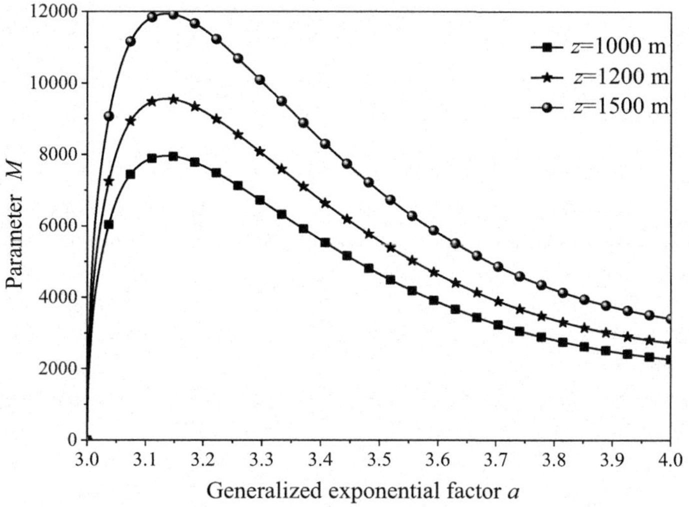

Fig. 3.12 Variation curve of M with α at different transmission distances z [20]

3.4.8 Turbulence Spectrum with Height Variation

Thus far, there are two models for the atmospheric turbulence structure in the Earth's atmosphere: (a) a two-layer model, where turbulence in the Earth's atmosphere is composed of Kolmogorov turbulence in the troposphere and non-Kolmogorov turbulence in the stratosphere; and (b) a three-layer model, where the turbulence in the Earth's atmosphere consists of Kolmogorov turbulence in the boundary layer, non-Kolmogorov turbulence in the free atmosphere, and non-Kolmogorov turbulence in the stratosphere. They have the following forms, respectively:

$$\Phi_{nB}(\kappa, z) = 0.033C_n^2(z)\kappa^{-11/3} \tag{3.28}$$

$$\Phi_{nF}(\kappa, z) = 0.015\tilde{C}_{nF}^2(z)\kappa^{-10/3} \tag{3.29}$$

$$\Phi_{nS}(\kappa, z) = 0.0024\tilde{C}_{nS}^2(z)\kappa^{-5}, \tag{3.30}$$

where z is the propagation distance, $C_n^2(z)$ is the structural parameter of the conventional Kolmogorov turbulent refractive index in the boundary layer ($\text{m}^{-2/3}$), $\tilde{C}_{nF}^2(z)$ is the structural parameter of the non-Kolmogorov refractive index in the free troposphere ($\text{m}^{-1/3}$), and $\tilde{C}_{nS}^2(z)$ is the structural parameter of the non-Kolmogorov refractive index in the stratosphere (m^{-2}). The above three parameters all depend on the altitude.

The three-dimensional power spectrum of the refractive-index fluctuations can be defined as follows [12]:

$$\Phi_n(\kappa, a) = A(\alpha) \cdot C_n^2 \cdot K^{-\alpha}, \tag{3.31}$$

where C_n^2 is the general refractive-index structure constant ($m^{3-\alpha}$) and $A(\alpha)$ is a constant that maintains consistency between the structure function and its power spectrum.

Arkadi found experimentally that the turbulence power spectrum in the free troposphere and stratosphere may exhibit non-Kolmogorov characteristics and that the spectral behavior of refractive-index fluctuations varies with altitude, which makes general models such as those in Eqs. (3.15)–(3.24) very unsuitable.

Based on the three-layer model of the refractive-index fluctuation spectrum in the free atmosphere (boundary layer (2–3 km), free troposphere (8–10 km), and the stratosphere above them), and according to experimental and theoretical data of the turbulence profile in the troposphere and lower stratosphere, Arkadi proposed a turbulence spectrum model that changes with height [13]:

$$\alpha(z) = \frac{\alpha_1}{1 + (z/H_1)^{b_1}} + \frac{\alpha_2 (z/H_1)^{b_1}}{1 + (z/H_1)^{b_1}} \times \frac{1}{1 + (z/H_2)^{b_2}} + \frac{\alpha_3 (z/H_2)^{b_2}}{1 + (z/H_2)^{b_2}}, \tag{3.32}$$

where $\alpha_1 = 11/3$, $\alpha_2 = -10/3$, $\alpha_3 = -5$, z is the height from the ground, H_1 and H_2 are the characteristic heights, and b_1 and b_2 are the fitting coefficients.

The experimental results showed that the turbulence characteristics varied unevenly with height. The height dependence of α is expressed by a three-layer height model, in which the first layer corresponds to the boundary layer with Kolmogorov turbulence ($\alpha = 11/3$). The second corresponds to a free troposphere with non-Kolmogorov turbulence ($\alpha = -10/3$). The third case is the stratosphere with non-Kolmogorov turbulence ($\alpha = -5$).

Figure 3.13 shows the power-profile distribution of the turbulent power spectrum under the following two conditions: $b_1 = 8$, $b_2 = 10$ (solid line), and $b_1 = 15$, $b_2 = 20$ (dashed line).

3.5 Source Noise Models

3.5.1 *Laser Relative-Intensity Noise Model*

In communication systems, the inherent noise of the laser itself is the main source of noise, and the relative-intensity noise (RIN) can be used to describe the intensity of the inherent noise of the laser itself. The average power of the emitted beam is P, the response is ρ, and the bandwidth is Δf. When the detector receives the beam, the average current detected by the detector is the product of the responsiveness and average power ρP. Therefore, the mean-square value of the current is [28]

$$\overline{i_{NL}^2} = RIN(\rho P)^2 \Delta f. \tag{3.33}$$

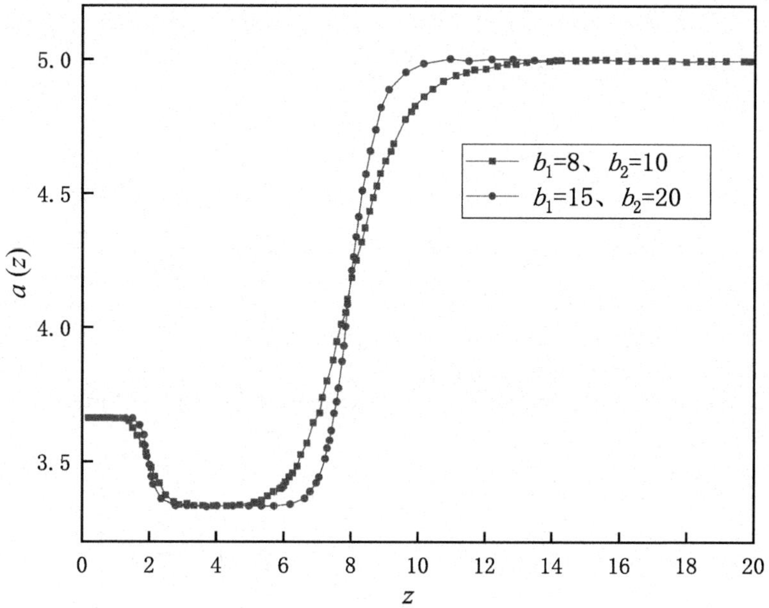

Fig. 3.13 Power-exponential profile distribution of a turbulent power spectrum [13]

Therefore, the average power caused by the fluctuation of the laser intensity is

$$\overline{P_{NL}^2} = \frac{\overline{i_{NL}^2}}{\rho^2}. \tag{3.34}$$

The relative-intensity noise of the laser can be obtained as

$$\text{RIN} = \frac{\overline{P_{NL}^2}}{P^2 \Delta f}, \tag{3.35}$$

where $\overline{i_{NL}^2}$ is the mean-square value of the current and $\overline{P_{NL}^2}$ is the average power, indicating that the laser noise is mainly related to the average power of the beam and the detector bandwidth.

3.5.2 Laser Phase-Noise Model

Spontaneous emission noise can cause phase fluctuations in the laser, resulting in a nonzero spectral linewidth Δv. In semiconductor lasers, because only a small number of photons are trapped in small cavities and the linewidth-enhancement factor cannot be ignored, the linewidth Δv is generally relatively large and can reach the megahertz range. The $\Gamma_{EE}(t)$ relationship between the spectrum of the radiated light field $E(t) = \sqrt{P}\exp(j\phi)$ and the field autocorrelation function is [5]

$$S(\omega) = \int_{-\infty}^{\infty} \Gamma_{EE}(t) e^{-j(\omega-\omega_0)\tau} d\tau \tag{3.36}$$

$$\Gamma_{EE}(t) = \langle E^*(t)E(t+\tau)\rangle = \langle \exp(j\Delta\phi(t))\rangle = \exp(-\langle\Delta\phi^2(t)\rangle/2), \tag{3.37}$$

where $\Delta\phi(t) = \phi(t+\tau) - \phi(t)$. E represents the field, t represents time, τ represents the time interval, j represents an imaginary number, ω represents the angular frequency, and ϕ represents the phase.

The Wiener process can be used to describe the laser phase-noise process as follows:

$$S_{EE}(v) = \frac{P}{2\pi\,\Delta v}\left[\frac{1}{1+\left(\frac{2(v+v_0)}{\Delta v}\right)^2} + \frac{1}{1+\left(\frac{2(v-v_0)}{\Delta v}\right)^2}\right], \tag{3.38}$$

where P is the optical power, v is the frequency, v_0 is the center frequency, and Δv is the laser linewidth.

3.6 Channel-Noise Models

Owing to the influence of atmospheric turbulence, when a laser beam is transmitted through the atmosphere, the light intensity at the receiving end undergoes random fluctuations, known as the intensity-scintillation effect, which can lead to signal fading at the receiver and reduce the quality of the received signal. Therefore, to improve the performance of free-space optical (FSO) systems, it is essential to establish channel models that describe different atmospheric-turbulence conditions. Table 3.2 shows the classification diagrams of the various atmospheric-turbulence channel models.

3.6.1 Log-Normal Distribution

The log-normal-distribution probability-density function describes the scintillation and fading statistics in the weak turbulence region. Considering the log-normal model, the probability-density function of the received optical signal is given by the following formula [1]:

Table. 3.2 Classification diagram of atmospheric-turbulence channel models

Channel model	Probability distribution function	Adapted to turbulence range
Log-normal distribution	$P(I)=\frac{1}{\sqrt{2\pi\sigma_l^2}I}\exp\left\{-\frac{(\ln(I/I_0)-E[I])^2}{2\sigma_l^2}\right\}$	Weak turbulence
Gamma–gamma distribution	$p(I)=\int_0^{+\infty}p_y(I\|x)p_x(x)dx=\frac{2(\alpha\beta)^{(\alpha+\beta)/2}}{\Gamma(\alpha)\Gamma(\beta)}I^{(\alpha+\beta)/2-1}K_{\alpha\_\beta}\left[2(\alpha\beta I)^{1/2}\right]$	Medium and strong turbulence
Negative-exponential distribution	$p(I)=\frac{1}{I_0}\exp\left(-\frac{I}{I_0}\right)$	Strong turbulence
K distribution	$P(I)=\frac{2a^{(a+1)/2}}{\Gamma(a)}I^{(a-1)/2}K_{a-1}(2\sqrt{aI})$	Medium and strong turbulence
Malaga distribution	$P(I)=A\sum_{k=1}^{\infty}a_kI^{\frac{\alpha+k}{2}-1}K_{\alpha-k}\left(2\sqrt{\frac{\alpha I}{\gamma\beta+\Omega'}}\right)$	Weak, medium, and strong turbulence
Ke distribution	$P(I)=A_1\times\frac{1}{\sqrt{2\pi}\times\sigma_1}\times e^{\left[-\frac{(I-\mu_1)^2}{2\times\sigma_1^2}\right]}+A_2\times\frac{1}{\sqrt{2\pi}\times\sigma_2}\times e^{\left[-\frac{(I-\mu_2)^2}{2\times\sigma_2^2}\right]}$	Weak turbulence

$$P(I)=\frac{1}{\sqrt{2\pi\sigma_l^2}I}\exp\left\{-\frac{(\ln(I/I_0)-E[I])^2}{2\sigma_l^2}\right\}, \tag{3.39}$$

where σ_l^2 is the logarithmic light-intensity variance and I is the normalized light-intensity value reaching each receiver.

Figure 3.14 shows the log-normal-distribution probability-density function curves σ_l^2 under different conditions. As shown in Fig. 3.14, as the turbulence intensity σ_l^2 increases, the curve gradually deviates from the mean and the tail becomes more severe, which is gradually inconsistent with the log-normal distribution. The log-normal distribution is only applicable to weak turbulence conditions (i.e., $\sigma_l^2 < 1$) because the scattering effect is also considered as the turbulence intensity increases.

3.6.2 Gamma–Gamma Distribution

Because the log-normal model is limited to weak turbulence, Shah derived another model based on the modified Rytov theory, called the gamma–gamma channel model, where the light-intensity fluctuation is developed as the product of two stochastic processes, the gamma–gamma distribution [1]:

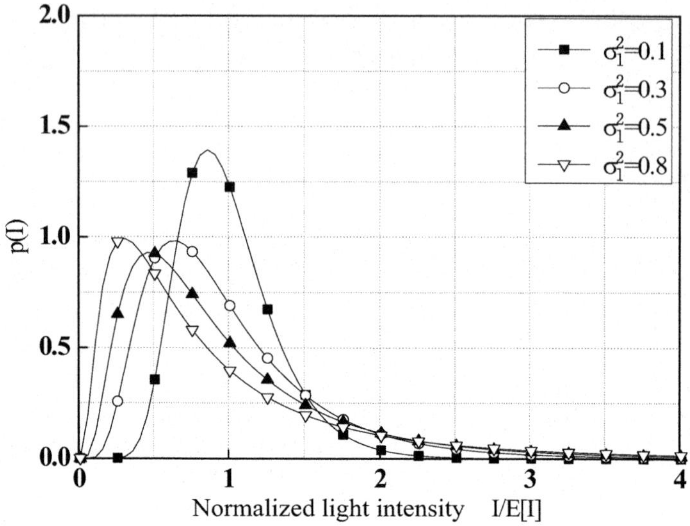

Fig. 3.14 Log-normal distribution probability-density function curves

$$p(I) = \int_0^{+\infty} p_y(I|x)p_x(x)dx = \frac{2(\alpha\beta)^{(\alpha+\beta)/2}}{\Gamma(\alpha)\Gamma(\beta)} I^{(\alpha+\beta)/2-1} K_{\alpha\_\beta}\left[2(\alpha\beta I)^{1/2}\right]. \quad (3.40)$$

The second type of *Bessel* correction function, $K_n(\cdot)$ with order n in the equation, is the gamma function $\Gamma(\cdot)$. α and β, the parameters used to measure the characteristics of large-scale and small-scale eddies in atmospheric turbulence, can generally be calculated using the following equation:

$$\begin{cases} \alpha = \left(\exp\left(\dfrac{0.49\sigma_R^2}{\left(1+1.11\sigma_R^{12/5}\right)^{7/6}}\right) - 1\right)^{-1} \\ \beta = \left(\exp\left(\dfrac{0.51\sigma_R^2}{\left(1+0.69\sigma_R^{12/5}\right)^{5/6}}\right) - 1\right)^{-1} \end{cases}. \quad (3.41)$$

Figure 3.15 shows the gamma–gamma distribution probability-density function for different turbulence intensities. It can be seen from the figure that the gamma–gamma distribution has a wider application scope than the log-normal distribution and can more accurately describe the statistical characteristics of light-intensity fluctuations in weak, medium, and strong fluctuation areas.

As can be seen from Fig. 3.16, when the size of σ_I^2 is small, $\beta \gg 1$, $\alpha \gg 1$ indicates that there are many effective numbers of large and small-scale scatterers.

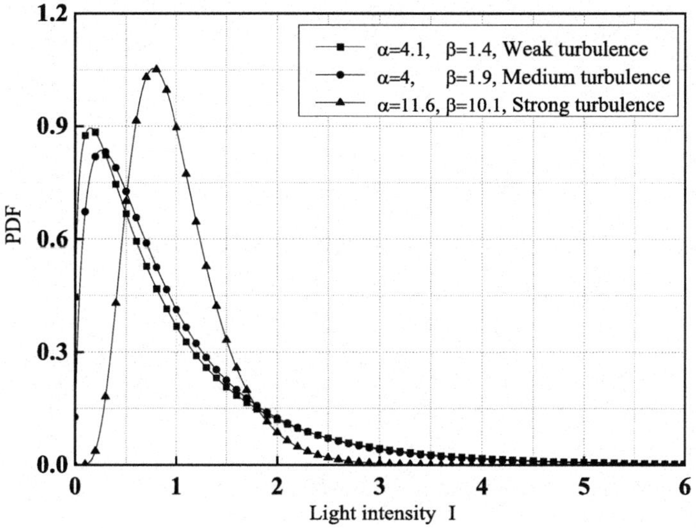

Fig. 3.15 Gamma–gamma distribution light-intensity probability-density functions

When the number σ_l^2 increases gradually, α and β decreases rapidly. When the number σ_l^2 increases from 1 to 100, α increases, but β decreases and approaches 1, indicating that the number of small-scale scatterers is determined by the horizontal spatial-coherence radius.

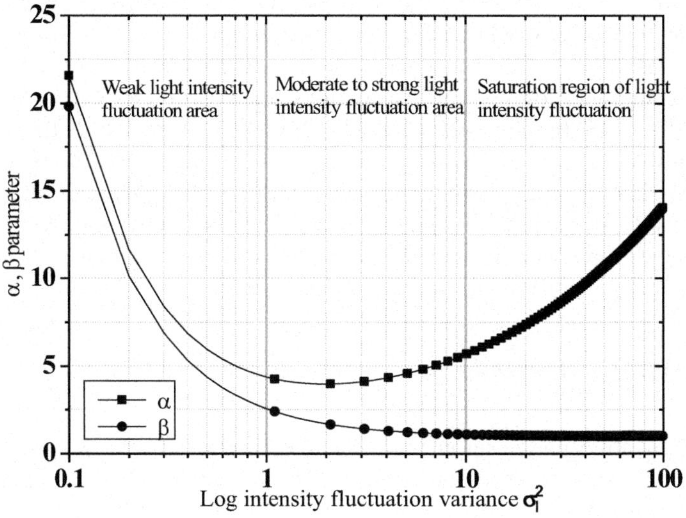

Fig. 3.16 Sum of σ_l^2 values under different β conditions

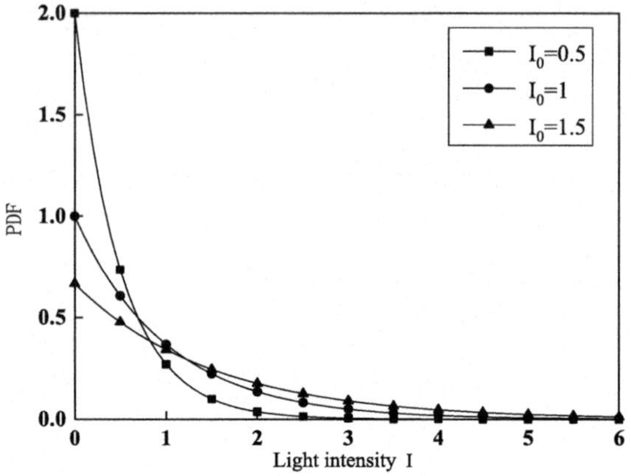

Fig. 3.17 Probability-density function of negative exponential-distribution light-intensity fluctuation

3.6.3 Negative Exponential Distribution

Variance-saturation scintillation regions may occur for strong-turbulence or long-distance transmissions. Under strongly fluctuating atmospheric turbulence, the light intensity at the receiving end follows a negative exponential distribution [1]:

$$p(I) = \frac{1}{I_0}\exp\left(-\frac{I}{I_0}\right), \quad I > 0, \tag{3.42}$$

where $I_0 = E[I]$ denotes the average light intensity. In the saturated state, the scintillation index $\to$ 1. The probability-density function of the negative exponential-distribution light-intensity fluctuation is shown in Fig. 3.17.

The probability distribution curves of the light-intensity fluctuations for the gamma–gamma and log-normal distributions under different Rytov variances are shown in Fig. 3.18. In the weak turbulent region, the log-normal distribution model is more in line with the experimental data; in the strong turbulent region, the log-normal distribution model exhibits a severe tailing phenomenon, which deviates significantly from the experimental data. In weak, medium, and strong turbulent regions, the gamma–gamma distribution model is more consistent.

3.6.4 Malaga Distribution

The Malaga atmospheric-turbulence channel model is a universal model that describes various turbulence distributions. The probability-density function of the Malaga distribution is expressed as [20]

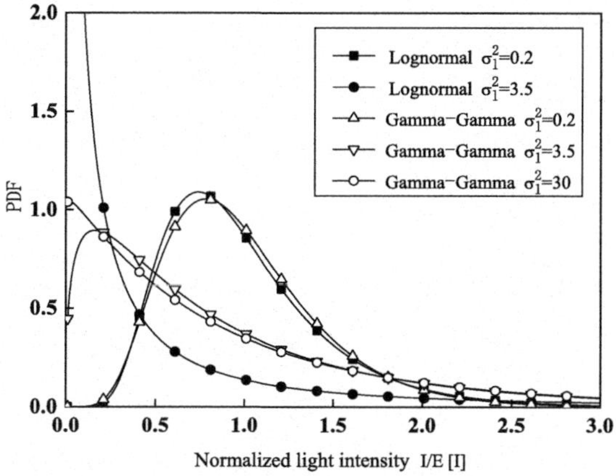

Fig. 3.18 Probability-density function curve of light-intensity distributions

$$f_I(I) = A\sum_{k=1}^{\infty} a_k I^{\frac{\alpha+k}{2}-1} K_{\alpha-k}\left(2\sqrt{\frac{\alpha I}{\gamma\beta+\Omega'}}\right) \tag{3.43}$$

$$\begin{cases} A = \dfrac{2\alpha^{\frac{\alpha}{2}}}{\gamma^{1+\frac{\alpha}{2}}\Gamma(\alpha)}\left(\dfrac{\gamma\beta}{\gamma\beta+\Omega'}\right)^{\beta} \\ a_k = \dfrac{(\beta)_{k-1}(\alpha\gamma)^{\frac{k}{2}}}{[(k-1)!]^2\gamma^{k-1}(\Omega'+\gamma\beta)^{k-1}}, \end{cases} \tag{3.44}$$

where a and β represent positive real parameters and fading parameters, respectively, and the Ω' parameters are the coherent average optical power. φ_A represents the phase of the light-wave line-of-sight propagation component and φ_B represents the phase of the light-wave line-of-sight propagation coupling component.

As shown in Fig. 3.19a, as the value of ρ increases, the value of f_I(I) inside the elliptical box increases accordingly. In the case of strong turbulence, compared with the gamma–gamma distribution, the declining trend of f_I(I) in the K distribution is relatively slow, and the tailing is relatively severe. As shown in Fig. 3.19b, under weak-turbulence conditions, as the ρ value increases, the f_I(I) curve gradually shifts to the right, while the peak value increases and the tailing phenomenon weakens [23].

3.6.5 *Ke's Distribution*

For visible-light communication in the Internet of Vehicles, Ke proposed a night background-light noise model (Ke's model) fitted with a double-Gaussian function called the Kirschner model [24]:

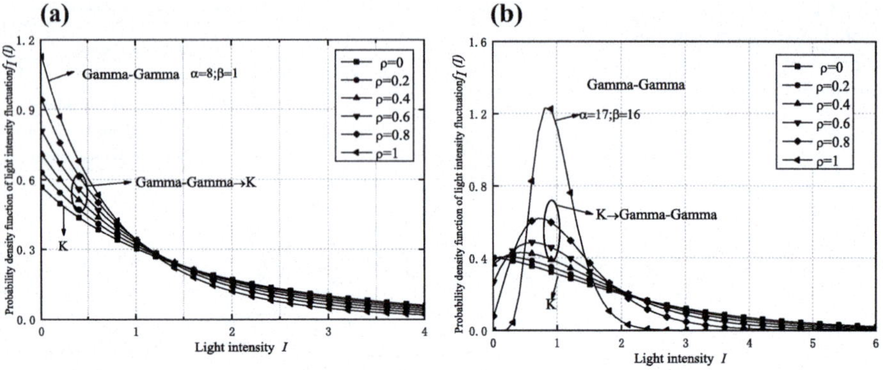

Fig. 3.19 ρ curve of the Malaga-distribution light-intensity fluctuation probability-density function changing with values of α and β under different conditions [23]. **a** $\alpha = 8, \beta = 1$. **b** $\alpha = 17, \beta = 16$

$$f(I) = A_1 \times \frac{1}{\sqrt{2\pi} \times \sigma_1} \times e^{\left[-\frac{(I-\mu_1)^2}{2\times\sigma_1^2}\right]} + A_2 \times \frac{1}{\sqrt{2\pi} \times \sigma_2} \times e^{\left[-\frac{(I-\mu_2)^2}{2\times\sigma_2^2}\right]}. \quad (3.45)$$

Figure 3.20 shows the Coriolis distribution model. The two Gaussian curves in the figure correspond to the two Gaussian functions of Eq. (3.45), and the corresponding variance and mean are different. In Gaussian curve 1, the background noise mainly includes the current background-light component of street lamps, with a weak light intensity and small light-intensity fluctuation variance. In Gaussian curve 2, the background noise mainly includes components such as front and rear headlights, extremely bright billboards, and traffic lights, with strong light intensities and a large light-intensity fluctuation variance [24].

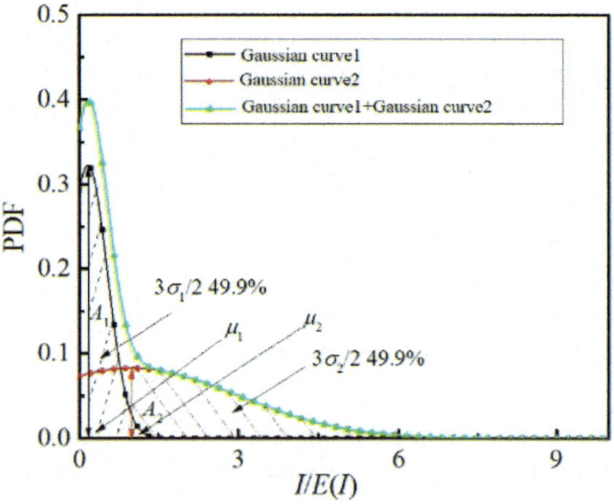

Fig. 3.20 Coriolis distribution model [24]

Figure 3.21 shows the probability distribution of background noise in nighttime visible-light communication under different scenarios. According to Fig. 3.21, the following can be seen:

1. The Coriolis distribution model can accurately describe the visible-light communication environment-noise model;
2. The probability distribution on rainy days does not differ significantly from that on sunny days; and
3. The intensity fluctuations generated by current streetlights are significantly smaller than those generated by other light sources.

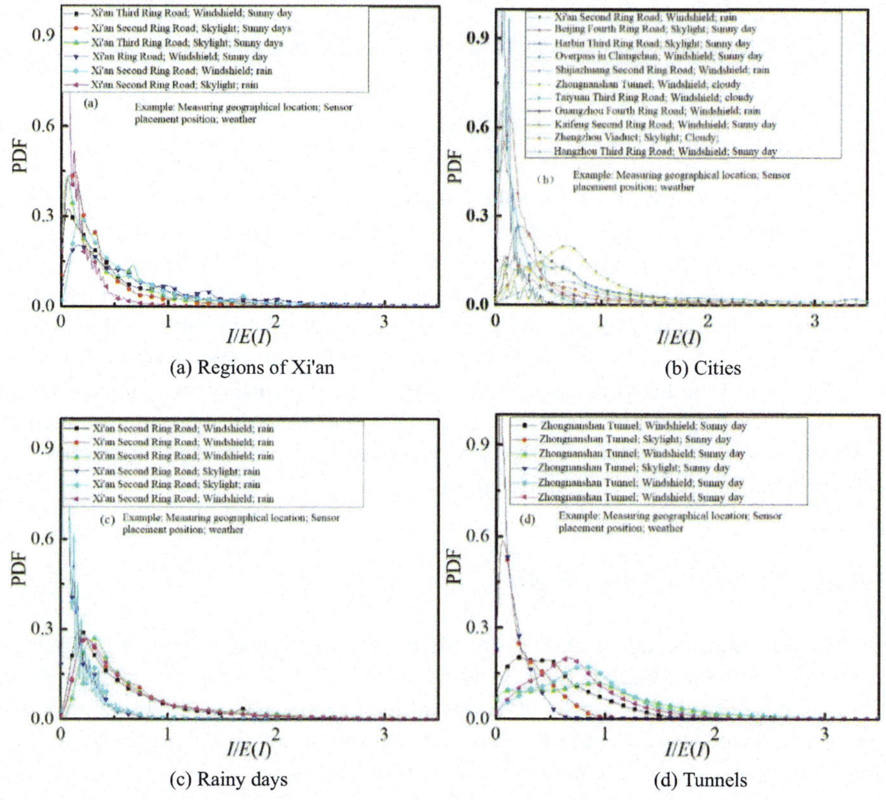

(a) Regions of Xi'an

(b) Cities

(c) Rainy days

(d) Tunnels

Fig. 3.21 Probability distribution of background noise in nighttime visible-light communication under different scenarios [24]

3.7 Signaling-Noise Models

3.7.1 Poisson-Noise Model

Poisson noise, also known as photon-counting noise or photon-jitter noise, mainly originates from the discreteness of photons. When an optical signal passes through a detector (such as an avalanche photodiode), the randomness of the photon number produces Poisson noise in the photocurrent. Poisson noise is statistical noise that describes the randomness of the photon-counting process.

Poisson noise is related to the light intensity, and the fluctuation in the photon number increases with increasing light intensity, resulting in more severe Poisson noise. Therefore, the Poisson distribution can be used for modeling [35]:

$$\mathrm{P}(\mathrm{X} = k) = \frac{e^{-\lambda}\lambda^k}{k!}, \tag{3.46}$$

where X represents a random variable, k is the number of occurrences per unit time, ! represents a factorial, and λ is the average occurrence rate per unit of time.

When designing optical-wireless communication systems, the impact of Poisson noise on the signal-to-noise ratio (SNR) and bit-error rate should be considered, and corresponding measures should be taken to reduce damage to the system performance. These measures include selecting high-performance photodetectors, adopting digital signal-processing technology, and optimizing the optical-system design. Poisson-noise models have important applications in many fields, such as medical imaging. For example, Poisson-noise models can be used to describe the absorption of X-rays through human tissues.

3.7.1.1 Photon-Counting Model

Photodiodes and avalanche diodes are often used as detection devices in optical-wireless communication receiving systems. When photons strike the diode surface, the avalanche effect they cause generates a high current, thus completing a photon count. In the receiver, the relationship between the input optical signal S_i and the output electrical signal S_0 can be represented as follows:

$$S_0 = S_i \cdot C_P \cdot M + S_n, \tag{3.47}$$

where C_p is the photon-detection efficiency, M is the device gain, and S_n is the noise signal.

To detect the amplitude of the output-pulse response, a judgment threshold can be set. Its probability distribution should satisfy the following requirements:

$$P(n, \lambda) = \frac{\lambda^n \exp(-\lambda)}{n!}, \tag{3.48}$$

where n and λ represent the number of photoelectrons and average pulse amplitude, respectively.

3.7.2 Multiplying Noise Model

In optical-wireless communication systems, when using devices with multiplication characteristics, such as avalanche photodiodes (APDs), PN junctions generate electron–hole pairs under the action of an external field. These electrons and holes move towards the N and P regions, respectively, under the action of an electric field. In the high-electric-field region (avalanche region), electrons and holes can obtain sufficient energy and form more electron–hole pairs through collisions and dissociations. This process is known as avalanche doubling.

The avalanche-doubling process is random because it depends on the path, velocity, and interaction with the lattice of electrons and holes within the avalanche region. This randomness leads to fluctuations in the multiplication factor, which results in avalanche-multiplication noise. Avalanche-doubling noise is superimposed on the detected optocurrent signal, reducing the signal-to-noise ratio, and thus affecting the performance of the communication system.

To evaluate the impact of avalanche-doubling noise on the system performance, this noise can be quantified by calculating the avalanche noise factor (F_n). The avalanche-multiplier noise factor F_n is a function of the gain (M) of the APD and is usually calculated using the following formula:

$$F_n = k \cdot M + \frac{1-k}{M}, \tag{3.49}$$

where M is the gain of the APD. k is the ionization coefficient ratio of the APD and represents the relative strength of the avalanche-doubling effect generated by the electrons and holes. The value of k is typically between 0 and 1, and is related to the material and structure of the APD.

When designing optical-wireless communication systems, it is necessary to balance the relationship between the APD gain and avalanche-multiplication noise. By optimizing the working voltage of the APD, selecting the appropriate gain, and adopting digital signal-processing technology, the effect of avalanche-multiplication noise on the system performance can be reduced.

3.7.3 Thermal-Noise Model

Thermal noise is mainly caused by the current generated by the random thermal motion of free electrons and ions inside a material. The characteristics of this noise can be determined by calculating its equivalent bandwidth and noise coefficients.

The thermal-motion direction of a single carrier exhibits a high degree of randomness, and over time, its average current tends toward zero. For a large number of carriers, the number of carriers that move instantaneously in one direction over a period of time is relatively large, whereas the number of carriers that move in another direction at another moment is also relatively large. However, their opportunities in all directions are equal. Therefore, the total current effect should be zero over a period of time.

At a specific moment, the current is not always zero and there may be a brief accumulation of local charges, resulting in fluctuating currents and voltages. This thermal noise in the load resistor is further amplified by a front-end amplifier. The thermal-noise power of the load resistor R_L can be defined as

$$S_t(f) = \frac{2k_B T}{R_L}, \tag{3.50}$$

where T represents the absolute temperature, R_L represents the load resistance, and k_B represents the Boltzmann constant, $k_B = 1.380649 \times 10^{-23}$ J/K.

Therefore, the power-spectral density of the thermal noise is related to the temperature, amplifier gain, and other factors; moreover, it is white noise. The noise figure can be used to describe the noise level of a device.

3.7.4 Photodetector-Amplification Noise Model

In photodetectors, the amplifier noise is an important source of noise caused by the thermal motion of electrons in resistors and capacitors. The magnitude of this noise is related to the amplification factors of the capacitance, resistance, photocurrent, and gain of the device.

In optical receivers, in addition to detectors, other devices generate a large amount of thermal noise, particularly the front high-frequency amplifiers. Owing to the high total gain of the amplifier group and the priority of high-frequency amplification, the noise of the high-frequency amplifier is gradually amplified in subsequent stages. Therefore, it is necessary to use a noise coefficient to describe the degree of SNR deterioration after the signal passes through an amplification circuit [6]. The noise coefficient is

$$N_F = 101g\frac{\text{Input signal noise power ratio}}{\text{Output signal noise power ratio}}(\text{dB}), \tag{3.51}$$

where N_F refers to the magnification of the thermal noise using a high-frequency amplifier. It can be seen that the smaller the N_F, the lower the additional noise caused by the amplifier itself. Owing to the thermal noise caused by both the transistors and resistive components in the circuit, the noise figure is usually greater than 1.

The amplified noise of the photodetector is white noise, which is independent of the frequency. In this case, the noise can be reduced by improving the internal structure of the amplifier. Therefore, establishing and analyzing the amplified noise model of photodetectors will help optimize their structure and performance and enhance their noise resistance and accuracy in practical applications.

3.7.5 Scattering-Noise Model

When light collides with the detector surface, a large number of electrons are generated. When electrons move at the output end of the detector, they trigger a current pulse effect, which can affect external monitors or sensors. This process can be described as follows:

$$x(t) = \sum_{j=1}^{K(v)} h(1 - Z_j), \tag{3.52}$$

where t represents time, v represents the volume of the detection site, j represents the jth electron, $h(t)$ represents the pulse response of a single electron, Z_j represents the release time of the jth electron, and $K(v)$ is a parameter used for counting. This count represents the number of electrons generated in the volume V of the detector area over the entire time interval, which is called the scattering-noise process.

3.7.6 Phase-Noise Model

The main factors contributing to phase noise are the laser phase jitter, time-varying fluctuations in the output phase of the local oscillator at the receiving end, nonlinear phase noise caused by the Kerr effect, and phase fluctuations caused by atmospheric turbulence and refractive-index fluctuations.

In a communication system, a photodetector converts the optical signal at the receiving end into an electrical signal, which is then multiplied by the signal output by the local oscillator to obtain the intermediate-frequency signal, and coherent demodulation is performed. However, with the passage of time, the phase of the output signal of the local oscillator may fluctuate, and the noise caused by this fluctuation makes it difficult to demodulate the subcarrier signal, causing the system performance to deteriorate.

The phase noise generated by the local oscillator φ follows the Tikhonov distribution [18] and its probability-density function is

$$P(\varphi) = \frac{\exp\left(\frac{\cos(\varphi)}{\sigma_\varphi^2}\right)}{2\pi I_0\left(\frac{1}{\sigma_\varphi^2}\right)}, |\varphi| \leq \pi, \tag{3.53}$$

where σ_φ^2 is the phase noise variance [21, 22] and $I_n()$ is the first type of n-order deformed Bessel function.

The Jacobi–Anger formula can convert the molecules of Eq. (3.53) into

$$\exp\left(\frac{\cos(\varphi)}{\sigma_\varphi^2}\right) = I_0\left(\frac{1}{\sigma_\varphi^2}\right) + 2\sum_{n=1}^{\infty} I_n\left(\frac{1}{\sigma_\varphi^2}\right) \cos n\varphi. \tag{3.54}$$

Substituting Eq. (3.54) into Eq. (3.53), the probability-density function of the phase noise can be expressed by a Fourier series as

$$p(\varphi) = \frac{1}{2\pi} + \sum_{n=1}^{\infty} c_n \cos(n\varphi), \tag{3.55}$$

where $c_n = \frac{I_n\left(\frac{1}{\sigma_\varphi^2}\right)}{\pi I_0\left(\frac{1}{\sigma_\varphi^2}\right)}$ is the Fourier coefficient of the probability-density function of Tikhonov phase noise.

3.7.7 1/f Noise Model

1/f noise is a type of random noise with long-range correlation, self-similarity, and nonstationarity. It is a general fluctuation phenomenon in the collective motion of particles, and reflects the internal characteristics of the system. It is largely caused by the interaction between impurities and defects in the device and carriers and can sensitively reflect various potential defects in the device.

The basic feature of *1/f* noise is that the noise power-spectral density is inversely proportional to the frequency. The $1/f$ current noise power spectrum is

$$S(f) = \frac{AI^\beta}{f^\gamma}, \tag{3.56}$$

where I is the current passing through the device, f is the frequency, A is determined by the structural characteristics of the device, and γ and β are specific coefficients.

The impact of *1/f* noise on optical communication systems may lead to data-transmission errors and reduce the system performance. Therefore, the designers of optical communication systems must take measures to minimize the impact of *1/f* noise as much as possible.

3.8 Transmission Characteristics of Optical Signals in Atmospheric Turbulence

3.8.1 Impact of Atmospheric Attenuation on Light Transmission

3.8.1.1 Atmospheric Absorption

In atmospheric channels, light waves interact with various molecules in the atmosphere, leading to molecular polarization and forced vibrations at the frequency of the incident light. By overcoming the resistance of atmospheric molecules and consuming energy, light waves are absorbed by the atmospheric molecules, resulting in laser power attenuation [25]. The absorption of a laser by atmospheric molecules depends on the different structures of the various molecules, and different absorbing molecules have completely different absorption lines and spectral absorption characteristics. Table 3.3 presents the absorption spectra of several main absorption molecules for visible and near-infrared light waves [25].

Domestic and foreign scholars have confirmed through extensive research [25] that water and other molecules in the atmosphere have corresponding absorption effects on light waves of different wavelengths. This absorption has a certain degree of selectivity: it exhibits strong absorption for certain wavelengths of light waves, which are almost impossible to pass through. For certain bands, they exhibit weak absorption and higher light-wave transmittance. We refer to these bands with higher transmittance as "atmospheric windows" [25].

Based on the distribution data of global atmospheric parameters with altitude, the absorption effect of a laser in the Earth's satellite atmospheric channel mainly occurs in the troposphere. As the altitude increases, the atmospheric-absorption effect on the laser transmission significantly weakens.

Table. 3.3 Main absorption spectra in the visible and near-infrared regions

Absorbing molecules	Central wavelength of the main absorption spectral line (μm)
H_2O	0.72 0.82 0.93 1.13 1.38 1.46 1.87 2.66 3.15 6.26 11.7 12.6 13.5 14.3
CO_2	1.4 1.6 2.05 4.3 5.2 9.4 10.4
O_2	4.7 9.6

For a laser beam with an incident power of $P(\lambda, 0)$, after transmission path L, the laser beam power decays, owing to the absorption of atmospheric molecules, as follows:

$$P(\lambda, L) = P(\lambda, 0) \exp[-k(\lambda)L], \tag{3.57}$$

where $P(\lambda, L)$ represents the laser power transmitted after passing a distance L; $k(\lambda)$ represents the absorption coefficient.

Research on aerosol particles has shown that, owing to their relatively large diameter, their light absorption is weak. Therefore, the absorption of light signals is not particularly obvious, but rather, they tend to reflect and scatter light.

3.8.1.2 Atmospheric Scattering

In atmospheric optics, regardless of the refraction or reflection of light, light propagates in a specific direction and has an intensity of zero in other directions. However, in an optically nonuniform medium, light can be seen in other directions; this optical phenomenon is called scattering. Atmospheric scattering refers to the influence of atmospheric particles on laser transmissions. Various particles with different diameters exist in the atmosphere, among which 0.1–10 μm-diameter particles have the greatest scattering effect on laser transmissions. The single-particle scattering theory uses Rayleigh scattering and Mie scattering to approximate the scattering effect of the atmosphere. Scattering does not decrease the laser energy, but it can affect the distribution characteristics of the light intensity and the shape of the light spots.

Molecular Scattering

Because of the small radius of atmospheric molecules, which is only on the order of 10^{-8} cm, the radiation wavelength is always much larger than the molecular radius in the visible and near-infrared bands. Scattering under this condition is usually called Rayleigh scattering. Rayleigh-scattering particles are mainly gas molecules; hence, it is also known as molecular scattering. The empirical formula for the Rayleigh scattering coefficient is [25]

$$\sigma_m = 0.827 \times N \times A^3 / \lambda^4, \tag{3.58}$$

where A is the cross-sectional area of the scattering element (cm^2); N is the number of molecules per unit volume (cm^{-3}); and λ is the wavelength of the light wave (μm).

Equation (3.58) shows that the molecular-scattering coefficient is directly proportional to the molecular density and molecular radius, and inversely proportional to the fourth power of the wavelength of the light wave [25]. As the radius of the scattering molecules increases, the scattering increases, and as the wavelength increases, the scattering weakens.

It can be inferred that visible light scatters more strongly than infrared light, whereas blue light scatters more strongly than infrared light. In a clear sky with relatively few larger particles, molecular scattering plays the main role, and because blue-light scattering is the strongest, a clear sky appears blue.

The volume-scattering coefficient of Rayleigh scattering is [26]

$$\sigma_m = \frac{8\pi^2\left(n^2-1\right)^2}{3N_S^2\lambda^4}\frac{6+3\delta_p}{6-7\delta_p} = \frac{8}{3}\left[\frac{\pi^2\left(n^2-1\right)^2}{N_S^2\lambda^4}\right], \tag{3.59}$$

where n is the refractive index of the particle and N_s is the density of the scattering elements.

The empirical formula for the scattering coefficient in air with low water–vapor content is

$$\sigma_m = 1.09 \times 10^{-3}\lambda^{-0.45}\,\mathrm{km}^{-1}. \tag{3.60}$$

In general, for particles with radius $r < 0.03\ \mu\mathrm{m}$, if the wavelength of the light wave is 1 μm, the calculation error of the Rayleigh scattering coefficient near m is $n < 1\%$ [27], which can accurately describe the intensity of molecular scattering in the atmosphere.

Aerosol Scattering

An aerosol is a colloidal particle formed by small solid or liquid particles dispersed and suspended in a gaseous medium. When the diameter of aerosol particles in the air is greater than or equal to the wavelength of the incident light, the distribution after the particles scatter the incident light is complex and asymmetric, and Rayleigh scattering is no longer applicable. The Mie scattering theory can be used to analyze such scattering phenomena.

The main characteristics of Mie scattering are that the distribution of the scattered light intensity with the angle becomes very complex, and the larger the particle size compared to the wavelength size, the more complex the distribution. As the particle size further increases, the angle of the scattered-light concentration becomes increasingly narrower.

In an actual communication process, the size and laser wavelength of aerosol particles, such as clouds, fog, rain, and snow, are equivalent or even larger. Therefore, for aerosol particles, the effect of Rayleigh scattering is relatively weak and can be ignored, whereas the impact of Mie scattering is mainly considered. The Mie scattering theory is a good approximation of aerosol scattering. The coefficient of Mie scattering is determined using Eq. (3.61) [28]:

$$\sigma_n = N(r)\pi r^2 Q_s(X_r, m), \tag{3.61}$$

where r is the particle radius, $N(r)$ is the number of particles per unit volume, and Q_s is the scattering efficiency (ratio of the energy scattered by the particles to the energy incident on the geometric cross section of the particles).

3.8.1.3 Atmospheric Attenuation

As a laser propagates in the atmosphere, the atmospheric-absorption effect causes the energy of the light to decrease with distance. The combined effect of atmospheric absorption and scattering manifests as atmospheric attenuation of laser transmissions measured by atmospheric transmittance. The atmospheric transmittance of monochromatic waves $T_{atm}(\lambda)$ can be expressed as follows [29]:

Horizontal uniform-path transmission:

$$T_{atm}(\lambda) = \exp[-k_e(\lambda)L]. \tag{3.62}$$

Oblique-path transmission:

$$T_{atm}(\lambda) = \exp\left(-\sec\varphi \int_0^Z k_e(\lambda, r)dr\right), \tag{3.63}$$

where $\tau(\lambda)$ is the atmospheric transmittance at wavelength λ, L is the horizontal transmission distance, Z is the vertical height of the slanted path, φ is the zenith angle of the slanted path, and $k_e(\lambda) = k_s + k_a$ is the atmospheric-extinction coefficient.

It should be noted that for oblique propagation, the atmospheric-extinction coefficient $k_e(\lambda)$ varies with height, so calculating the transmittance requires integrating the path. Assuming that the initial luminous flux is I_0, the luminous flux $I(L)$ after the transmission distance L is

$$I(L) = I_0 T_{atm}(\lambda). \tag{3.64}$$

If τ is defined as the optical thickness of atmospheric optical-transmission channels, it can be used to characterize channel-attenuation characteristics:

$$\tau = \tau_a + \tau_s, \tag{3.65}$$

where τ_a is the scattering optical thickness and τ_s is the absorption optical thickness.

The atmospheric-extinction coefficient $k_e(\lambda)$ varies significantly in different weather conditions, owing to atmospheric uncertainty. The influence of molecular scattering in the near-ground atmosphere is small, and the attenuation of light energy is mainly caused by the Mie scattering of large-scale particles. When designing atmospheric optical-wireless communication links, the wavelength within the atmospheric window is generally chosen; therefore, atmospheric absorption has a relatively small

Table. 3.4 International visibility levels [31]

Grade	Weather state	Visibility (km)	Scattering coefficient	Grade	Weather state	Visibility (km)	Scattering coefficient
0	Extremely dense fog	< 0.05	> 78.2	5	Haze	2–4	1.960–0.954
1	Thick fog	0.05–0.2	78.2–19.6	6	Light haze	4–10	0.954–0.391
2	Medium fog	0.2–0.5	19.6–7.82	7	Sunny	10–20	0.391–0.196
3	Mist	0.5–1	7.82–3.91	8	Very sunny	20–50	0.196–0.078
4	Light mist	1–2	3.91–1.96	9	Extremely clear	> 50	0.0141

impact on optical-energy attenuation, and atmospheric attenuation is mainly affected by atmospheric scattering.

Owing to the complexity of the number and scale distribution of various scattering particles in the atmosphere, it is difficult to directly calculate the atmospheric scattering and absorption coefficients. Therefore, atmospheric attenuation caused by near-ground laser atmospheric transmission can be calculated using visibility.

The empirical formula between the atmospheric-extinction coefficient and visibility is [30]

$$k_e(\lambda, R_v) = \frac{3.912}{R_v}\left(\frac{550}{\lambda}\right)^q \tag{3.66}$$

$$q = \begin{cases} 1.6 & R_v \geq 50\,\text{km} \\ 1.3 & 6\,\text{km} \leq R_v < 50\,\text{km} \\ 0.585R_v^{1/3} & R_v < 6\,\text{km} \end{cases}, \tag{3.67}$$

where R_v is the atmospheric visibility (km) and λ is the laser wavelength (nm). $k_e(\lambda, R_v)$ is the extinction coefficient (km^{-1}). Visibility is the main indicator used to measure the attenuation effect of the atmosphere on visible light. During the daytime, it refers to the farthest distance that the human eye can see, with the horizontal sky as the background. At night, it refers to the distance at which a medium-intensity unfocused light source can be seen. Meteorology generally divides visibility into ten levels based on meteorological conditions, as shown in Table 3.4.

3.8.2 *Impact of Atmospheric Turbulence on Laser Transmissions*

Owing to the different physical properties of different parts of the atmosphere, coupled with the influence of heat and wind, the atmosphere flows continuously, resulting in the formation of different airflow vortices from temperature, pressure, density, flow velocity, and size. These vortices undergo continuous movement changes, and their movements affect each other and are superimposed on each other to form random turbulence movements, that is, atmospheric turbulence.

The turbulent movement causes the atmospheric refractive index to undulate, resulting in random fluctuations of the amplitude and phase parameters of the light waves, and the flickering, twisting, division, diffusion, spatial coherence, and polarization changes of the beam. This influence is illustrated in Fig. 1.1.

3.8.2.1 Light-Intensity Fluctuation

As shown in Fig. 3.22, when the beam diameter is much larger than the turbulent scale, the beam cross section will contain multiple turbulent vortices, each of which generates independent scattering and diffraction effects of the beam illuminated on it, causing the detector to receive a light intensity I that changes over time and fluctuates around the average value $\langle I \rangle$, which is known as light-intensity scintillation. The size of the flicker is directly related to the strength of the turbulence, and the flicker effect of the light intensity can increase the bit-error rate of communication systems [32].

According to Rytov's classic light-intensity scintillation theory, the logarithmic variance of the light-intensity fluctuations in Kolmogorov's locally uniform and isotropic weak-turbulent region $\sigma_{\ln I}^2$ can be expressed as [32]

$$\sigma_{\ln I}^2 = C_0 C_n^2 k_0^{7/6} r^{11/6}, \tag{3.68}$$

where r represents the transmission distance; C_0 is a constant, where $C_0 = 1.23$ for plane waves and $C_0 = 0.496$ [85] for spherical waves; C_n^2 is the atmospheric refractive-index structure constant; and $k_0 = 2\pi / \lambda$ represents the number of waves.

$$\sigma_{\ln I}^2 = 1.23 C_n^2 k_0^{7/6} r^{11/6} (\text{plane wave}) \tag{3.69}$$

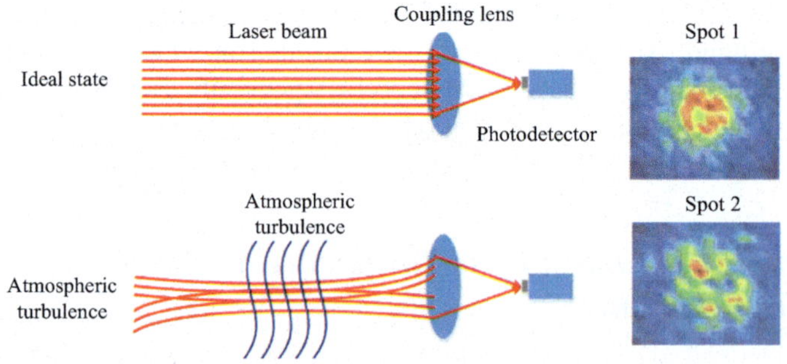

Fig. 3.22 Impact of light-intensity scintillation on an atmospheric laser-communication system [32]

$$\sigma_{\ln I}^2 = 0.496 C_n^2 k_0^{7/6} r^{11/6} (\text{spherical wave}) \tag{3.70}$$

The Rytov theory better reflects the strength of atmospheric scintillation. When the turbulence intensity is small, $\sigma_{\ln I}^2 < 1$. As C_n^2 and r increases, the logarithmic intensity variance of $\sigma_{\ln I}^2$ gradually increases, with a numerical value greater than 1. After reaching the maximum value, multiple scattering in the atmosphere begins to play a major role, $\sigma_{\ln I}^2$ does not increase with C_n^2 and r but gradually decreases and eventually approaches 1, which is the scintillation-saturation effect.

When $\sigma_{\ln I}^2 > 1$, it indicates the occurrence of a scintillation-saturation effect (strong fluctuation region). The empirical formula for $\sigma_{\ln I}^2$ at this time is [32]

$$\sigma_{\ln I}^2 = 1 + 0.87 \left(1.23 C_n^2 k_0^{7/6} r^{11/6}\right)^{-2/3}. \tag{3.71}$$

The scintillation characteristics of the laser spot intensity can be expressed not only by the logarithmic-intensity variance $\sigma_{\ln I}^2$, but also by the logarithmic amplitude variance σ_χ^2 or the normalized intensity variance σ_l^2.

For low-intensity turbulence, the expression for the logarithmic-amplitude variance σ_χ^2 is [32]

$$\sigma_\chi^2 = 0.252 \sigma_l^2 = 0.31 C_n^2 k_0^{7/6} r^{11/6} (\text{plane wave}) \tag{3.72}$$

$$\sigma_\chi^2 = 0.0954 \sigma_l^2 = 0.125 C_n^2 k_0^{7/6} r^{11/6} (\text{spherical wave}). \tag{3.73}$$

In the weakly turbulent region, σ_χ^2 satisfies the Gaussian distribution form.

3.8.2.2 Beam Drift

As shown in Fig. 3.23, a wireless laser-communication system consists of two parts: a transmitting antenna and a receiving antenna. The transmitting and receiving antennas are typically selected from a Cassegrain telescope, and their optical structure can be simplified to a single-lens system. From Fig. 3.23, the following can be seen.

Ignoring Atmospheric Turbulence

When atmospheric turbulence is ignored, the emitted beam is transmitted to the receiving surface of the receiving optical antenna, and its main optical axis coincides with the viewing axis of the receiving antenna. The receiving optical system converges the beam focus on the receiving end of the communication detector. A splitter prism is placed in the internal optical path of the optical system, dividing the

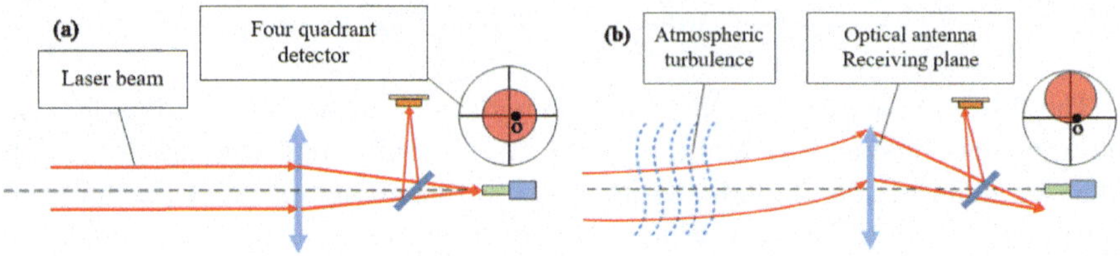

Fig. 3.23 Impact of spot drift on optical-wireless communication systems [9]. **a** Ignoring atmospheric turbulence; **b** considering atmospheric turbulence

receiving optical path into two: one beam is used for communication detection and the other is used for spot-position detection.

The detection sensor is a four-quadrant detector, and the position of the light spot on the detector's receiving surface can be used to determine the center position of the light spot in the four-quadrant detector. The center of the light spot coincides with the center of the four-quadrant detector [33].

Considering Atmospheric Turbulence

When considering atmospheric turbulence, the beam drifts, owing to the influence of atmospheric turbulence during transmission, resulting in the random movement of the beam perpendicular to the optical axis and a deviation from the receiving surface. At this point, the optical axis of the transmitting beam no longer coincides with the axis of view of the receiving antenna; an angle causes the focus of the beam converging through the optical antenna to shift, and the shifted focus is no longer incident on the receiving surface of the communication detector. At this point, the beam power received by the communication detector decreases significantly, adversely affecting the signal detection of the communication system.

The offset of the centroid of the light spot relative to the center position of the four-quadrant detector can be detected through four-quadrant detection. With further turbulence development, the transmitting beam may completely drift out of the receiving surface of the receiving antenna. At this point, the receiving antenna cannot converge with the transmitting beam, which interrupts the communication link [33].

Usually, the drift of a light spot is described by a change in the center position of the light point. The centroid is defined as follows [33]:

$$\rho_0 = \frac{\iint \rho I(\rho) d\rho}{\iint I(\rho) d\rho} \tag{3.74}$$

$$x_0 = \frac{\iint x I(x, y) dx dy}{I(x, y) dx dy} \tag{3.75}$$

$$y_0 = \frac{\iint yI(x, y)dxdy}{I(x, y)dxdy}, \tag{3.76}$$

where ρ_O is the centroid coordinate and x_O and y_O are the centroid coordinates in the horizontal and vertical directions, respectively. $\iint$ represents double integrals.

The offset of the center of the light spot can be expressed using the following formula [9]:

$$\sigma_\rho^2 = \langle \rho_0^2 \rangle = \iint \iint \left(\frac{(\rho_1 \cdot \rho_2)I(\rho_1)I(\rho_2)d\rho_1 d\rho_2}{\left[\iint I(\rho)d\rho\right]^2} \right), \tag{3.77}$$

where ρ_0 is the centroid coordinate, $\iint$ represents double integrals, $\langle \cdot \rangle$ represents the ensemble average, and $\iiiint$ represents quadruple integrals.

Assuming that the drifts in the horizontal and vertical directions are not interdependent, the overall drift variance of the laser spot can be expressed using the following formula:

$$\sigma_\rho^2 = \sigma_x^2 + \sigma_y^2. \tag{3.78}$$

In both cases, σ_x is the moving average variance in the horizontal direction and σ_y is the moving average variance in the vertical direction.

When z is 0–L, the deviation of the laser beam in the $z = L$ direction is caused by a turbulent disturbance in the z direction. Thus, the offset variance in the acceptance plane (L-z) can be expressed as [9]

$$\sigma_\rho^2 = 6.08D^{-1/3} \left[L^2 \int_0^L C_n^2(z)dz - 2L \int_0^L C_n^2(z)zdz + \int_0^L z^2 C_n^2(z)dz \right]. \tag{3.79}$$

Assuming that the changes in turbulence intensity are consistent during transportation, [9]

$$\sigma_\rho^2 = 2.03 C_n^2 L^3 D^{-\frac{1}{3}}. \tag{3.80}$$

When the aperture of the transmitting antenna of the focused beam is D, the following equation applies [9]:

$$\sigma_\rho^2 = 6.08 D^{\frac{-1}{3}} L^2 \int_0^L C_n^2(z)\left(1 - \frac{z}{L}\right)^{\frac{11}{3}} dz. \tag{3.81}$$

Assuming that the variation in turbulence intensity is constant during transmission [9],

$$\sigma_\rho^2 = 1.305C_n^2L^3D^{-\frac{1}{3}}. \tag{3.82}$$

3.8.2.3 Beam Extension

As shown in Fig. 3.24, if a very short observation time is taken on the receiving plane, the distance of a widened spot with radius r_1 deviates from the origin r_2, which is the beam-drift amount. If the time taken is sufficiently long, a large spot with a mean-square radius of r_3 is generated by the random walk of the spot. The former is called short-term beam expansion, whereas the latter is called long-term beam expansion [34].

Beam expansion refers to the change in the radius or area of a light spot on the receiving plane caused by turbulent vortices. When light passes through small-scale vortices, diffraction effects occur, leading to beam expansion. When light passes through large-scale vortices, refraction occurs, causing beam drift [34].

If the emitting laser beam is a Gaussian beam and it is in an ideal channel, there is no turbulence effect. The received light intensity I and radius of the received spot ω_f can be calculated using the following formula [34]:

$$I(s,\rho) = I_0\left(\frac{\omega_0}{\omega_f}\right)^2 \exp\left[-\frac{2\rho^2}{\omega_f^2}\right] \tag{3.83}$$

$$\omega_f = \sqrt{\omega_0^2 + \left(\frac{2s}{k\omega_0^2}\right)}, \tag{3.84}$$

where I_0 is the emission intensity, $2\omega_0$ is the diameter of the laser output aperture, and s is the transmission distance.

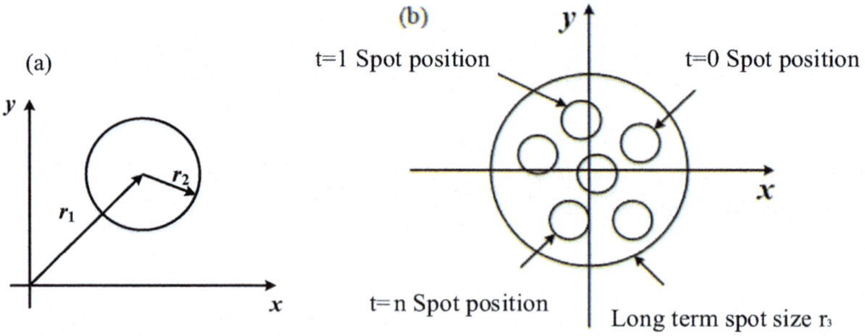

Fig. 3.24 Schematic diagram of the spot observed on the receiving plane. **a** Short-term observed image on the receiving surface; **b** long-term observed image on the receiving surface

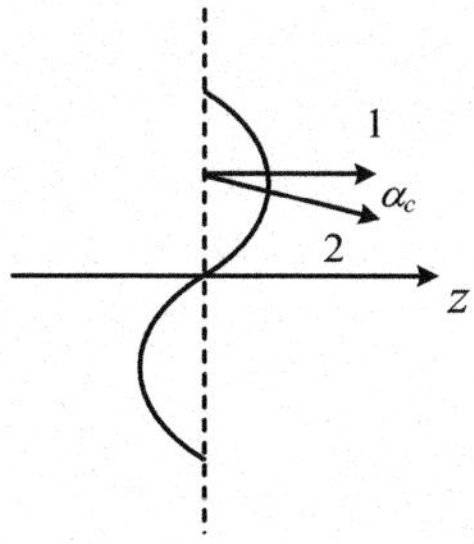

Fig. 3.25 Schematic diagram of the local arrival angle. (1) Uniform phase front normal; (2) equal phase normal

Considering the beam-spreading effect caused by turbulence in actual atmospheric channels, the radius of the transmitted laser spot is

$$\omega_t = (\omega_f^2 + 4.38 C_n^2 l_0^{-1/3} s^3)^{1/3}. \tag{3.85}$$

According to Eqs. (3.84) and (3.85), the change in the spot radius can be calculated as

$$\omega_{t2}^2 = 4.38 C_n^2 l_0^{-1/3} s^3. \tag{3.86}$$

When the variation ω_{t2}^2 of the spot radius ranges from 0.1 to 10 mm, the attenuation of the laser intensity can reach 3 dB.

3.8.2.4 Arrival-Angle Fluctuation

As shown in Fig. 3.25, laser propagation in a uniform medium has a uniform wavefront. When propagating in a turbulent atmosphere, the fluctuation of the atmospheric refractive index in different parts of the beam cross section results in different phase changes in different parts of the beam wavefront [35].

3.8.2.5 Aiming Error

In optical-wireless communication links, the random shift of a beam caused by changes in the atmospheric refractive index results in signal attenuation and aiming errors at the detector. The width of a beam propagating through atmospheric turbulence at distance z can be approximated as ω_z [36]:

$$\omega_z \approx \omega_0 \left[1 + \varepsilon \left(\frac{\lambda z}{\pi \omega_0^2} \right)^2 \right]^{1/2}, \tag{3.87}$$

where ω_0 is the beam width at the transmitting end, $\varepsilon = \left(1 + 2\omega_0^2/p_0^2(z)\right)$, $p_0(z) = \left(0.55C_n^2k^2z\right)^{-3/5}$ is the coherence length, C_n^2 is the atmospheric structure constant, $k = 2\pi/\lambda$ represents the number of light waves, and z is the propagation distance.

The aiming error can be approximated as [36]

$$h_p(r) \approx A_0 \exp\left(-\frac{2r^2}{\omega_{zeq}^2}\right), \tag{3.88}$$

where r is the spot drift, $A_0 = \left[erf(v)\right]^2$, $\omega_{zep} = \omega_z\left[\sqrt{\pi}erf(v)/\left(2v\exp(-v^2)\right)\right]^{1/2}$, equivalent beam radius $v = \sqrt{\pi/2}R_a/\omega_z$, R_a is the detector radius, and A_0 is the optical power when the beam offset $r = 0$.

Assuming that the offsets in the horizontal and vertical directions are the same, a Rayleigh distributed beam-offset model is used:

$$f_r(r) = \frac{r}{\sigma_s^2}\exp\left(-\frac{r^2}{2\sigma_s^2}\right), \tag{3.89}$$

where σ_s^2 is the jitter variance received by the detector.

The probability distribution of the aiming error can be expressed as [36]

$$f_p(h_p) = \frac{\eta^2}{A_0^{\eta^2}}h_p^{\eta^2-1}, \tag{3.90}$$

where $\eta = \omega_{zeq}/2\sigma_s$ is the ratio of the equivalent beam radius at the detector to the pointing-displacement error.

3.9 Summary

This chapter presented research on the transmission characteristics of laser signals in the atmosphere. First, the cause of atmospheric turbulence was analyzed, followed by a theoretical analysis of the impact of the influence of atmospheric absorption, atmospheric scattering, atmospheric turbulence, and aiming errors on optical-wireless communication systems. It also focused on the advantages and disadvantages of different atmospheric-turbulence power-spectrum models and provided a theoretical basis for further research and analysis.

References

1. Ke XZ, Deng LJ (2016) Wireless laser communication. Science Press, Beijing, pp 1–169
2. Monin AS (1977) Statistical fluid mechanics: the mechanics of turbulence. Am J Phys 318(10):202–220
3. Rosa R (2006) Turbulence theories. IEEE Trans Antennas Propag 45(42):295–303
4. Wu XQ (2006) Atmospheric optics turbulence, models and measurement technology. J Anhui Normal Univ 29(2):104–107
5. Tofsted DH (1992) Outer-scale effects on beam-wander and angle-of-arrival variances. Appl Opt 31(27):5865–5870
6. Hufnagel RE (1974) Topical meeting on optical propagation through turbulence. Boulder in Proc IEEE 10(51):45–52
7. ITU—R (2001) Propagation data and prediction methods required for the design of earth-space telecommunication systems. Radio Commun Study Group Meet 206(3):618–624
8. Lee SY (1986) Theory of multidimensional wave packet propagation. Chem Phys 108(3):451–459
9. Ke XZ, Wu JL, Yang SJ (2021) Research progress and prospects of atmospheric turbulence for optical-wireless communication. J Radio Sci 36(03):323–339
10. Kolmogorov AN (1941) The local structure of turbulence in an incompressible viscous fluid for very large Reynolds numbers. Sov Phys Usp 30(42):301–305
11. Karman VT (1948) Progress in the statistical theory of turbulence. Proc Natl Acad Sci U S A 34(11):530–539
12. Tatarski VI (1961) Wave propagation in a turbulent medium. Hill Pergamon Press, New York, pp 69–78
13. Karman VT (1963) From low-speed aerodynamics to astronautics. Pergamon Press, New York, pp 152–168
14. Greenwood DP, Tarazano D (1974) A proposed form for the atmospheric microtemperature in the input range. USAF Rome Air Development Center. Technical Report RADC, 42(16):442–456
15. Hill RJ (1978) Modified spectrum of atmospheric temperature fluctuations and its application to optical propagation. J Opt Soc Am 68(04):892–897
16. Andrews LC (1992) An analytical model for the refractive index power spectrum and its application to optical scintillations in the atmosphere. J Mod Opt 39(9):1849–1853
17. Lipatack C (2004) Optical wave propagation through non-Kolmogorov atmospheric turbulence. University of Central Florida, pp 152–159
18. Zilberman A, Golbraikh E, Kopeika NS (2008) Propagation of electromagnetic waves in Kolmogorov and non-Kolmogorov atmospheric turbulence: three-layer altitude model. Appl Opt 47(34):6385–6391
19. Toselli I, Andrews LC, Phillips RL et al (2008) Free space optical system performance for laser beam propagation through non Kolmogorov turbulence for uplink and downlink paths. Opt Eng 47(2):115–122
20. Ke XZ, Tian XC (2015) Modeling and simulation of two-dimensional correlated K-distribution turbulent channels. J Opt 35(04):41–49
21. Wang CH (2019) Research on phase noise characteristics and compensation techniques of optical-wireless subcarrier modulation. Xi'an University of Technology, Xi'an, pp 34–42
22. Ke XZ, Wang CH, Chen D (2018) Phase noise analysis of subcarrier modulation system in Malaga atmospheric turbulence channel. J Commun 39(11):80–86
23. Wang S (1979) Optical beam propagation for a partially coherent source in the turbulent atmosphere. JOSA 69(9):319–322
24. Ke XZ, Qin HH, Yang SJ, et al (2021) Night background light noise model of Vehicle-to-everything visible light communication system. J Radio Sci 36(06):986–990
25. Xu CY, Xie DL, Yang H (1999) Analysis of laser atmospheric transmission transmissivity. Optoelectronics 26(18):7–11

26. Zhou BH, Chen SR (1984) Laser principle. Arms industry Press, Beijing, pp 81–88
27. Zhong XH (2004) Fundamentals of modern optics. Peking University Press, Beijing, pp 467–472
28. Jakeman E, Pusey PN (2002) The significance of K-distributions in scattering experiments. Phys Rev Lett 14(40):546–550
29. Isaac IK, Harel H, Prasanna A et al (1990) Scintillation reduction using multiple transmitters. SPIE 29(90):102–113
30. Deng DZ (2004) Impact of atmospheric random channel on wireless laser communication. Southwest Jiaotong University, Chengdu, pp 51–53
31. Kruse PW (1962) Elements of infrared technology: generation transmission and detection. John Wiley & Sons. Inc., New York, pp 151–155
32. Song ZF, Zhang YX, Gong ZB (1986) Discussion on several issues of beam drift in turbulent atmosphere. J Quantum Electron 10(01):57–64
33. Jiao Y (2012) Current status and future development trends of laser communication technology. Inf Commun 10(5):206–207
34. Steart DA, Essenwanger OMA (1982) Survey of fog and related optical propagation characteristics. Rev Geophys Space Phys 20(3):481–495
35. Zhang WT, Zhu BH (2007) Research on the influence of atmospheric turbulence on laser signal transmission. J Univ Electron Sci Technol China 36(4):784–787
36. Cui LY, Fei W, Xue BD (2018) Theoretical investigations of infrared optical wave modulation transfer function models in anisotropic marine turbulence. J Opt Soc Am 35(9):1496–1503

Chapter 4
Atmospheric-Turbulence Noise-Measurement Experiment

This chapter first introduces the meaning of atmospheric-turbulence noise, derives the atmospheric-turbulence noise model, conducts a detailed study of the noise-distribution model, and analyzes the adaptation of the turbulence areas of various noise-distribution models. Considering the system performance in terms of the signal-to-noise ratio, bit-error rate, fading probability, and interrupt probability, the impact of atmospheric-turbulence noise on the performance of a optical-wireless communication system was analyzed.

4.1 Atmospheric-Turbulence Noise

When a signal beam travels through air, it is affected by the refraction of the air medium. This occurs in optical phenomena, such as strong scintillation, light-spot drift, and beam expansion. These phenomena affect the receivers in optical-wireless communication systems. Because these effects are caused by various factors related to turbulence in the atmospheric channel, the noise that affects optical-wireless transmissions is referred to as atmospheric-turbulence noise.

Turbulence noise is not additive noise. It acts on the signal light in the form of multiplication or convolution, which is equivalent to modulating the signal light through the physical characteristics of the atmospheric channel. Frequently, only two things are considered to be turbulence noise: the refraction caused by random temperatures changes in the atmosphere and interference that sends light flashes to the receiver. In this chapter, in addition to the previous elements, scattering elements are also considered to be turbulence. Air-soluble beams and droplets that move through the atmosphere cause light to scatter, and the scattered light causes light and random attenuation.

X. Ke and C. Ke, *Noise Models in Optical-Wireless Communication Systems*,
Optical Wireless Communication Theory and Technology,
https://doi.org/10.1007/978-981-97-7550-7_4

4.1.1 Theoretical Derivation

The propagation of light beams through atmospheric turbulence involves relatively complex physical processes, including optical phenomena, such as light-intensity attenuation, intensity flickering, and spot drift. The received light-field strength can be expressed using the following formula [1]:

$$u_0(r) = A_0(r)\exp(i\phi_0(r)), \tag{4.1}$$

where the independent variable r is the propagation-path vector of the light, and $u_0(r)$ is the field strength of the received light, which can be expressed as a combination of the field-strength amplitude $A_0(r)$ and phase $\phi_0(r)$.

Considering that propagation is affected by atmospheric turbulence, the received light field can be described as [2]

$$u(r) = A(r)\exp(i\phi(r)), \tag{4.2}$$

where $u(r)$ is the received light field affected by atmospheric turbulence, which can also be expressed as a combination of the amplitude $A(r)$ and phase $\phi(r)$.

Rytov's theory indicates that $u_0(r)$ and $u(r)$ satisfy the following relationship [2]:

$$u(r) = u_0(r)\exp(\Phi), \tag{4.3}$$

where Φ is an exponential factor that characterizes atmospheric turbulence. Substituting Eqs. (4.1) and (4.2) into Eq. (4.3) and taking the logarithm, we obtain

$$\Phi = \ln\left[\frac{u(r)}{u_0(r)}\right] = \ln\left[\frac{A(r)}{A_0(r)}\right] + i[\phi(r) - \phi_0(r)] \triangleq \chi + iS, \tag{4.4}$$

where χ is the logarithm of the ratio of amplitude A to A_0, and S is the difference between phases $\phi(r)$ and $\phi_0(r)$.

Atmospheric turbulence affects the magnitude and phase of the received light-field intensity. For optical-wireless communication systems that directly detect light-intensity modulation, the phase term S can be ignored; hence, Φ can be expressed as

$$\Phi = \ln\left[\frac{A(\mathbf{r})}{A_0(\mathbf{r})}\right] = \chi. \tag{4.5}$$

After the beam propagates through atmospheric turbulence, if background-light noise is considered, the received signal of the detector can be expressed as

$$A(t) = A_0(t)\exp(\chi) + e_n, \tag{4.6}$$

where $A_0(t)$ is the light-intensity signal transmitted through an ideal atmospheric medium without disturbance and χ is the turbulent multiplicative-noise factor. e_n is expressed as turbulent additive noise in the beam transmission [1], which includes background-light noise, detector electrical noise, shot noise, resistance thermal noise, $1/f$ noise, etc.

Background-light noise refers to light emitted or reflected by light sources other than the signal light received by the optical-wireless receiver. Background light sources include ambient-temperature radiation, direct sunlight, reflection of sunlight by objects in the field of view, and atmospheric scattering of sunlight. Atmospheric turbulence also affects background lighting.

4.1.2 Influence of Atmospheric-Turbulence Noise on Optical-Communication Performance

4.1.2.1 Signal-to-Noise Ratio

Based on Rytov's theory, the propagation formula of a plane wave in a free medium is [1]

$$E_0(\vec{r}) = A_0(\vec{r}) \exp(i\phi_0(\vec{r})), \tag{4.7}$$

where $E_0(\vec{r})$ is the electromagnetic-field distribution of the beam, $A_0\left(\vec{r}\right)$ is the amplitude of the laser beam without turbulence, and ϕ_0 is the phase of the light wave.

Beam expansion and drift significantly weaken the signal at the receiving end. The fluctuation rate of the beam deflection is less than 1 or 2 kHz [2], and the beam-expansion influence is also small. Therefore, the light-intensity fluctuation caused by a change in the atmospheric-turbulence refractive index is the most important factor of signal attenuation. A change in the atmospheric refractive index causes a change in the laser-beam distribution. This wave equation can be expressed as [2]

$$E(\vec{r}) = A(\vec{r}) \exp(i\phi_0(\vec{r})) = E_0(\vec{r}) \exp(\Phi), \tag{4.8}$$

where $A_0\left(\vec{r}\right)$ represents the amplitude of the laser beam transmitted in a turbulent flow. According to Eq. (4.4),

$$\Phi = \ln\left[\frac{A(\vec{r})}{A_0(\vec{r})}\right] + i(\phi(\vec{r}) - \phi_0(\vec{r})) = \sigma_\chi + iS, \tag{4.9}$$

where i is the imaginary unit. σ_χ^2 represents the real part of the logarithmic amplitude fluctuation caused by atmospheric turbulence and S is the imaginary part of the light-wave phase fluctuation. As shown in Eq. (4.9), the intensity and phase fluctuations of the received signal caused by atmospheric turbulence are the main factors affecting the signal-to-noise ratio (SNR) and bit-error rate (BER).

For plane waves, when the turbulence intensity is weak, the refractive-index structure constant C_n^2 is symmetrical about the optical path, and the logarithmic light-intensity fluctuation variance is [2]

$$\sigma_{\ln I}^2 = \left\langle (\ln I - \langle \ln I \rangle)^2 \right\rangle = 1.23 C_n^2 k_0^{7/6} r^{11/6}. \tag{4.10}$$

Under weakly turbulent conditions, the variance in the logarithmic amplitude during a horizontal transmission is [2]

$$\sigma_\chi^2 = 0.31 C_n^2 k_0^{7/6} r^{11/6}. \tag{4.11}$$

Assuming that only the noise caused by atmospheric turbulence is considered and the influence of other noises is ignored, σ_χ^2 [2] can be obtained as follows:

$$\sigma_\chi^2 = \ln\left[\frac{A(\vec{r})}{A_0(\vec{r})}\right] = \ln\left[\frac{A_0(\vec{r}) + A_n(\vec{r})}{A_0(\vec{r})}\right] = \ln(1+\varepsilon), \tag{4.12}$$

where $A_n(\vec{r})$ denotes the noise amplitude. If $\varepsilon = \frac{A_n(\vec{r})}{A_0(\vec{r})}$, when the signal strength is I_0 and the noise strength is $\langle I_n \rangle$, the SNR caused by atmospheric turbulence is [3]

$$SNR = \frac{I_0}{\langle I_n \rangle} = \frac{\left\langle A_0^2(\vec{r}) \right\rangle}{\left\langle A_n^2(\vec{r}) \right\rangle} = \left[\langle \varepsilon^2 \rangle\right]^{-1}. \tag{4.13}$$

For weak turbulence, ε is very small, so $\sigma_\chi = \ln(1+\varepsilon) \approx \varepsilon$; therefore [3],

$$SNR = \left(\sigma_\chi^2\right)^{-1} = \left(0.31 C_n^2 k_0^{7/6} r^{11/6}\right)^{-1}. \tag{4.14}$$

In Eq. (4.13), $\sigma_{\ln I}^2 = 1.23 C_n^2 k_0^{7/6} r^{11/6}$. Combining Eqs. (4.13) and (4.14), the relationship between the SNR and logarithmic-intensity variance can be obtained as follows:

$$SNR = \left[\frac{31}{123} \sigma_{\ln I}^2\right]^{-1}. \tag{4.15}$$

Figure 4.1 shows the relationship between the SNR and transmission distance using lasers of different wavelengths for transmission, with an atmospheric-turbulence structure constant of $C_n^2 = 10^{-14}\ \mathrm{m}^{-2/3}$.

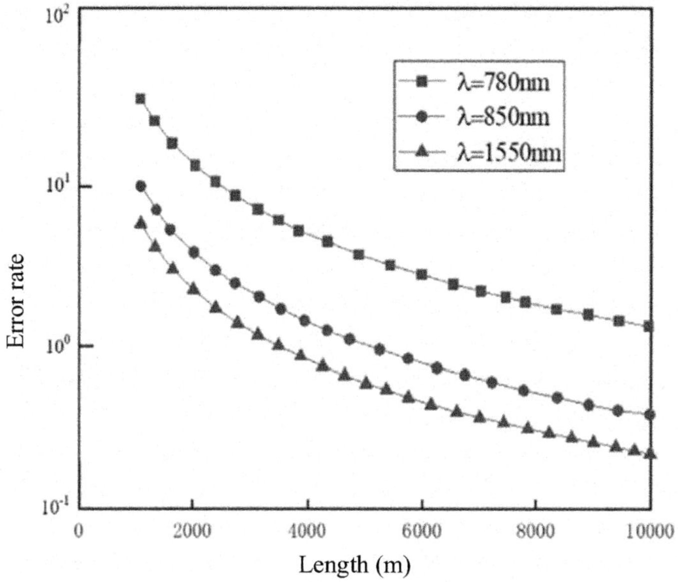

Fig. 4.1 Relationship between the SNR and transmission distance at different wavelengths λ

As shown in Fig. 4.1, the SNR decreases with an increase of the transmission distance. For the same transmission distance, the SNR of the system differs when the laser wavelength is different. The signal-to-noise ratio is better when using long-wavelength laser transmissions than when using short-wavelength laser transmissions. When the transmission power is the same, the signal-to-noise ratio attenuates to 1 when a laser with a 1550-nm wavelength transmits at a distance of 5000 m. A laser with an 850-nm wavelength attenuates to 1 by the time it reaches 3500 m.

The signal-to-noise ratios of lasers with 780-nm and 850-nm wavelengths are slightly different. When the SNR attenuates to 1, the transmission-distance difference between the two is approximately 100 m. The SNR of the laser with a 780-nm wavelength drops to 0.4679. It is completely submerged in the noise and the signal cannot be identified.

Figure 4.2 shows the relationship between the SNR and transmission distance when the laser wavelength $\lambda = 1550$ nm under different atmospheric-turbulence structure constants. It can be seen from Fig. 4.2 that the SNR decreases with the increase of the transmission distance. The SNR of the system also differs when the transmission distance is the same and the atmospheric-turbulence intensity is different.

When using a 1550-nm wavelength for transmission, under weak-turbulence conditions ($C_n^2 = 10^{-15}\,\text{m}^{-2/3}$), the SNR is significantly better than that of strong turbulence. When the transmission distance is the same, the SNR of a system

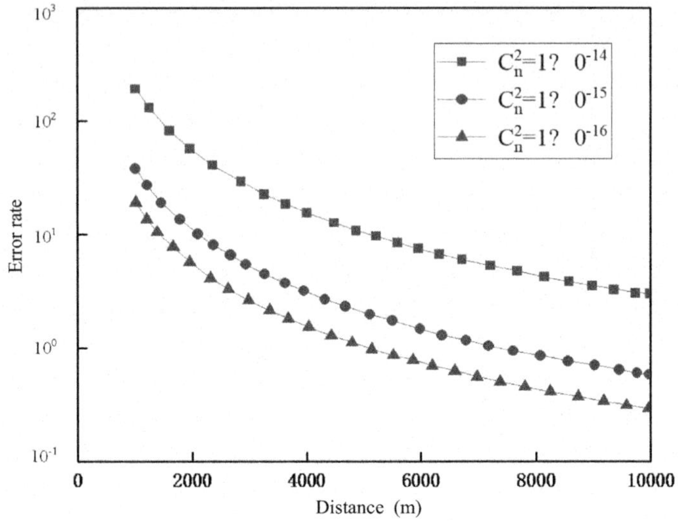

Fig. 4.2 Relationship between the SNR and transmission distance for different C_n^2

with different turbulence intensities is also different. Moreover, the atmospheric-turbulence intensity has a great influence on the signal-to-noise ratio. Under weak turbulent conditions, a laser can travel up to 10 km while the SNR is still greater than 1

4.1.2.2 Bit-Error Rate

Atmospheric turbulence changes the amplitudes of light waves. When not considering other noises, but only the impact of atmospheric turbulence on the bit-error rate of a communication system, we can approximate the amplitude change as being caused by atmospheric-turbulence noise. The logarithmic light-intensity fluctuation variance expression is [4]

$$\ln\frac{I(r,t)}{I_0} = 2\ln\frac{A(r)}{A_0(r)} = 2\frac{A_0(r)+A_i(r)}{A_0(r)} = 2\ln(1+\varepsilon), \tag{4.16}$$

where $A_i(r)$ is the noise amplitude and $\varepsilon = A_i(r)/A_0(r)$ is the noise-to-signal amplitude ratio. When ε is very small, we approximate that $\chi = \ln(1+\varepsilon) \approx \varepsilon$. Assuming that the signal strength is I_0 and the noise strength is $\langle I_n \rangle$, the SNR caused by atmospheric turbulence can be expressed as [4]

$$SNR = \frac{I_0}{\langle I_n \rangle} = \left\langle \frac{A_0^2(r)}{A_i^2(r)} \right\rangle = \frac{1}{\langle \varepsilon^2 \rangle} = \frac{1}{\langle \chi^2 \rangle}, \tag{4.17}$$

where I_0 is the signal strength, $\langle I_n \rangle$ is the noise strength, $A_i(r)$ is the noise amplitude, and $\varepsilon = A_i(r)/A_0(r)$ is the amplitude ratio of the noise and signal.

Under weakly turbulent conditions, the logarithmic light-intensity fluctuation variance and logarithmic amplitude-fluctuation variance have the following relationship:

$$\sigma_I^2 = 4\chi^2. \tag{4.18}$$

Under strong turbulence, the signal-to-noise ratio simplified by the Taylor series can be approximated as [4]

$$SNR = \frac{1}{\langle \chi^2 + \chi^3 + \cdots \rangle} \approx \frac{1}{\alpha \langle \chi^2 \rangle}, \quad 1 \leq \alpha \leq 2, \tag{4.19}$$

where α is the flicker-intensity factor.

In an actual laser communication system, the laser emitted by the transmitter can be regarded as a plane wave after being collimated by an optical lens. Therefore, we analyzed it as a plane wave. For a plane wave, the logarithmic light-intensity fluctuation under weakly turbulent conditions is

$$\sigma_I^2 = 1.23 C_n^2 k^{7/6} L^{11/6}, \tag{4.20}$$

where C_n^2 is the atmospheric-turbulence refractive-index structure constant, $k = 2\pi/\lambda$ is the wavenumber, and L is the propagation distance. For a digital laser-communication system, the bit-error rate when the optical receiver receives an optical signal is

$$BER = \frac{1}{2}\left[erfc\left(\frac{SNR}{\sqrt{2}} \right) \right]. \tag{4.21}$$

Combining this with Eq. (4.14), we determine that the relationship between the bit-error rate and logarithmic light-intensity fluctuation variance is

$$BER = \frac{1}{2}\left[erfc\left(\frac{4}{\sqrt{2}\sigma_I^2} \right) \right] = \frac{1}{2}\left[erfc\left(\frac{4}{\sqrt{2} \times 1.23 C_n^2 k^{7/6} L^{11/6}} \right) \right]. \tag{4.22}$$

Figure 4.3 shows the relationship between the BER and the transmission distance using transmission lasers of different wavelengths with an atmospheric-turbulence structure constant of $C_n^2 = 10^{-14}\,\mathrm{m}^{-2/3}$.

It can be seen from Fig. 4.3 that under the same atmospheric-turbulence conditions, the BER increases with an increase in the transmission distance. When the transmission distance is the same and the transmission wavelength differs, the bit-error rate also differs. When the transmission wavelength is different, the bit-error rate is better when using long-wavelength transmissions than when using short-wavelength transmissions.

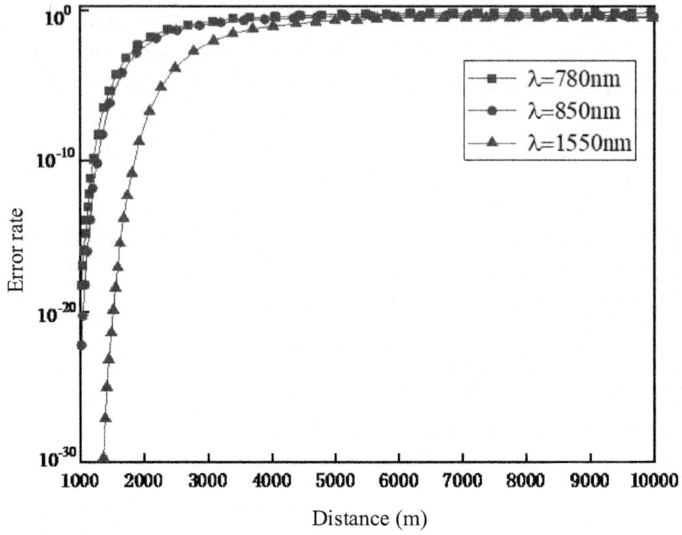

Fig. 4.3 Relationship between the bit-error rate and transmission distance at different wavelengths λ.

When the bit-error rate of the system is required to be less than 10^{-9}, a laser with a 1550-nm wavelength corresponds to a transmission distance of approximately 2000 m, and a laser with an 850-nm wavelength corresponds to a transmission distance of approximately 1200 m. At this time, the transmission distances of the lasers with 780-nm and 850-nm wavelengths are slightly different; the transmission-distance difference between the two is approximately 100 m.

The relationship between the BER and the transmission distance was simulated under different atmospheric-turbulence structure constants with a fixed 1550-nm wavelength. The simulation results are shown in Fig. 4.4.

It can be seen from Fig. 4.4 that for the same laser wavelength, the atmospheric-turbulence structure constant has a great influence on the BER. When using 1550-nm wavelengths for transmission, under weak turbulence conditions ($C_n^2 = 10^{-14}\,\mathrm{m}^{-2/3}$), the BER is significantly better than that in a strong turbulent flow. In a weakly turbulent flow, when the BER is less than 10^{-9}, the laser can travel 7 km.

4.1.2.3 Fading Probability

The fading probability indicates the possibility that the received light intensity will be lower than a given light-intensity threshold. When the signal-to-noise ratio output by the detector is sufficiently high, the fading probability P_{fa} can be given by the cumulative-distribution function of the light-intensity signal [5]:

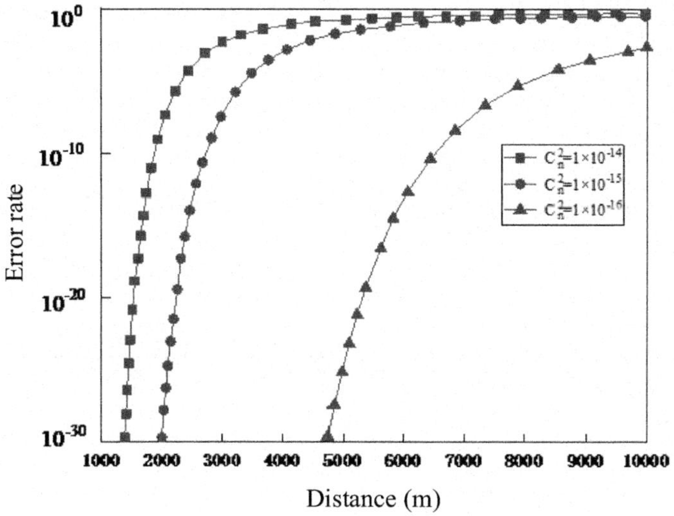

Fig. 4.4 Relationship between the bit-error rate and transmission distance under different C_n^2 conditions

$$P_{\mathrm{fa}} = \Pr(I \le I_T) = \int_0^{I_T} p_I(I)dI, \tag{4.23}$$

where I_{T} represents the light-intensity threshold level and $p_I(I)$ is the probability-density function of the light-intensity fluctuation. The fading-threshold parameter F_T is defined in [5] to indicate a fading threshold lower than the average received light intensity. The unit is dB and the expression is [5]

$$F_T = 10 \lg\left[\frac{\langle I(0, L)\rangle}{I_T}\right], \tag{4.24}$$

where I_{T} is the average light intensity along the Gaussian-beam axis.

Under weak fluctuation, $p_I(I)$ in Eq. (4.24) is a log-normal distribution, and the corresponding fading-probability model is

$$P_{\mathrm{fa}} = \frac{1}{2}\left[1 + erf\left(\frac{\frac{1}{2}\sigma_{\ln I}^2 - 0.23F_T}{\sqrt{2}\sigma_{\ln I}}\right)\right], \tag{4.25}$$

where $\sigma_{\ln I}^2$ represents the variance of the logarithmic light intensity, and $erf(\cdot)$ represents the error function. Under medium and strong turbulence, $p_I(I)$ in Eq. (4.23) follows a gamma–gamma distribution. The corresponding fading-probability model is

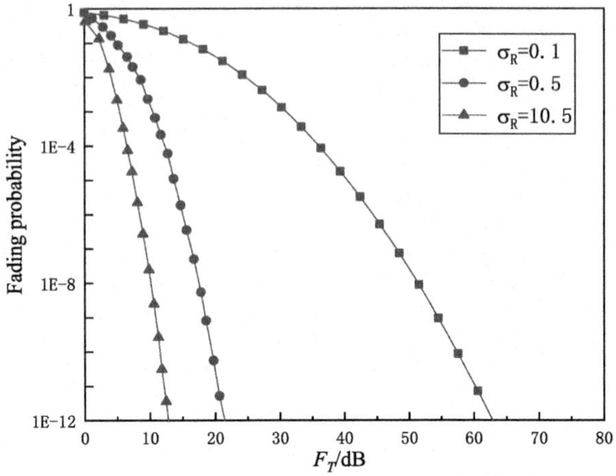

Fig. 4.5 Variation curve of the fading probability with a threshold value under different turbulence intensities

$$P_{\text{fa}} = \frac{\pi}{\Gamma(\alpha)\Gamma(\beta)\sin[\pi(\alpha-\beta)]} \times \left\{ \begin{array}{l} \dfrac{(\alpha\beta)^{\beta}}{\beta\Gamma(\beta-\alpha-1)}\exp(-0.23\beta F_T) \\ \times {}_1F_2[\beta;\beta+1,\beta-\alpha-1;\alpha\beta]\exp(-0.23\beta F_T) \\ -\dfrac{(\alpha\beta)^{\beta}}{\alpha\Gamma(\beta-\alpha-1)}\exp(-0.23\alpha F_T) \\ \times {}_1F_2\big[\alpha;\alpha+1,\alpha-\beta-1;\alpha\beta\exp(-0.23\beta F_T)\big] \end{array} \right\}, \tag{4.26}$$

where $\sigma_{\ln I}^2$ represents the variance of the logarithmic light intensity and $erf(\cdot)$ represents the error function.

Figure 4.5 shows the variation curve of the fading probability with the threshold under different turbulence intensities. It can be observed from the figure that with a continuous increase in the threshold parameter F_T, the fading probability begins to decrease gradually, which shows that the optical-wireless communication-system performance has improved to varying degrees. However, owing to the difference in turbulence intensity, there are certain differences in the fading-probability characteristics with the threshold parameters. As the turbulence intensity increases, the fading probability of the noise signal increases, and the performance of the communication system is more severely degraded.

4.1.2.4 Channel Outage Probability

Assuming that the signal strength received by the wireless communication system is $\langle I \rangle$, the flicker signal-to-noise ratio can be defined as [6]

$$S/N = \eta^2 I^2 / N_0, \tag{4.27}$$

where η is the photoelectric-conversion efficiency and N_0 is the additive Gaussian white-noise power spectrum. Then, the outage probability can be expressed as [6]

$$\begin{aligned} P_{out} &= P(S/N \le \mu) = P\left(\eta^2 I^2 / N_0 \le \mu\right) \\ &= P\left(I \le \frac{\sqrt{\mu N_0}}{\eta}\right) = \int_0^{\frac{\sqrt{\mu N_0}}{\eta}} f(I) dI, \end{aligned} \tag{4.28}$$

where μ is the threshold value of the signal-to-noise ratio and I is the received light-intensity value. Substituting Eqs. (4.7) and (4.8) under the two turbulence forms into Eq. (4.29), the outage probability P_{out} in the cases of weak turbulence and medium–strong turbulence, respectively, can be obtained.

In the case of weak turbulence, P_{out} is

$$P_{out} = 1 - \frac{1}{2} erfc\left[\frac{\left(\ln \mu + \sigma_0^2/2\right)}{\sigma^2 \sqrt{2}}\right]. \tag{4.29}$$

In the case of medium and strong turbulence, P_{out} *is*

$$P_{out} = \frac{1}{\Gamma(\alpha)\Gamma(\beta)} \times G_{1,3}^{2,1}\left[\alpha\beta\mu\left[\begin{matrix}1 \\ \alpha, \beta, 0\end{matrix}\right]\right]. \tag{4.30}$$

$G_{p,q}^{m,n}$ [ ] is a Meijer G-function.

The channel-outage probability and transmission distance were simulated under different turbulence-intensity conditions. In the simulation, the laser wavelength $\lambda = 1.55$ μm, and the atmospheric refractive-index structure constant is $C_n^2 = 5.02 \times 10^{-15}$ m$^{-2/3}$ (weak turbulence), $C_n^2 = 8.04 \times 10^{-14}$ m$^{-2/3}$ (medium turbulence), and $C_n^2 = 1.26 \times 10^{-12}$ m$^{-2/3}$ (strong turbulence). The transmission distances were set to $L = 500$ and 1000 m.

The channel-outage probability adopts a log-normal channel distribution under weak turbulence and uses the gamma–gamma channel distribution under medium and strong turbulence. Figure 4.6 shows the simulation results of the outage probability under different turbulence intensities and transmission distances.

As shown in Fig. 4.6, the outage probability of the two channel models increases with an increase in the normalized-threshold SNR, indicating that the optical-communication reliability decreases. When the system transmission distance $L =$ 500 and 1000 m, the outage probability increases with an increase in the turbulence intensity under the same normalized-threshold signal-to-noise ratio.

When SNR = 0.5 dB and $L = 500$ m, the channel-outage probabilities for weak, medium, and strong turbulence are 0.09, 0.21, and 0.41, respectively. When SNR

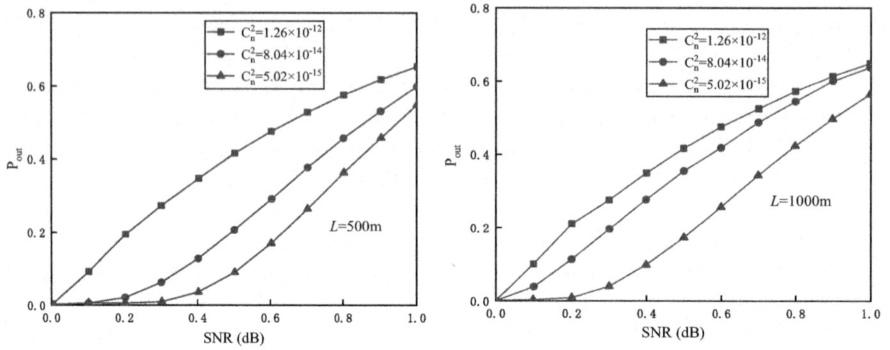

Fig. 4.6 Channel-outage probability under different turbulence intensities

= 0.5 dB and L = 1000 m, the outage probabilities are 0.17, 0.35, and 0.4, respectively. The simulation results show that when the communication distance is short, the change in the reliability of the optical-wireless communication system is significantly affected by the atmospheric refractive-index structure constant. When the communication distance increases, the outage probability is less affected by a change in C_n^2.

4.2 Experimental Method and Measurement Links

The impact of atmospheric-turbulence noise on laser transmissions significantly decreases the accuracy and transmission distance of laser communications, even interrupting the communication in severe cases. We established three measured links to analyze the probability-density distribution of actual atmospheric-turbulence noise samples under different weather and transmission-distance conditions. The atmospheric refractive-index structure constant, partial slope and steepness, fading characteristics, and interrupt probability were observed.

4.2.1 Experimental Method

The experimental system for transmitting through the atmosphere is shown in Fig. 4.7. The wavelength of the transmitting end was 525 nm, and a semiconductor laser with a 200-mW power and a 5-mm beam diameter was used as the light source. After the laser emission was expanded by focusing the lens (focal length of 10 mm), it was shot from a MACRO-style telescope with a 105-mm diameter.

After the atmospheric-channel transmission, the laser signal was focused by a Maksutov–Cassegrain (Mak-Cass) optical telescope at the receiving end before

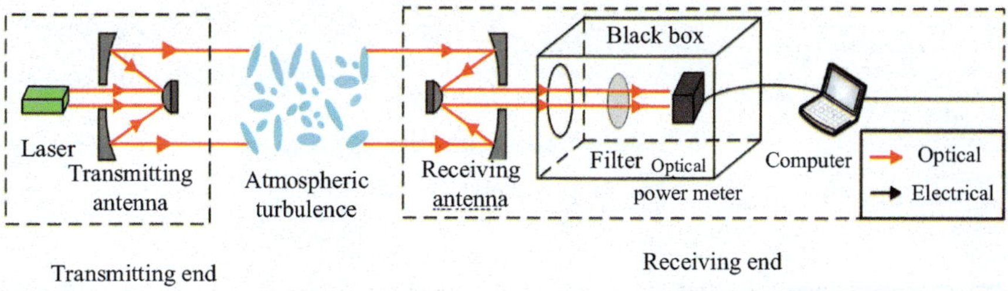

Fig. 4.7 Experimental schematic diagram

measuring the optical power. In addition, the meteorological parameters during the experiment were recorded. To minimize the impact of background light, filters were placed after the receiving-end Mak-Cass optical telescope, and the signal-receiving device was placed in a homemade dark box.

4.2.2 Measurement Links

From August 2021 to April 2023, we established three laser-transmission measurement links in Xi'an, Shaanxi Province, China, and conducted multiple measurement experiments. The three links are as follows:

Link 1: The transmission end was located on the eighth floor of Building No. 2 of the Jinhua Campus of Xi'an University of Technology. The receiving end was located on the 11th floor of the sixth building of the Jinhua Campus of the Xi'an University of Technology. The link passed through terrain including buildings and trees.

Link 2: The transmission end was located in the Kisen District of Xi'an. The receiving end was located on the 8th floor of the sixth building of the Jinhua Campus of Xi'an University of Technology. The link passed through areas with streets and buildings.

Link 3: The transmission end was located in Xiaojiazhai Village, Bailuyuan, Xi'an. The receiving end was located on the 12th floor of the sixth building of the Jinhua Campus of Xi'an University of Technology. The link passed through relatively complicated terrain, mainly buildings, woods, and streets. The complex terrain disrupted the atmospheric state, which had a certain impact on the experimental measurements (Figs. 4.8, 4.9 and 4.10).

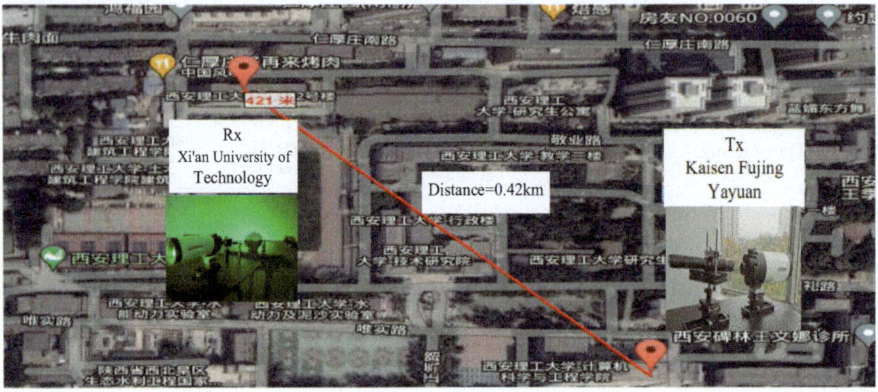

Fig. 4.8 0.42-km wireless laser-communication link diagram

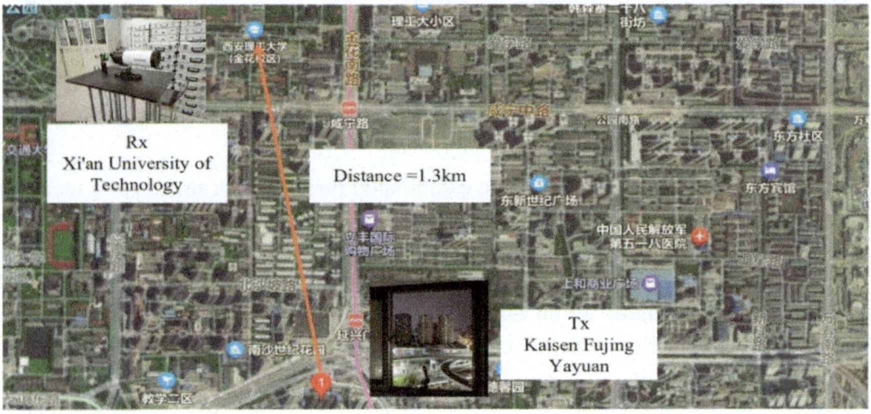

Fig. 4.9 1.32-km wireless laser-communication link diagram

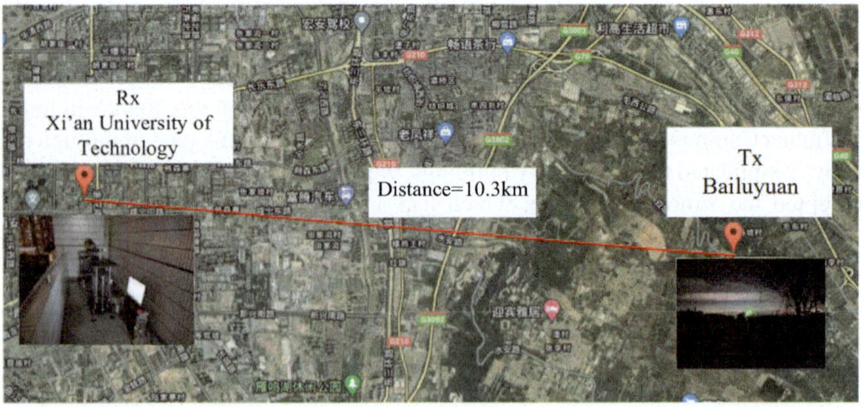

Fig. 4.10 10.3-km wireless laser-communication link diagram

4.3 Impact of Atmospheric Turbulence on Laser Transmissions

4.3.1 Light Intensity Probability-Density Distribution

During turbulent atmospheric transmissions, the laser intensity undulates randomly, owing to the random field of the atmospheric refractive index. Theory and experiments have proven that the uniform orientation process is the same and has a stable increase. The intensity fluctuation follows the normal distribution under weak-turbulence conditions and follows the gamma–gamma distribution in strong turbulence [7]. However, turbulence in an actual atmosphere not only varies between weak and strong fluctuations, but does not strictly follow the distribution with stable increments (especially in complex terrain). Figure 4.11 shows the probability-density distribution diagrams of different weather-measurement samples.

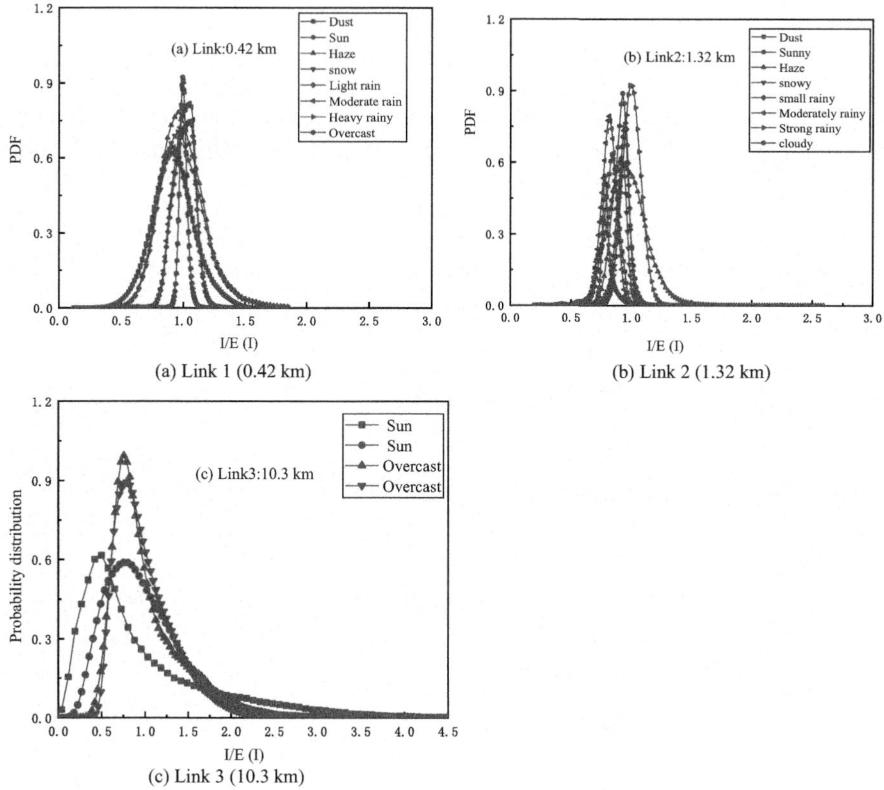

(a) Link 1 (0.42 km)

(b) Link 2 (1.32 km)

(c) Link 3 (10.3 km)

Fig. 4.11 Probability-density distribution of the light intensity of measured samples

The interval of each data group was divided into 600 equal areas according to [min (I), max (I)]. The frequency count of i in the area constitutes another sequence, $y = (y_1, y_2, \ldots, y_n)$. After drawing the fitting curve, each value of i corresponds to the y value on the fitting curve, and we obtain the third sequence, $Y = (Y_1, Y_2, \ldots Y_n)$. R can be defined as [8]

$$R = \frac{\langle Y \cdot y \rangle - \langle Y \rangle \cdot \langle y \rangle}{\sqrt{DY \cdot Dy}}, \tag{4.31}$$

where DY and Dy are the differences between Y and y, respectively, and < > indicates the comprehensive average. In Tables 4.1, 4.2 and 4.3, R^2 ($0 < R^2 < 1$) represents correlation coefficient, and ${R_1}^2$ and ${R_2}^2$ are the superior fitting values of the positive-state distribution. In the gamma–gamma distribution, R^2 is closer to 1, indicating that the result is closer to the ideal.

Considering Fig. 4.11 and Tables 4.1, 4.2 and 4.3, it can be seen that the measurement distances of the Link 1 and Link 2 samples are 0.42 and 1.32 km, the turbulence strength is in a weak-medium turbulence area, light intensity fluctuation degree is small.

In Table 4.2, ${R_1}^2$ under various weather conditions is higher than ${R_2}^2$, and the probability-density distribution conforms to a positive-state distribution. The order of heavy rain and cloudy days weakens, resulting in the peak probability value of the probability-density function, and the distribution interval of the probability-density function gradually decreases.

Table 4.1 R^2 fitting parameters for Link 1 (0.42 km)

Trial (Link 1)	Dust	Sun	Fog	Snow	Drizzle	Moderate rain	Heavy rain	Overcast
${R_1}^2$	0.9948	0.9965	0.9961	0.9955	0.9987	0.9912	0.9989	0.9914
${R_2}^2$	0.9719	0.9817	0.9726	0.9717	0.9837	0.9815	0.9836	0.9788

Table 4.2 R^2 fitting parameters for Link 2 (1.32 km)

Trial (Link 2)	Dust	Sun	Fog	Snow	Light rain	Moderate rain	Heavy rain	Overcast
${R_1}^2$	0.9831	0.983	0.9912	0.9946	0.9954	0.9961	0.9968	0.9981
${R_2}^2$	0.9742	0.9851	0.9826	0.9811	0.9823	0.9817	0.9891	0.9911

Table 4.3 R^2 fitting parameters for Link 3 (10.3 km)

Trial (Link 3)	Sunny	Sunny	Overcast	Overcast
${R_1}^2$	0.9517	0.9582	0.9617	0.9641
${R_2}^2$	0.9898	0.9902	0.9918	0.9916

The distance in Fig. 4.11c is 10.3 km, which is far greater than the measurement distances of Link 1 and Link 2. Moreover, it is in a medium–strong turbulence area. R_2^2 in Table 4.3 is higher than R_1^2. Rather than following the normal distribution, it follows the gamma–gamma distribution. Compared with the overcast measurement sample, the peak probability value of the probability-density distribution measured on sunny days was low, the distribution interval was larger, and the tail phenomenon was obvious.

4.3.2 Skewness and Steepness

The skewness and steepness are statistics describing the symmetry and steepness degree of the distribution of a sample sequence, respectively. This method is used to determine whether the sample-sequence distribution follows a normal distribution. A partial slope is used to describe the distribution symmetry centered on the statistical average of the sample sequence.

A negative bias coefficient indicates that the sample sequence is more widely distributed on the left side of the average. A positive slope coefficient indicates that the sample sequence has a broader distribution on the right side of the average value. If the slope coefficient is zero, then the sample sequence follows a Gaussian distribution or other symmetrical distribution.

The steepness coefficient is used to describe the statistics of the sample sequence relative to the normal distribution flatness. A high steepness coefficient indicates that the sample sequence is steeply distributed near the average. A low steepness coefficient indicates that the sample sequence is flat near the average.

The partial slope s and steepness degree K of the sample sequence x [9] are calculated as follows:

$$S = E\left[\left(\frac{X - E(X)}{D(X)}\right)^3\right] \tag{4.32}$$

$$K = E\left[\left(\frac{X - E(X)}{D(X)}\right)^4\right], \tag{4.33}$$

where E(X) and D(X) represent the expectations and squares of the sample sequence, respectively.

Figure 4.12 shows the relationship between the slope of the measuring sample, the steepness, and the Rytov index. From Fig. 4.12a, b, it can be seen that the noise in various types of weather conditions increases with the Rytov index, and the absolute value of the partial slope and steepness increase. The slope and steepness degree of the noise samples of the hazy, dusty, and sunny days are significantly higher than the slope and steepness of the noise samples of rainy and cloudy days. The skewness degree of the noise signal and its statistical distribution in the samples of link 1 and link 2 decreased in the order of dust day, sunny day, haze day, snow day, light rain, moderate rain, heavy rain and cloudy day, and the steepness was also the same. The skewness and steepness of the noise signal and its statistical distribution in the measurement sample of Link 3 decrease in the order of sunny and cloudy days.

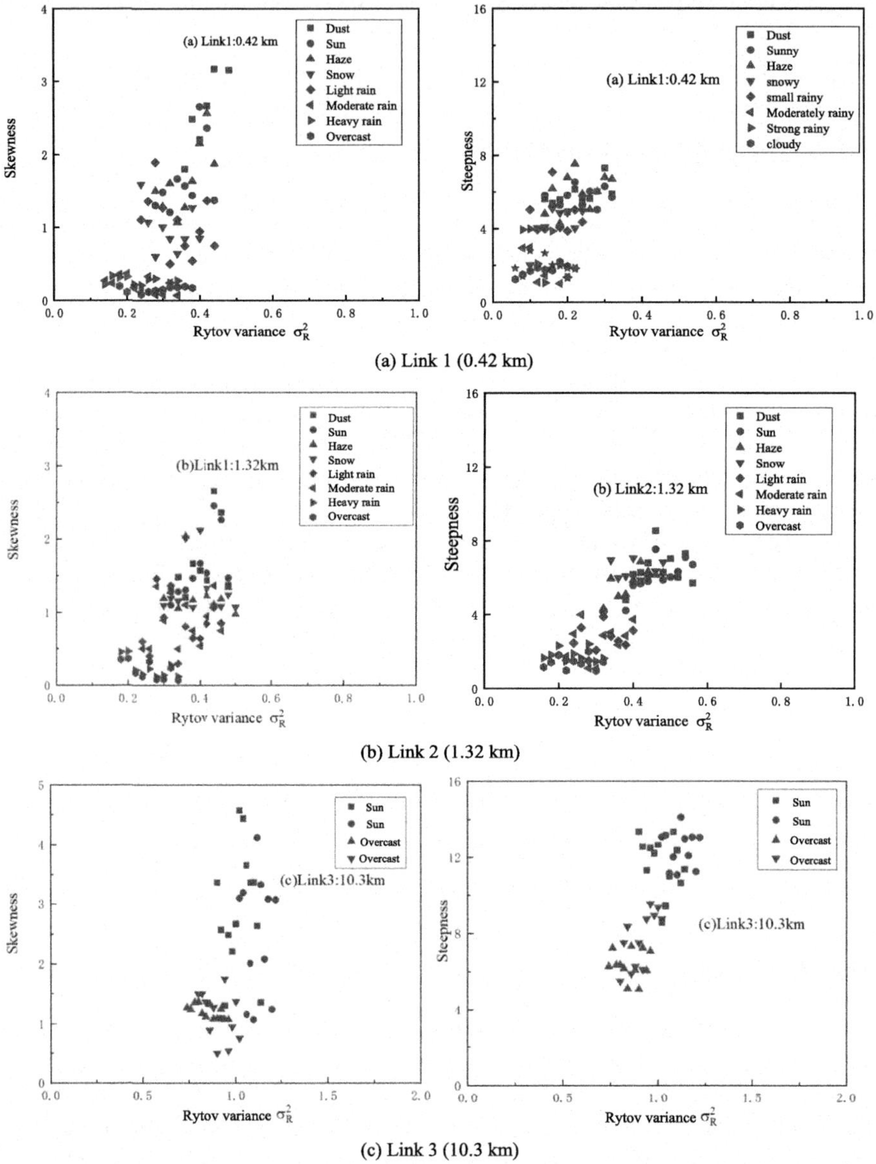

Fig. 4.12 Relationship between the skewness and steepness of the measurement samples and Rytov exponent

In Fig. 4.12a, the Rytov index of the tested sample data is in the overall range of 0.01–0.4, which is a weak turbulence area. In Fig. 4.12b the Rytov index of the sample data is 0.1–0.6, which is in the weak to medium turbulence area. In Fig. 4.12c, the Rytov index of the sample data in the test is 0.7–1.3, which is in the medium–strong turbulence area. This shows that the Rytov index increases with the increase of the link distance. The turbulence strength increases accordingly, and the degree of skewness and steepness of the noise sample also increases accordingly.

4.3.3 Atmospheric Refractive-Index Structure Constant

Optical turbulence is mainly caused by random changes in the temperature. The atmospheric refractive-index structure constant is an important parameter that reflects the characteristics of atmospheric optical turbulence, and quantitatively describes the optical turbulence strength [7]. According to the Davis turbulence theory, the structure constant can be divided into three categories:

1. $C_n^2 \gg 2.5 \times 10^{-13}\,\text{m}^{-2/3}$ indicates strong turbulence.
2. $6.4 \times 10^{-17}\,\text{m}^{-2/3} \ll C_n^2 < 2.5 \times 10^{-13}\,\text{m}^{-2/3}$ indicates medium turbulence.
3. $C_n^2 \ll 6.4 \times 10^{-17}\,\text{m}^{-2/3}$ indicates weak turbulence [7].

The constant's value changes depending on the weather conditions, geographical location, and underlying surface conditions. The atmospheric refractive-index structure constant quantitatively describes the intensity of the optical turbulence. This section uses the scintillation method to measure C_n^2 [7], and its expression is

$$C_n^2 = 4.48 C D_t^{7/3} L^{-3} \sigma_X^2, \tag{4.34}$$

where L is the link length, $C = D_r/D_t$, where D_r and D_t are the sending aperture size and receiving aperture size, respectively. σ_X^2 is the logarithmic amplitude variance, which can be expressed as [7]

$$\sigma_X^2 = 0.25 \ln\left[1 + \left(\frac{s}{\langle I \rangle}\right)^2\right], \tag{4.35}$$

where $\langle I \rangle$ is the mean and s is the standard deviation of the light intensity.

The experimental measured light intensity values were processed to obtain the mean light intensity and the standard deviation of light intensity. Then, the experimental analysis data were calculated using Eqs. (4.34) and (4.35) to obtain the refractive-index structure constant of the measurement sample.

Considering Fig. 4.13 and Table 4.4, it can be seen that the main concentration of C_n^2 in the 0.42-km-link measurement samples, the main concentration is between 1×10^{-18} m$^{-2/3}$ and 1×10^{-16} m$^{-2/3}$, and only a small part is near 1×10^{-16} m$^{-2/3}$, and only a small portion is in the range of 1×10^{-18} m$^{-2/3}$–1×10^{-16} m$^{-2/3}$, with a maximum C_n^2 value of 1.06×10^{-16} m$^{-2/3}$. The majority of the 1.32-km link measurement samples are concentrated within the 1×10^{-17} m$^{-2/3}$–1×10^{-15} m$^{-2/3}$ interval range, with a maximum value of 2.02×10^{-15} m$^{-2/3}$. The 10.3-km-link measurement samples are all between 1×10^{-14} m$^{-2/3}$ and 1×10^{-12} m$^{-2/3}$. Within the range of 1×10^{-14} m$^{-2/3}$–1×10^{-12} m$^{-2/3}$, the maximum value is 3.15×10^{-12} m$^{-2/3}$, indicating that the turbulence intensity increases with the link distance.

In the noise samples measured in both Link 1 and Link 2, the values of dust, haze, sunny days, snowy days, and light rainy days are significantly higher than those of heavy rain, moderate rain, and cloudy days. In the samples tested in Link 3, the values for the sunny day are significantly higher than those for the cloudy day.

The intensity of the atmospheric turbulence during the testing period was divided according to the Davis inequality [10]. Among the measurement samples in Link 1, 95 values belonged to the moderate-turbulence category (1.13% of the total sample size) and 8305 values belonged to the weak-turbulence category (98.87% of the total sample size), with no strong-turbulence samples. In the measurement samples of Link 2, 1848 values belonged to the medium-turbulence category (22.12% of the total sample size) and 6552 values belonged to the weak-turbulence category (77.88% of the total sample size), with no strong turbulence observed. In the Link-3 measurement

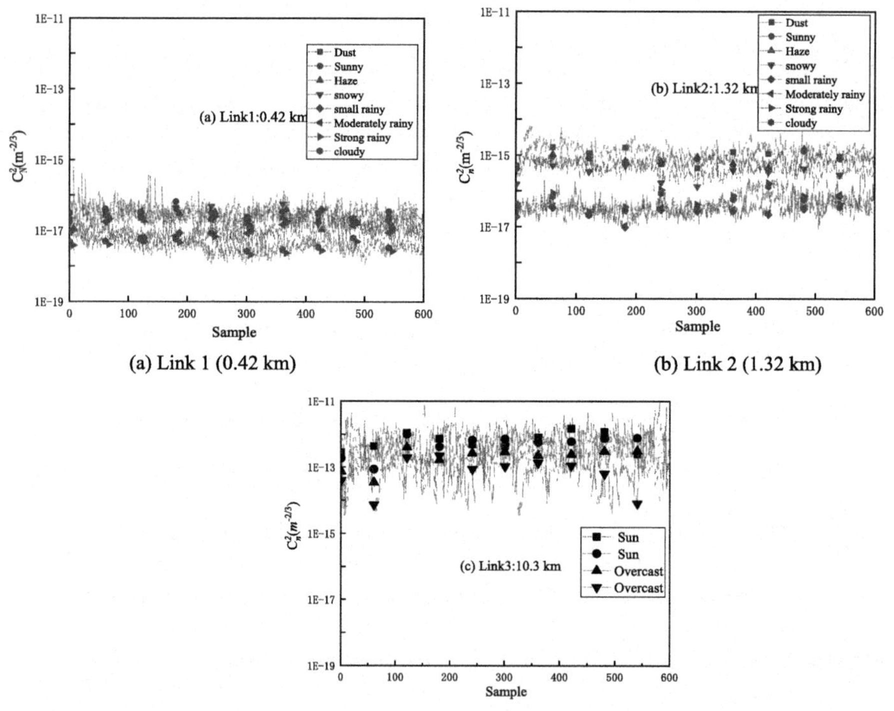

(a) Link 1 (0.42 km)

(b) Link 2 (1.32 km)

(c) Link 3 (10.3 km)

Fig. 4.13 Refractive-index structure constants of measured samples

Table 4.4 Statistical distribution of the C_n^2 of the measurement samples

Trial	Mean (C_n^2)	Statistical distribution (C_n^2)	Max (C_n^2)	Min (C_n^2)
T1	4.39×10^{-17}	3.75×10^{-17}	1.07×10^{-16}	5.39×10^{-18}
T2	4.25×10^{-17}	2.76×10^{-17}	1.36×10^{-16}	6.28×10^{-18}
T3	3.10×10^{-17}	1.76×10^{-17}	1.06×10^{-16}	7.75×10^{-18}
T4	3.01×10^{-17}	1.38×10^{-17}	1.12×10^{-16}	9.75×10^{-18}
T5	1.15×10^{-17}	7.93×10^{-18}	4.76×10^{-17}	2.84×10^{-18}
T6	1.02×10^{-17}	6.41×10^{-18}	3.96×10^{-17}	2.22×10^{-18}
T7	8.98×10^{-18}	5.38×10^{-18}	2.10×10^{-17}	1.35×10^{-18}
T8	8.75×10^{-18}	5.89×10^{-18}	2.58×10^{-17}	8.84×10^{-19}
T9	8.45×10^{-18}	5.24×10^{-18}	1.81×10^{-17}	2.33×10^{-19}
T10	7.26×10^{-18}	5.23×10^{-18}	3.67×10^{-17}	2.55×10^{-19}
T11	6.78×10^{-18}	1.02×10^{-18}	9.17×10^{-18}	1.01×10^{-19}
T12	6.66×10^{-18}	8.41×10^{-18}	3.23×10^{-18}	1.16×10^{-19}
T13	5.83×10^{-18}	3.41×10^{-18}	3.67×10^{-18}	2.55×10^{-19}
T14	4.83×10^{-18}	3.41×10^{-18}	3.67×10^{-18}	2.55×10^{-19}
T15	8.83×10^{-15}	3.41×10^{-16}	5.71×10^{-15}	8.76×10^{-16}
T16	9.86×10^{-15}	4.25×10^{-16}	3.67×10^{-15}	6.12×10^{-16}
T17	5.93×10^{-15}	3.41×10^{-16}	2.88×10^{-15}	5.55×10^{-16}
T18	6.05×10^{-15}	3.23×10^{-16}	2.68×10^{-15}	6.21×10^{-16}
T19	4.13×10^{-15}	2.88×10^{-16}	1.81×10^{-15}	4.55×10^{-16}
T20	4.83×10^{-15}	2.64×10^{-16}	1.62×10^{-15}	4.31×10^{-16}
T21	8.83×10^{-16}	8.98×10^{-17}	3.67×10^{-15}	2.55×10^{-16}
T22	9.16×10^{-16}	1.41×10^{-16}	3.67×10^{-15}	3.02×10^{-16}
T23	7.39×10^{-16}	6.41×10^{-17}	1.81×10^{-15}	2.33×10^{-16}
T24	7.12×10^{-16}	5.56×10^{-17}	1.80×10^{-15}	1.62×10^{-16}
T25	4.57×10^{-16}	5.88×10^{-17}	2.02×10^{-15}	9.03×10^{-17}
T26	4.39×10^{-17}	3.75×10^{-17}	3.72×10^{-16}	9.67×10^{-18}
T27	5.74×10^{-17}	3.96×10^{-17}	2.41×10^{-16}	1.42×10^{-17}
T28	6.30×10^{-17}	4.11×10^{-17}	3.15×10^{-17}	1.26×10^{-17}
T29	1.63×10^{-13}	1.51×10^{-13}	1.89×10^{-12}	1.11×10^{-14}
T30	6.30×10^{-13}	4.11×10^{-13}	3.15×10^{-12}	1.98×10^{-14}
T31	1.03×10^{-14}	2.45×10^{-15}	1.29×10^{-12}	9.41×10^{-14}
T32	5.33×10^{-14}	3.96×10^{-15}	2.89×10^{-13}	1.30×10^{-14}

sample, 792 values belonged to the moderate-turbulence category (accounting for 33.17% of the total sample size), and 1608 values belonged to the strong-turbulence category (66.83% of the total sample size), with no weak turbulence observed.

4.4 Effect of Atmospheric Turbulence on Optical Communication Performance

4.4.1 *Interrupt Probability*

The interrupt probability used to measure the performance of a communication system. When the real transmission rate of a communication system is lower than its minimum rate, normal communication will be interrupted. This seriously affects the quality of the communication line.

The laser wavelength $\lambda = 525$ nm and the transmission distances are 0.42, 1.32, and 10.3 km. We substitute the atmospheric refractive-index structure constants C_n^2 and Rytov variances obtained from the experimental-measurement samples in Sect. 4.2 into Eqs. (4.40)–(4.42) to calculate their interruption probability.

Figure 4.14 shows the interruption probability of the measurement samples. As shown in the figure, as the signal-to-noise ratio increases, the interruption probability of the system continuously decreases. When P_{out} is 0.01, the SNRs in the 0.42-km link under different weather conditions measurement samples are 5.23 dB, 5.21 dB, 5.18 dB, 5.16 dB, 5.15 dB, 5.10 dB, and 5.08 dB, respectively. The SNRs in the 1.32-km link measurement samples are 6.46 dB, 6.32 dB, 6.24 dB, 6.13 dB, 5.98 dB, 5.64 dB, and 5.31 dB, respectively. The SNRs in the 10.3-km link measurement samples are 13.57 dB, 11.44 dB, 9.18 dB, and 8.32 dB, respectively. Based on Table 4.4, it can be seen that under the same signal-to-noise ratio, the interruption probability decreases in the order of dusty days, sunny days, hazy days, snowy days, light rain, moderate rain, heavy rain, and cloudy days.

4.4.2 *Fading Probability*

The fading probability refers to the probability that the received signal strength will decrease to a specific level, owing to the influence of fading effects during the signal-propagation process. For wireless communication systems, the communication-channel quality directly affects the reliability of the signal transmission; therefore, it is necessary to optimize the channel characteristics by accurately measuring the fading probability. For the experiment, the laser wavelength was set to 525 nm with transmission distances of 0.42, 1.32, and 10.3 km. The atmospheric refractive-index

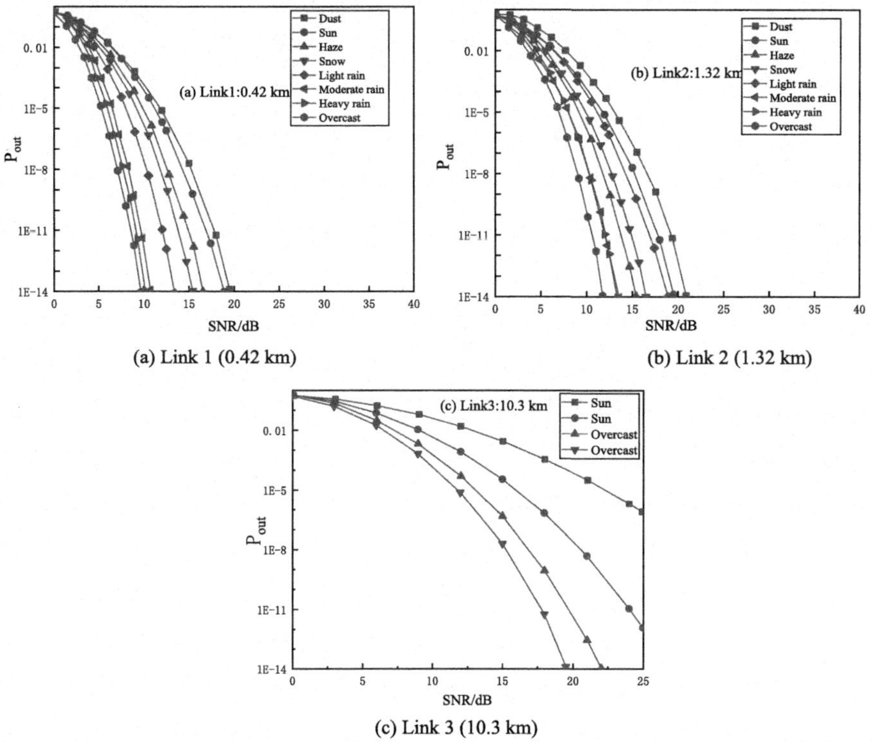

Fig. 4.14 Interruption probability of measurement samples. **a** Link 1 (0.42 km). **b** Link 2 (1.32 km). **c** Link 3 (10.3 km)

structure constants C_n^2 and Rytov variances obtained from the experimental measurement samples in Sect. 4.3.3 are respectively substituted into Eqs. (4.36) to (4.38) to calculate the fading probability of the measurement samples.

From the overall trend of the curves, as the threshold parameter F_T continues to increase, the fading probability begins to gradually decrease, indicating that the performance of the optical-wireless communication system has improved to varying degrees. However, there are certain differences in the characteristics of the fading probability with changes in the threshold parameters, depending on the fluctuation of the signal light intensity.

From Fig. 4.15, it can be observed that the decline rate of the measured-sample fading-probability curve gradually slows, indicating that as the scintillation index increases, the fading probability of the noise signal increases and the performance of the communication system decreases more severely. The threshold parameters required in the different transmission environments differ significantly when the fading probability is maintained at a certain level. For example, in the 10.3-km link measurement sample, owing to the severe fluctuation of the signal light intensity, to maintain the fading probability at the 10^{-10} level, the specific values taken by F_T

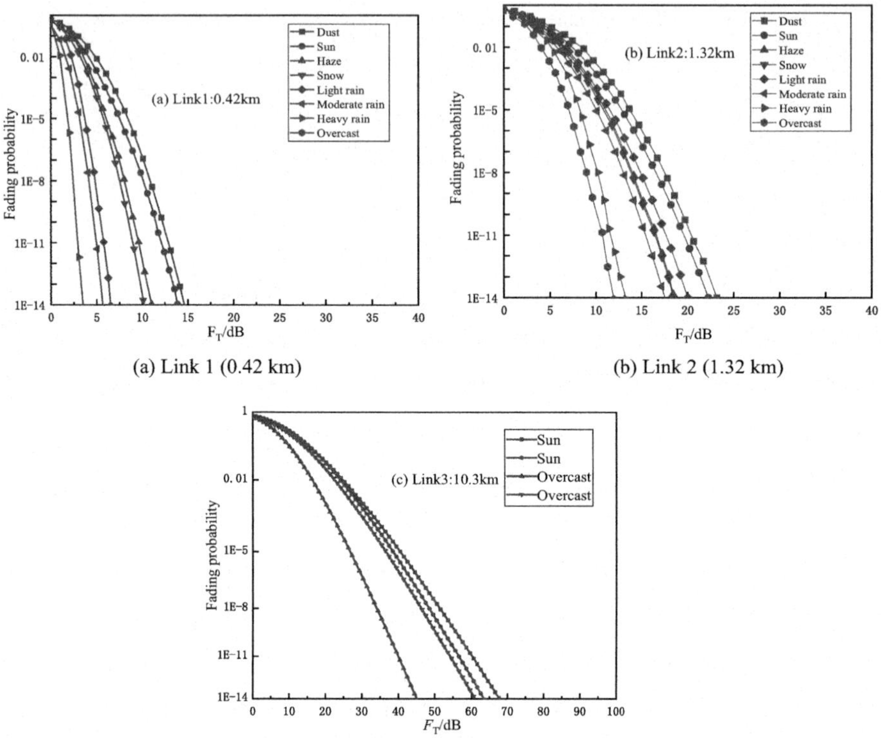

(a) Link 1 (0.42 km)

(b) Link 2 (1.32 km)

(c) Link 3 (10.3 km)

Fig. 4.15 Measurement of the fading probability of the samples

must be relatively large, approximately 38.40 dB, 52.30 dB, 56.70 dB, and 58.10 dB, respectively.

4.5 Experiments on the Aperture-Smoothing Effect

The impact of atmospheric turbulence on a laser transmission can affect the communication-system performance and even interrupt the communication. The large-aperture receiving method uses the aperture-smoothing effect to reduce fluctuations in the received light intensity, thereby reducing the impact of atmospheric-turbulence noise on the communication-system performance.

This section analyzes the theory of the aperture-smoothing effect. The measurement results show that, as the receiving aperture increases, the scintillation index will decrease and decrease greatly. Therefore, smoothing the aperture can improve the atmospheric laser-communication performance.

4.5.1 Aperture-Smoothing Effect

Large-aperture receiving technology can effectively reduce the scintillation effect caused by atmospheric turbulence in atmospheric channels. Large-aperture receiving technology reduces atmospheric scintillation at the receiving end by increasing the receiving aperture; this is called the aperture-smoothing effect. The aperture-smoothing factor is a characteristic of the aperture-smoothing effect. In the actual measurements, when different sizes of receiving apertures were used for reception, the impact on the final-system flicker-effect analysis results varied.

In long-distance atmospheric laser communications, the light-field area that reaches the plane of the receiving end is significantly larger than that of the receiving aperture. At this time, the wavefront of the received laser beam can be regarded as a plane wave. The uncertainty in the light energy received by the photodetector affects the reliability of the laser-communication link, and the variance of the received light-intensity fluctuation is represented by the scintillation index.

The comprehensive aperture-smoothing effect is related to the inner and outer diameters of the atmospheric turbulence. The received light-intensity fluctuation variance of a space laser-communication system with receiving-antenna aperture D_G is [11]

$$\sigma_I^2(D_G) = \frac{\langle I^2 \rangle - \langle I \rangle^2}{\langle I \rangle^2}$$

$$= \frac{16}{\pi D_G^2} \int_0^{D_G} \rho B_I(\rho, L) \left[\cos^{-1}\left(\frac{\rho}{D_G}\right) - \left(\frac{\rho}{D_G}\right) \sqrt{1 - \frac{\rho^2}{D_G^2}} \right] d\rho, \quad (4.36)$$

where I is the received optical power, D_G is the receiving aperture, and $B_I(\rho, L)$ is the light-intensity covariance function on the aperture plane. Using a variable substitution of $x = \rho/D_G$, Eq. (4.37) can be replaced by [10]

$$\sigma_I^2(D_G) = \frac{16}{\pi} \int_0^1 x B_I(x D_G, L) \left[\cos^{-1} x - x\sqrt{1 - x^2} \right] dx. \quad (4.37)$$

Generally, the aperture-smoothing effect is described by aperture-smoothing factor A, which is defined as the ratio of the normalized intensity-fluctuation variance of the optical signal received by the receiver, whose aperture is D_G, and the normalized intensity-fluctuation variance of the optical signal received by the point receiver. The signal is normalized by the square of its mean value, and the general form of the aperture-smoothing factor is [12]

$$A = \sigma_I^2(D_G)/\sigma_I^2(0), \quad (4.38)$$

where $\sigma_I^2(0) = B_I(0, L)$ represents the scintillation index of the spot aperture ($D_G = 0$). In this case, we can write the aperture-smoothing factor as [11]

$$A = \frac{\sigma_I^2(D_G)}{\sigma_I^2(0)} = \frac{16}{\pi}\int_0^1 x b_I(xD_G, L)\left[\cos^{-1}x - x\sqrt{1-x^2}\right]dx. \tag{4.39}$$

$b_I(\rho, L) = B_I(\rho, L)/B_I(0, L)$ is the normalized covariance function.

In a weakly fluctuating turbulent region, the light-intensity fluctuation follows a logarithmic-normal distribution. When the transverse-coherence length at the receiving surface is larger than the Fresnel-zone scale $\sqrt{\lambda L}$, the turbulent flow is considered to be weakly fluctuating. For isotropic and uniform turbulence, the covariance function can be written as [11]

$$B_I(\rho) = 16\pi^2 k^2 \int_0^\infty d\kappa \cdot \Phi(\kappa) \int_0^L dz \cdot J_0(\kappa\rho) \cdot \sin^2\left(\frac{\kappa^2(L-z)}{2k}\right), \tag{4.40}$$

where k is the wavenumber, $\Phi(\kappa)$ is the refractive-index fluctuation power spectrum, L is the transmission distance, and J_0 is the zero-order Bessel function of the first type.

For plane waves, the small internal scale is $l_0 \ll (L - lk)^{1/2}$; the Kolmogorov spectrum can be used and the Rytov variance is

$$B_I(\rho = 0) = 1.23k^{7/6}L^{11/6}C_n^2 = \sigma_I^2. \tag{4.41}$$

The aperture-smoothing factor can be approximately written as follows [13]:

$$A = \left[1 + 1.07\left(\frac{kD^2}{4L}\right)^{7/6}\right]^{-1}. \tag{4.42}$$

Churnside [12] calculated the circular aperture-smoothing factor of diameter D for plane and spherical waves under different transmission conditions. The results are summarized in Tables 4.5 and 4.6.

It can be seen from Fig. 4.16 that in the simulation analysis of three different receiving apertures, the overall change trend of the scintillation index is consistent. When σ_R is less than 2.4, the scintillation index σ_I^2 increases with the increase of σ_R. When σ_R is equal to 2.4, the scintillation index σ_I^2 reaches the saturation value. The scintillation index σ_I^2 decreases gradually with the increase of σ_R.

Table 4.5 Aperture-averaging factor of a plane wave

Fluctuation condition	Aperture-averaging factor
Weak fluctuation condition $l_0 \ll \sqrt{L\lambda} \ll r_0$	$A = \left[1 + 1.812\left(\frac{D^2}{L\lambda}\right)^{7/6}\right]^{-1}$
Weak fluctuation condition $\sqrt{L\lambda} < \min[r_0, l_0]$	$A = \left[1 + 2.21\left(\frac{D}{l_0}\right)^{7/3}\right]^{-1}$
Strong fluctuation condition $l_0 < r_0 < \sqrt{L\lambda}$	$\beta_l^2 = 1 + 1.373\left(\frac{r_0^2}{L\lambda}\right)^{1/3}$ $A = \frac{\beta_l^2 + 1}{2\beta_l^2}\left[1 + \left(\frac{D}{r_0}\right)^2\right]^{-1} + \frac{\beta_l^2 - 1}{2\beta_l^2}\left[1 + 0.415\left(\frac{r_0 D}{L\lambda}\right)^{7/3}\right]^{-1}$
Strong fluctuation condition $r_c < \min\left[\sqrt{L\lambda}, l_c\right]$	$r_c = 2.1\left[1.20k^2 \int_0^L C_n^2(z)\right]\left(0.545k^2 C_n^2 L l_0^{-1/5}\right)^{-1/2}$ $\beta_l^2 = 1 + 1.744\left(\frac{r_0 l_0}{L\lambda}\right)^{1/3}$ $A = \frac{\beta_l^2 + 1}{2\beta_l^2}\left[1 + 1.10\left(\frac{D}{r_0}\right)^2\right]^{-1} + \frac{\beta_l^2 - 1}{2\beta_l^2}\left[1 + 3.251\left(\frac{r_0 D}{L\lambda}\right)^{7/3}\right]$

Table 4.6 Aperture-averaging factor of spherical waves

Fluctuation condition	Aperture-averaging factor
Weak fluctuation condition $(5/3)l_0 \ll \sqrt{L\lambda} \ll r_0$	$A = \left[1 + 0.3624\left(\frac{D^2}{L\lambda}\right)^{7/6}\right]$
Weak fluctuation condition $\sqrt{L\lambda} < \min[r_0, (5/3)L_0]$	$A = \left[1 + 0.109\left(\frac{D^2}{l_0}\right)^{7/3}\right]^{-1}$
Weak fluctuation condition $l_0 < r_0 < \sqrt{L\lambda}$	$\beta_l^2 = 1 + 4.343\left(\frac{r_0^2}{L\lambda}\right)^{1/3}$ $A = \frac{\beta_l^2 + 1}{2\beta_l^2}\left[1 + \left(\frac{D}{r_0}\right)^2\right]^{-1} + \frac{\beta_l^2 - 1}{2\beta_l^2}\left[1 + 2.560\left(\frac{r_0 D}{L\lambda}\right)^{7/3}\right]^{-1}$
Strong fluctuation condition $r_0 < \min[\sqrt{L\lambda}, l_0]$	$r_c = 2.1\left(0.545k^2 C_n^2 L l_0^{-1/5}\right)^{-1/2}$ $\beta_l^2 = 1 + 3.271\left(\frac{r_0 l_0}{L\lambda}\right)^{1/3}$ $A = \frac{\beta_l^2 + 1}{2\beta_l^2}\left[1 + \left(\frac{D}{r_0}\right)^2\right]^{-1} + \frac{\beta_l^2 - 1}{2\beta_l^2}\left[1 + 1.367\left(\frac{r_0 D}{L\lambda}\right)^{7/3}\right]^{-1}$

Note l_0 refers to the internal turbulence scale; $\sqrt{\lambda L}$ is the Fresnel scale; r_0 is the Fried parameter; D represents the aperture of the receiving antenna and β_l^2 is the variance of the light-intensity fluctuation

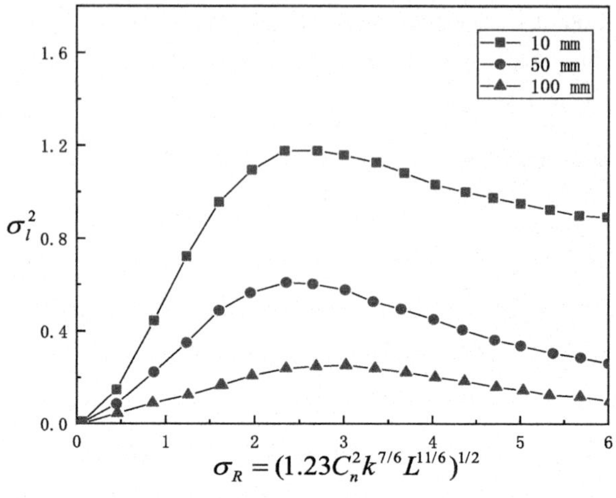

Fig. 4.16 Changes in the scintillation index with the turbulence intensity under different receiver apertures

Moreover, it can be found that when the receiving aperture is 10 mm, the value of the overall scintillation index is the largest, and the variation range is 0–1.2, indicating that the light-intensity fluctuation is the strongest in this case. When the receiving aperture is 100 mm, the value of the overall scintillation index is the smallest, and the range of change is 0–0.2; it shows that the light-intensity fluctuation is the weakest in this scenario. The light-intensity fluctuation degree decreases sequentially in the order of receiving apertures of 10, 50, and 100 mm. This shows that a large receiving aperture plays an important role in suppressing the light-intensity fluctuation.

Figure 4.17 shows the variation curve of the aperture-averaging factor A with a normalized aperture size under different turbulence conditions (weak turbulence $\sigma_R = 0.3$, medium turbulence $\sigma_R = 4$, and strong turbulence $\sigma_R = 30$). The normalized aperture size $\left(kD_G^2/4L\right)^{1/2}$ is defined as the ratio of the receiving-antenna aperture $D_G/2$ to the Fresnel region $(L/k)^{0.5}$.

From Fig. 4.17, it can be observed that under weak to moderate turbulence conditions, the aperture-averaging effect is only significant when aperture $D_G > (L/k)^{0.5}$. Under strong turbulence conditions, the aperture-averaging effect is only noticeable when the receiving-antenna aperture D_G is greater than the spatial coherence size. This is because the turbulence size is smaller under strong-turbulence conditions, so the aperture-averaging effect under strong-turbulence conditions is more significant than that under weak-turbulence conditions. By comparing the aperture-averaging factor A under different turbulence conditions, it can be concluded that the aperture-averaging effect increases with the increase in turbulence intensity.

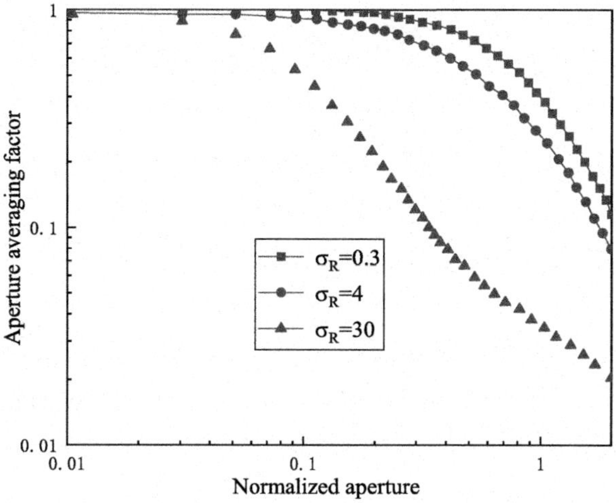

Fig. 4.17 Variation curve of the aperture-averaging factor A with a normalized aperture size under different turbulence conditions (beam radius $W_0 = 1$ cm, wavelength $\lambda = 0.633$ μm)

The fading probability describes the percentage of time that the received light intensity falls below a specified threshold, I_T. Given the probability-density function (PDF) model of the fluctuation of the light intensity $p_I(I)$, the fading probability can be obtained as a function of the threshold I_T from the cumulative probability:

$$P(I \leq I_T) = \int_0^{I_T} P_I(I)dI. \tag{4.43}$$

Under weak fluctuation conditions, $p_I(I)$ in Eq. (4.43) is a logarithmic-normal distribution. Its corresponding fading-probability model is

$$P_r(I \leq I_T) = \frac{1}{2}\left\{1 + erf\left[\frac{\frac{1}{2}\sigma_I^2(0, L + L_f) - 0.23F_T}{\sqrt{2}\sigma_I(0, L + L_f)}\right]\right\}, \tag{4.44}$$

where $erf(x)$ is the error function.

To obtain Eq. (4.44), we define the fading threshold F_T. The unit of F_T is decibels (dB) and it represents the ratio of the axial-average light intensity to the threshold.

$$F_T = 10\log_{10}\left(\frac{\langle I(0, L + L_f)\rangle}{I_T}\right).[dB] \tag{4.45}$$

Under moderate to strong fluctuation conditions, $p_I(I)$ in Eq. (4.43) is a dual-gamma distribution, and its corresponding fading-probability model is

$$P_r(I \le I_T) = \frac{\pi}{\Gamma(\alpha)\Gamma(\beta)\sin[\pi(\alpha-\beta)]} \times \left\{ \begin{array}{l} \frac{(\alpha\beta)^\beta}{\beta\Gamma(\beta-\alpha-1)}\exp(-0.23\beta F_T) \\ \times {}_1F_2[\beta;\beta+1,\beta-\alpha-1;\alpha\beta]\exp(-0.23\beta F_T) \\ -\frac{(\alpha\beta)^\beta}{\alpha\Gamma(\beta-\alpha-1)}\exp(-0.23\alpha F_T) \\ \times {}_1F_2\left[\alpha;\alpha+1,\alpha-\beta-1;\alpha\beta\exp(-0.23\beta F_T)\right] \end{array} \right\}. \tag{4.46}$$

For parameters α and β in the gamma–gamma distribution model, the definition in Eq. (4.8) is typically used. When considering the aperture-averaging effect, α and β can be rewritten as [11]

$$\alpha = \frac{1}{\exp\left[\sigma^2_{\ln X}(D_G)\right]-1},\ \beta = \frac{1}{\exp\left[\sigma^2_{\ln Y}(D_G)\right]-1}. \tag{4.47}$$

In addition, the variance of the light-intensity fluctuations can be written as

$$\sigma_I^2\left(0, L+L_f\right) = \exp\left[\sigma^2_{\ln X}(D_G)+\sigma^2_{\ln Y}(D_G)\right]-1. \tag{4.48}$$

In this formula, the large-scale and small-scale logarithmic fluctuation variances are [14]

$$\sigma^2_{\ln X}(D_G) = \frac{0.49\left(\frac{\Omega_G-\Lambda_1}{\Omega_G+\Lambda_1}\right)\sigma_B^2}{\left[1+\frac{0.4\left(2-\overline{\Theta}\right)(\sigma_B/\sigma_R)^{12/7}}{(\Omega_G+\Lambda_1)\left(\frac{1}{3}-\frac{1}{2}\overline{\Theta}_1+\frac{1}{5}\overline{\Theta}_1^2\right)^{6/7}}+0.56\left(1+\overline{\Theta}_1\right)\sigma_B^{12/5}\right]} \tag{4.49}$$

$$\sigma^2_{\ln Y}(D_G) = \frac{0.51\sigma_B^2(\Omega_G+\Lambda_1)\left(1+0.69\sigma_B^{12/5}\right)^{-5/6}}{\Omega_G+\Lambda_1+1.20(\sigma_R/\sigma_B)^{12/5}+0.83\sigma_R^{12/5}}. \tag{4.50}$$

The limited size of the lens aperture is characterized by the dimensionless Ω_G Fresnel parameter, Θ and Λ are the Gaussian-beam parameters of the wave incident on the lens, σ_R^2 is the Rytov variance of the plane wave, and σ_B^2 is the Rytov variance of the Gaussian beam [12]:

$$\sigma_B^2 \cong 3.86\sigma_R^2\left\{0.40\left[(1+2\Theta_1)^2+4\Lambda_1^2\right]^{5/12}\cos\left[\frac{5}{6}\tan^{-1}\frac{1+2\Theta_1}{2\Lambda_1}\right]-\frac{11}{16}\Lambda_1^{5/6}\right\}. \tag{4.51}$$

Figure 4.18 shows the variation curve of the fading probability of the communication-system link with the fading threshold F_T under different receiving

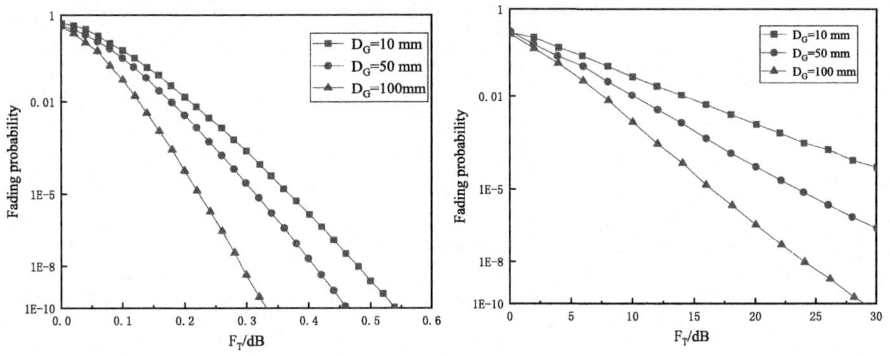

Fig. 4.18 Curve of the system fading probability with fading threshold F_T under different receiver apertures. **a** $C_n^2 = 1 \times 10^{-16}\,m^{-2/3}$, **b** $C_n^2 = 1 \times 10^{-13}\,m^{-2/3}$

apertures. The fading threshold $F_T = 10\lg(< I(r, L)) > /I_T)$ is defined as the additional loss that the system can receive. In Fig. 4.18a, if the fading threshold F_T is 0.3 dB and D_G is 10 mm, the link-fading probability is 1.44×10^{-4}. When D_G is 50 mm, the link-fading probability is 1.38×10^{-5}. When D_G is 100 mm, the link-fading probability is 3.57×10^{-9}.

In Fig. 4.18b, if the fading threshold F_T is 20 dB and D_G is 10 mm, the link-fading probability is 1.17×10^{-3}. When D_G is 50 mm, the link-fading probability is 6.24 $\times 10^{-5}$. When D_G is 100 mm, the link-fading probability is 4.2×10^{-7}. From the system link-fading probability curve, it can be observed that as the receiving aperture increases, the link-fading probability decreases, and the communication quality of the optical-wireless communication is enhanced.

4.5.2 Aperture and Diameter Smoothing Experiment

4.5.2.1 Experimental Device

Based on the schematic shown in Fig. 4.7, we fabricated three baffles with different round holes. The radii were 10, 20, and 50 mm. A baffle was placed on the receiving antenna. The size of the aperture was recorded and analyzed using experimental data for different apertures.

4.5.2.2 Scintillation Index

We changed the size of the receiving-terminal apertures for data collection and compared the scintillation indices of different sizes for receiving apertures of different sizes, as shown in Fig. 4.19.

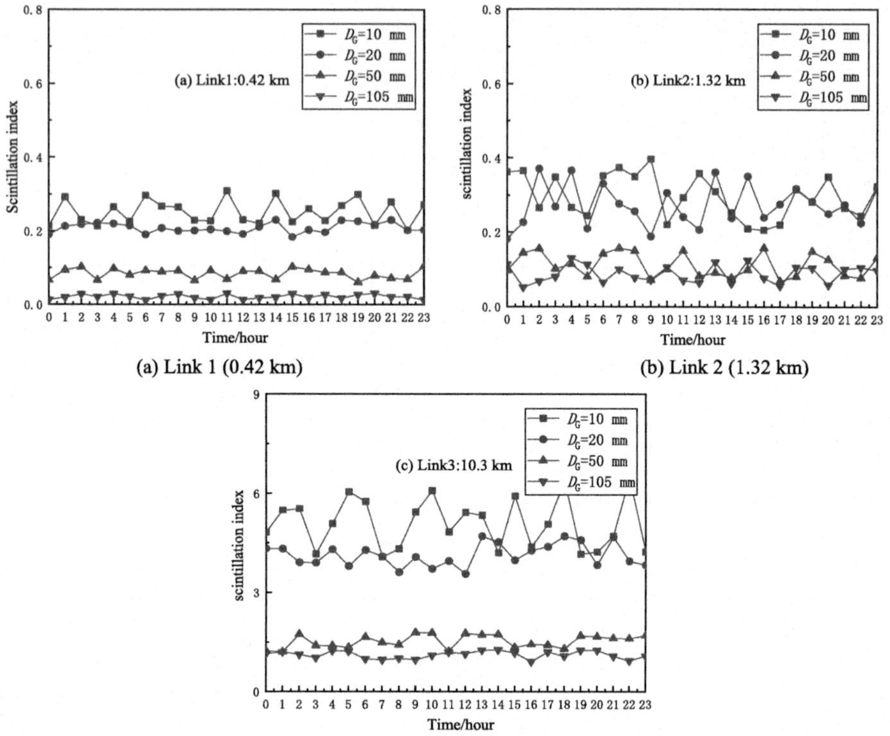

(a) Link 1 (0.42 km) (b) Link 2 (1.32 km) (c) Link 3 (10.3 km)

Fig. 4.19 Scintillation-index change curve under different aperture sizes

According to Fig. 4.19, it can be seen that the scintillation index decreases as the aperture increases for 10, 20, 50, and 105 mm. The scintillation index generated by the acceptance of the 105-mm receiving aperture is significantly reduced, and it is relatively stable. Moreover, the scintillation indices displayed in Fig. 4.19a–c were consistent with the decline. This law is consistent and reflects the accuracy of the measurement. Compared with the 50-mm and 105-mm receiving apertures, when choosing 10-mm and 20-mm receiving apertures for the measurement, the scintillation-index changes were large and unstable.

4.5.2.3 Fading Characteristics

The fading probability refers to the probability that the signal intensity received during the signal dissemination will drop to a certain level. For wireless communication systems, the quality of the communication channel directly affects the reliability of the signal transmission; therefore, it is necessary to optimize the channel characteristics by accurately measuring the fading probability.

The laser wavelength $\lambda = 525$ nm and the transmission distances are 0.42 km, 1.32 km, and 10.3 km. According to Eqs. (4.44)–(4.46), the atmospheric refractive-index structure constant C_n^2 and the Rytov variance of the experimental-measurement sample were used to calculate the fading probabilities.

Figure 4.20 shows the change in the decline of linkages of the communication system with the changing curve of the antenna aperture D_G and the fading threshold F_T. The fading threshold F_T is defined as $F_T = 10\lg(\langle I(r, L)\rangle)/I_T)$, which is the additional loss that the system can receive.

As shown in Fig. 4.20, as the threshold parameter F_T continues to increase, the fading probability gradually decreases, indicating that the performance of the optical-wireless communication system has been improved to varying degrees. Certain differences were observed in the changing characteristics. The larger the receiving aperture, the lower the speed of the sample-decay probability curve, which means that as the receiving aperture increases, the probability of the noise signal decreases.

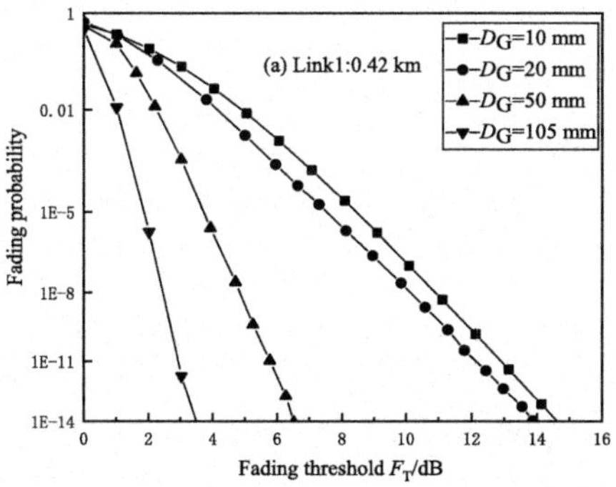

(a) Link 1 (0.42 km)

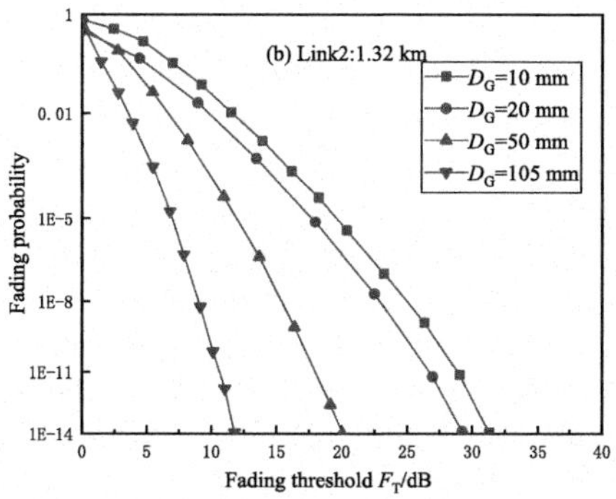

(b) Link 2 (1.32 km)

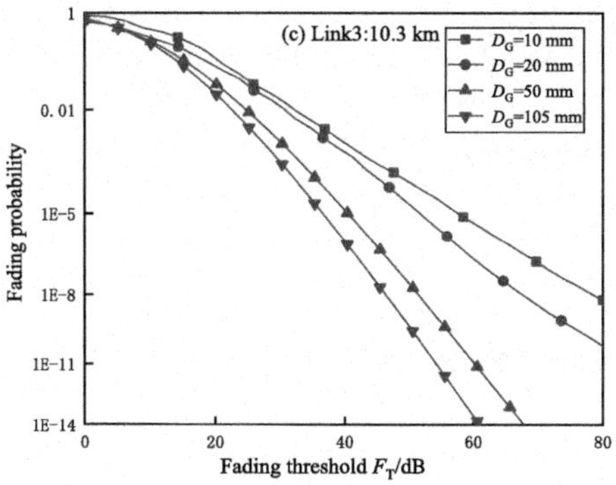

(c) Link 3 (10.3 km)

Fig. 4.20 Fading-probability variation curve under different aperture sizes

4.5.2.4 Atmospheric Refractive-Index Structure Constant

The atmospheric refractive-index structure constant C_n^2 is an important parameter that reflects the characteristics of atmospheric optical turbulence and quantitatively describes the optical turbulence strength. The scintillation method is used to measure C_n^2 [7], which is expressed as

$$C_n^2 = 4.48 C D_t^{7/3} L^{-3} \sigma_X^2, \tag{4.52}$$

where L is the link length and $C = D_r/D_t$, where D_r and D_t are the sending and receiving aperture sizes, respectively. σ_X^2 is the logarithmic-amplitude variance, which can be expressed as [7]

$$\sigma_X^2 = 0.25 \ln\left[1 + \left(\frac{s}{\langle I \rangle}\right)^2\right], \tag{4.53}$$

where $\langle I \rangle$ is the average light intensity, and s is the standard deviation of the light intensity. The distances are 0.42, 1.32, and 10.3 km. The light-intensity values obtained from the experimental measurements were processed to obtain the mean and standard deviation of the light intensity. Then, the experimental analysis data were substituted into Eqs. (4.52) and (4.53) to calculate the refractive-index structure constant of the measured sample.

According to Fig. 4.21, it can be seen that for the 0.42-km-link measurement sample, C_n^2 is mainly concentrated between 1×10^{-18} m$^{-2/3}$ and 1×10^{-16} m$^{-2/3}$. In the 1.32-km-link measurement sample, a large part of the C_n^2 concentration appears within the range of 1×10^{-17} m$^{-2/3}$–1×10^{-14} m$^{-2/3}$. In the 10.3-km-link measurement sample, C_n^2 is within the interval range 1×10^{-16} m$^{-2/3}$–1×10^{-12} m$^{-2/3}$, which shows that the turbulence intensity increases with the link distance.

This is consistent with the rules of the refractive index structure constant of the atmospheric-turbulence noise measurement samples in Sect. 4.3; however, when selecting different receiving apertures, the turbulence strength will weaken. The smoothing effect can inhibit the impact of atmospheric-turbulence noise on the communication-system performance.

4.5.2.5 Skewness and Steepness

Figure 4.22 shows the relationship between the skewness, steepness, and Rytov exponent of the measured samples for different receiving apertures. From Fig. 4.22a, b, it can be observed that the skewness of the noise signals and their statistical distributions in the samples of Link 1, Link 2, and Link 3 decreases in the order of $D_G = 10$ mm, $D_G = 20$ mm, $D_G = 50$ mm, and $\mathrm{D_G} = 105$ mm; the steepness is the same. This indicates that as the receiving aperture increases, the skewness and steepness of the sample will decrease.

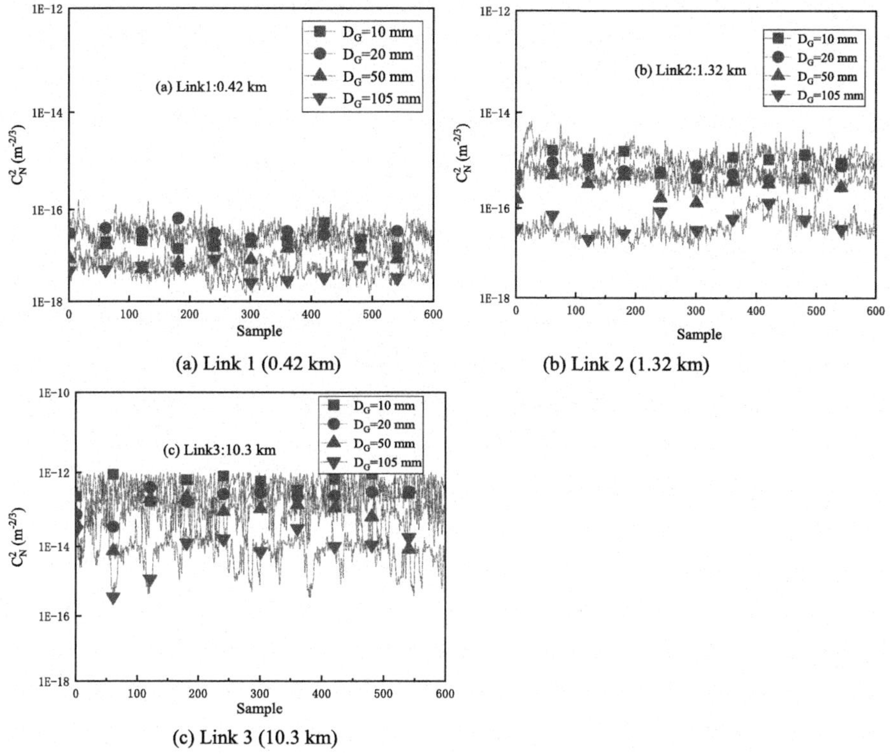

Fig. 4.21 Refractive-index structure constants of measured samples

The Rytov exponent of the sample data tested in Fig. 4.22a ranges from 0.01 to 0.4, indicating a weak-turbulence region. The Rytov exponent of the sample data tested in Fig. 4.22b ranges from 0.1 to 0.6; thus, it is located in a weak-to-moderate turbulence region. The Rytov exponent of the sample data tested in Fig. 4.22c ranges from 0.7 to 1.3, indicating a moderate-to-strong turbulence region. This indicates that the Rytov exponent increases with increasing link distance, and the turbulence intensity increases accordingly. The skewness and steepness of the noise samples also increase accordingly.

4.6 Summary

When a laser is transmitted through the atmosphere, it is affected by atmospheric turbulence. This seriously restricts the further development of atmospheric laser communication technology and is the main problem that must be solved for use in practical applications. Based on the principle of the aperture-smoothing effect, this chapter presented an experimental study on the effect of smoothing and turbulence

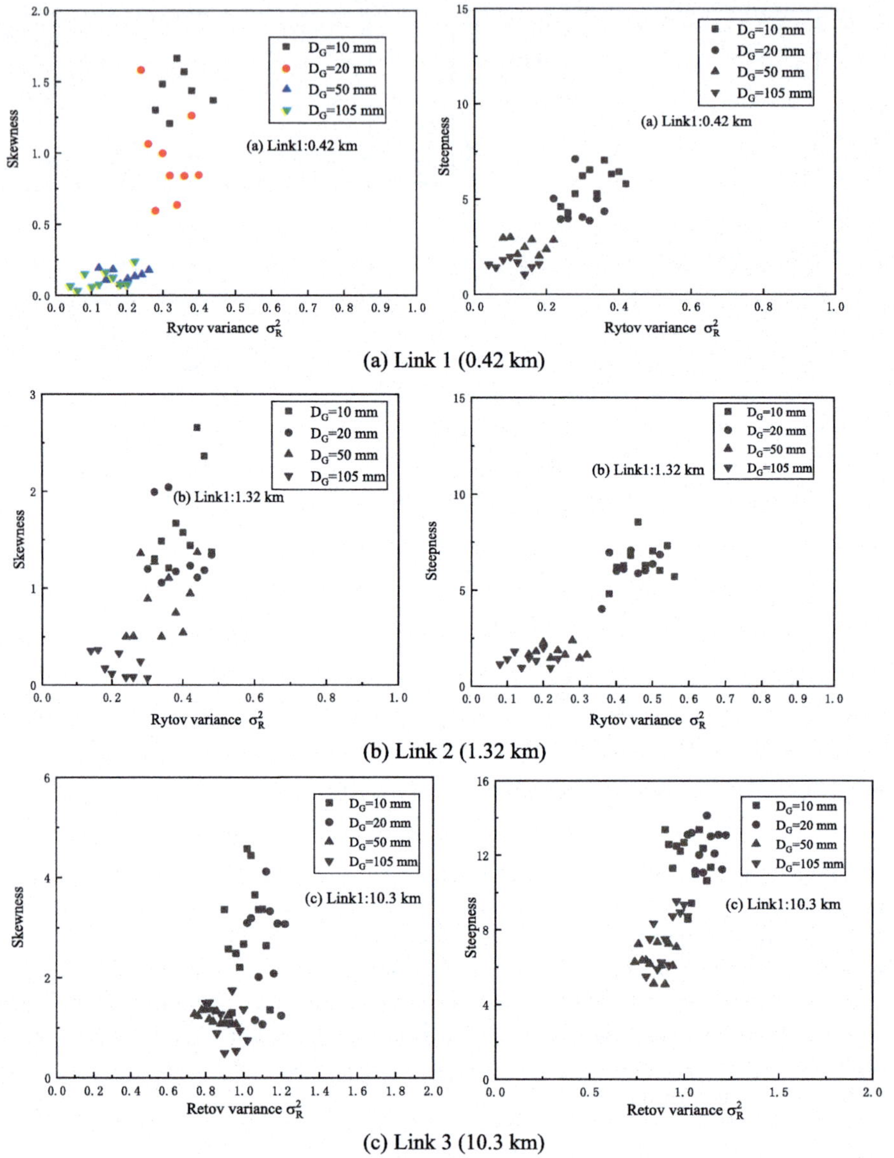

Fig. 4.22 Relationship between the skewness and steepness of the measurement samples and the Rytov exponent

on the aperture diameter, measured three different experimental links, and analyzed the experimental measurement data.

References

1. Kunkel KE, Walters DL (1983) Modeling the diurnal dependence of the optical refractive index structure parameter. J Geophys Res Oceans 88(15):10999–11004
2. Fante RL (1980) Some physical insights into beam propagation in strong turbulence. Radio Sci 15(4):757–762
3. Xie MQ, Ke XZ (2013) Research on the influence of atmospheric turbulence on the signal to noise ratio of optical-wireless communication systems. Prog Laser Optoelectron 50(11):65–71
4. Zhu X, Kahn JM (2002) Free-space optical communication through atmospheric turbulence channels. IEEE Trans Commun 50(8):1293–1300
5. Wu XJ, Wang HX, Li BF et al (2015) Analysis of the impact of atmospheric turbulence on the fading characteristics of optical-wireless communication in different transmission environments. China Laser 42(05):270–277
6. Han LQ, Wang Q, Xin TKG (2010) Analysis of outage probability of Free-space optical communication under atmospheric turbulence. Infrared Laser Eng 39(04):660–663
7. Wang HQ, Li Y, Hu Q et al (2018) Experimental study on the fluctuation characteristics of nighttime light intensity in Lanzhou region. Acta Photonica Sin 47(04):194–201
8. Ma XS, Zhu WY, Rao RZ (2007) The average aperture factor of circular aperture for light wave scintillation in turbulent atmosphere. J Opt 52(09):1543–1547
9. Frida SV, Cynthia Y (2007) Aperture averaging effects on the probability density of irradiance fluctuations in moderate-to-strong turbulence. Appl Opt 46(11):2099–2108
10. Wentao Z, Baohua Z (2007) Research on the influence of atmospheric turbulence on the transmission of laser signals. J Univ Electron Sci Technol 36(4):784–787
11. Lu HQ, Wang LH, Li Y et al (2012) Computational simulation of aperture averaging effect in space laser communication. Appl Opt 33(03):619–623
12. Dai TJ (2021) Research on long distance atmospheric laser communication system and atmospheric turbulence suppression technology. Changchun University of Science and Technology, Jilin Changchun, pp 51–60
13. Yang CQ, Jiang WH, Rao CH (2007) Influence of aperture averaging on bit error rate of Free-space optical communication. Acta Opt Sin 18(02):212–218
14. Wu XJ, Wang HX, Liu M et al (2012) Analysis of the influence of receiving aperture on the performance of optical-wireless communication bit error rate. Res Opt Commun 20(01):64–66

Chapter 5
Atmospheric-Turbulence Suppression Method

Optical-wireless signals are affected by atmospheric turbulence during their transmission through atmospheric channels [1]. Atmospheric turbulence can cause light-intensity fluctuations, beam expansion, beam drift, and phase fluctuations, causing the performance of optical-wireless communication systems to deteriorate and making it difficult to fully utilize their advantages. Therefore, suppressing the effects of atmospheric turbulence on optical communication is crucial [2]. This chapter introduces several key technical principles and the status of research on atmospheric-turbulence suppression.

5.1 Key Technologies for Suppressing Atmospheric Turbulence

To reduce the impact of atmospheric turbulence on optical-wireless communication, and further improve its system performance and communication quality, domestic and foreign scholars have proposed solutions and related technologies for suppressing atmospheric turbulence. These include large-aperture reception, adaptive-optics, diversity, and partially coherent optical technologies. This section introduces several key technological principles for suppressing atmospheric turbulence, and their research status.

5.1.1 Large-Aperture Receiving Technology

Large-aperture receiving technology has been proposed to suppress the impact of atmospheric turbulence on optical-wireless communication and further improve its system performance and communication quality.

X. Ke and C. Ke, *Noise Models in Optical-Wireless Communication Systems*,
Optical Wireless Communication Theory and Technology,
https://doi.org/10.1007/978-981-97-7550-7_5

5.1.1.1 Current Research Status

In 1973, experiments by Kerr et al. [3] showed that under the same turbulent conditions, the volatility of the received optical power decreased as the received aperture size increased; when the transmitting aperture remained constant, the volatility of the received optical power decreased as the receiving aperture increased.

In 2006, Yuksel et al. [4] proposed an optical-antenna receiver-aperture optimization method by analyzing scintillation spot images, providing the optimal aperture and proving that the aperture-smoothing factor was independent of the aperture shape and only related to the aperture area.

In 2008, Chen et al. [5] conducted a 16-km laser communication link experiment and analyzed the light-intensity fluctuation characteristics of the receiving link. They found that receiving through a large-aperture antenna for longer distance communication reduced the turbulence-induced light intensity fluctuation and enabled them to more easily analyze the light-intensity fluctuation distribution.

In 2009, Prokes [6] found that the signal-power attenuation by atmospheric turbulence increased with an increase in communication distance, and that the scintillation effect of atmospheric turbulence could be suppressed by appropriately increasing the receiving-aperture size using large-aperture reception techniques.

In 2009, Chen et al. [7] analyzed the relationship between the aperture-smoothing factor and the variance of light-intensity fluctuation in a large-aperture receiving system and determined the effect of different receiving-aperture sizes on the signal-to-noise ratio (SNR) of the system with a given bit-error rate (BER), which laid the foundation for selecting the appropriate receiving-aperture size.

In 2011, Lou et al. [8] conducted a near-earth laser-transmission experiment and analyzed the effect of different receiving-aperture sizes on the optical-intensity fluctuation; the results showed that large-aperture receivers could receive a larger spot area, which effectively suppressed the optical-intensity scintillation effect of turbulence.

In 2012, Lu et al. [9] analyzed the effect of the antenna aperture on the quality of free-space optical (FSO) communication and found that the signal scintillation index decreased with an increase in the receiving aperture, as did the fading probability of the communication link.

In 2014, Wang [10] conducted experiments on atmospheric laser communication using different receiving-aperture sizes and found that as the receiving-aperture size increased, the BER of the communication system decreased, suppressing the light-intensity fluctuation effect and improving the performance of the entire communication system.

In 2017, Li et al. [11] found that the scintillation index increased with an increase in turbulence intensity and conducted experiments on communication links at different turbulence intensities. They found that the scintillation index could be reduced by increasing the receiving-aperture size, the scintillation index could reach the saturation state, and an appropriate increase in the receiving-aperture size could effectively suppress the turbulence-scintillation effect.

In 2019, Soni et al. [12] studied the performance of a free-space optical communication system under different rainfall intensities using large-aperture reception, and analyzed the suppression of atmospheric turbulence using the aperture-smoothing effect. This led to an average signal-to-noise ratio improvement of 1.58 dB using large-aperture reception.

In 2021, Dai [13] analyzed the effect of different aperture sizes on the light-intensity scintillation index in a wireless laser-communication system, using a large-aperture Cassegrain antenna as the receiving system. It was found that the light-intensity scintillation index of the large-aperture reception was small and stable, effectively suppressing the scintillation effect of atmospheric turbulence.

5.1.1.2 Suppression Principle

Because of the light-intensity flickering caused by atmospheric turbulence, the received light power at the receiving end can fluctuate. Increasing the antenna size at the receiving end can reduce the fluctuation variance. Large-aperture reception technology increases the aperture size of the receiving end, enabling it to receive more light energy and suppress the atmospheric-turbulence effect.

Large-aperture reception technology is widely used in optical-wireless communication systems and its inhibitory effect on flickering is particularly significant. Large-aperture receiving technology reduces the scintillation effect caused by atmospheric turbulence by increasing the receiving aperture at the receiving end. This is called the aperture-smoothing effect. Under different atmospheric-turbulence intensities, the influence of different receiving-aperture sizes on the scintillation effect varies; therefore, the aperture-smoothing factor characterizes the aperture-smoothing effect. Based on the aperture-smoothing factor under different turbulence conditions, the optimal aperture of the receiving antenna can be determined to suppress the influence of turbulent scintillation.

The Cassegrain telescope system is composed of two mirrors, as shown in Fig. 5.1. The incident laser beam passes through the primary mirror with a parabolic surface and is output by the secondary mirror with a hyperbolic surface. By appropriately increasing the aperture of the Cassegrain antenna, the power of the received optical signal can be increased; it also has an aperture-averaging effect that weakens fluctuations, which can effectively suppress the influence of atmospheric turbulence on the beam.

Using large-aperture receiving technology can, to some extent, suppress the effect of light-intensity flickering on optical communication links. However, the aperture cannot be unlimited. Increasing the aperture implies an increase in the size and weight of the receiver, which limits applications with high portability requirements.

5.1.2 Partially Coherent Optical Technology

Partially coherent optical-transmission technology has been applied in optical communications. Coherent optical transmission technology is commonly used in

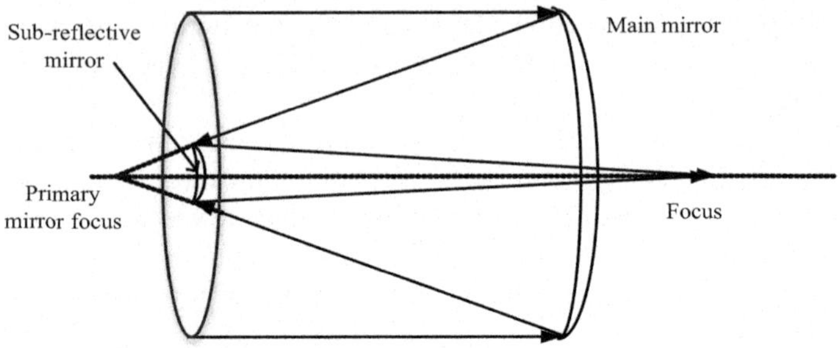

Fig. 5.1 Schematic diagram of a large-aperture Cassegrain telescope [14]

optical communication. Owing to various factors, such as atmospheric turbulence, the phase and amplitude of the signal change, decreasing the signal coherence. In partially coherent optical-transmission technology, the phase and amplitude of the optical signal can undergo certain changes, but still maintain a certain degree of coherence. Partially coherent optical-transmission technology can process and repair signals through modulation and demodulation, thereby improving the transmission reliability.

5.1.2.1 Current Research Status

In 1978, Wu et al. [15] discovered that partially coherent beams had a strong anti-turbulence effect by analyzing the transmission characteristics of partially coherent Gaussian–Schell-model (GSM) beams under atmospheric turbulence.

In 2003, Ricklin et al. [16] investigated whether the BER of free-space optical communication systems was affected by the turbulence strength and degree of beam coherence. The final results demonstrated that reducing the beam coherence reduced the degree of optical-power jitter at the receiver, which reduced the BER of the communication system.

In 2003, Aristide et al. [17] used a partially coherent light for transmission in free-space optical communication systems. It was shown that partially coherent light was less affected by atmospheric turbulence than fully coherent light, and the effect of turbulence could be better suppressed by appropriately reducing the beam coherence.

In 2010, Borah et al. [18] investigated optimizing the coherence length of a spatially partially coherent beam for free-space optical communication under weak-turbulence conditions. The effects of different parameters on the coherence length were discussed. The optimization of the coherence length reduced the turbulence effect and improved the system performance.

In 2011, Drexler et al. [19] improved the performance of free-space optical communication systems by controlling the coherence of the transmitted beams. It was demonstrated that using partially coherent beams reduced the scintillation values by 50%, compared to fully coherent beams, under different turbulence-intensity conditions.

In 2014, Zhang et al. [20] investigated the transmission characteristics of partially coherent Gaussian beams with different coherences and experimentally derived the scintillation index for beam transmission in atmospheric turbulence. The results showed that reducing the spatial coherence of the beam suppressed the light-intensity scintillation in turbulence.

In 2016, Zhang et al. [21] derived Rayleigh-interval and turbulence-distance analytic equations for a radial partially coherent GSM array beam during atmospheric-turbulence transmissions. They showed that the effect of atmospheric turbulence could be effectively suppressed by appropriately reducing the beam coherence.

In 2019, Zhang [22] investigated whether anisotropic atmospheric turbulence could lead to poor optical-signal quality, proposed a novel twisted rectangular Laguerre-Gaussian correlated Schell-model (TRLGCSM) beam generated by partially coherent light superposition. They verified that the beam could effectively suppress the effects of atmospheric turbulence.

In 2019, Yu et al. [23] found that part of a coherent beam was affected by atmospheric turbulence. Moreover, the beam-drift phenomenon caused by atmospheric turbulence could be reduced by changing the coherent structure of the optical field and reducing the beam coherence.

In 2019, Zhang [24] conducted indoor simulation experiments using a liquid–crystal spatial-light modulator (LC-SLM) to generate a partially coherent beam. Using the LC-SLM for atmospheric-turbulence simulation, they verified that the partially coherent beam generated by the LC-SLM had better turbulence resistance than the fully coherent beam.

In 2022, Yang et al. [25] analyzed the transmission characteristics of a partially coherent radially polarized twisted beam under atmospheric turbulence and derived its twist factor. They showed that the larger the twist factor of the beam, the better the turbulence resistance.

In recent years, Xi'an University of Technology conducted extensive research on partially coherent light transmissions in the atmosphere [26]. In 2014, the transport properties of partially coherent GSM beams in atmospheric turbulence were studied by Han et al. [27, 28]. They verified that partially coherent light had better turbulence resistance than fully coherent light. The experimental results showed that the beam was less affected by beam drift when the spatial coherence length of the beam was greater than 0.005 m within a transmission distance of 10 km.

In 2015, Zhang et al. [29] studied the scintillation effect of partially coherent light subjected to atmospheric turbulence, and analyzed the variance in the light-intensity fluctuation of partially coherent light under different atmospheric turbulences. The experiments showed that partially coherent light had a better anti-turbulence effect under the same turbulent conditions, and that the turbulence effect could be effectively suppressed by reducing the beam coherence.

In the same year, Wang et al. [30, 31] studied beam-expansion expressions for partially coherent GSM beams and discussed the transmission characteristics when the beams were transmitted obliquely and horizontally. They showed that the light-intensity distribution when the beam was normalized for horizontal transmission was the same as that for an oblique-range transmission. The beam was less affected by turbulence during the oblique transmission.

Wang et al. [32] studied the transmission characteristics of a partially coherent electromagnetic Schell-model beam in upstream and downstream links through atmospheric turbulence and demonstrated that the beam could transmit over longer distances in the downlink transmission.

In 2016, Wang et al. [33] studied the transmission characteristics of partially coherent vortex beams under atmospheric turbulence. It was demonstrated that the partial vortex beam had a better transmission performance than the full vortex beam. Zhang et al. [34] studied the transmission characteristics of a partially coherent GSM array beam under atmospheric turbulence. It was demonstrated that the array beam had a better turbulence-suppression ability than a single-beam. Zhang et al. [35] studied and analyzed a partially coherent Gaussian–Schell-model pulsed beam (GSMP) and verified that its expansion was less affected by turbulence than a fully coherent beam was affected by turbulence.

In 2017, Wang et al. [36] derived an expression for the average scattering radius of a GSM beam at its receiving end. It was verified that the larger the initial radius of the GSM beam or the smaller the coherence, the less its average scattering radius was affected by atmospheric turbulence. In the same year, Xue et al. [37, 38] investigated the transmission characteristics of GSM beams at various turbulence scales and intensities. This study demonstrated the ability of inclined-range transmissions to resist turbulence better than horizontal transmissions under the same turbulent conditions.

Wang et al. [39] studied the transmission characteristics of partially coherent Airy beams. They showed that the smaller the beam-truncation factor, the less it was affected by turbulence and the more complete the light-intensity distribution. Zhang et al. [40] studied the transmission characteristics of partially coherent Airy beams and showed that they had less beam drift than Gaussian beams under the same turbulent conditions and had strong turbulence-suppression and recovery properties. Wang et al. [41] studied the transmission characteristics of partially coherent off-axis vortex beams and derived an expression for the light-intensity distribution.

In 2019, Wang et al. [42, 43] investigated a partially coherent electromagnetic Gaussian–Schell-model vortex (EGSMV) beam and verified that the larger the topological charge of the beam in an atmospheric-turbulence transmission, the more information the receiver side received, and the larger the midline dark spot of the beam.

5.1.2.2 Suppression Principle

No completely coherent light source exists. A partially coherent light, which lies between fully coherent and incoherent light, is prevalent. A highly coherent laser light is prone to scattering in inhomogeneous media, especially in atmospheres where the

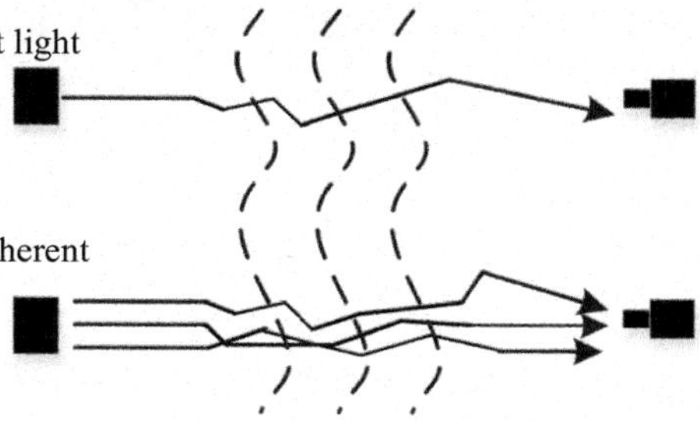

Fig. 5.2 Schematic diagram of partially coherent optical anti-turbulence [13]

transmission is affected by turbulence, resulting in an uneven light-intensity distribution. Therefore, the atmospheric-turbulence effect can be suppressed by appropriately reducing the optical coherence. Studies have shown that partially coherent light is less affected by turbulence in atmospheric transmissions, and the scintillation index is less than that of highly coherent light, which has better turbulence resistance [44].

Partially coherent light can suppress turbulent disturbances because of its polychromatic nature. Even under the influence of atmospheric turbulence, the transmission effect is better than that of single-frequency coherent light. Therefore, using partially coherent light for transmissions results in less flickering and can deliver more light energy at the receiving end. When coherent light propagates in turbulence, the beam refracts in various directions, resulting in only a portion of the light energy being received at the receiving end. A schematic diagram of partially coherent light suppressing turbulence is shown in Fig. 5.2.

GSM beams [45] are a typical spatially correlated structure of partially coherent light. For the evolution characteristics of such beams in atmospheric turbulence, it was found that GSM beams had better turbulence resistance than fully coherent beams. In addition, some partially coherent beams with other novel correlation structures, such as Hermite–Gaussian–Schell-model (HGSM) beams [46], cosine-Gaussian–Schell-model (CGSM) beam arrays [47], and multi-Gaussian–Schell model (MGSM) beams [48], were found to effectively suppress the negative atmospheric-turbulence effects for all beams transmitted in atmospheric turbulence.

However, it should be noted that during a partially coherent light transmission, the beams diverge faster, resulting in a correspondingly large signal-energy loss; thus, in practical applications, it is necessary to generate coherent light of a suitable coherence length.

5.1.3 Diversity Technology

Diversity technology is a technique that receives signals in parallel through multiple independent receiving channels and then merges and processes them. The main purpose is to improve the reliability of signal reception. In communication systems, diversity technology is commonly used to reduce factors, such as multipath fading, interference, and noise, during a signal transmission, thereby improving the signal quality of the system. It receives multiple independent signal paths simultaneously and processes them to reduce the signal distortion and signal-to-noise ratio degradation.

5.1.3.1 Current Research Status

In 1996, Ibrahim [49] hypothesized that the output signals of small-aperture receivers faded independently of each other and investigated a spatial-reception diversity technique using multiple reception apertures.

In 2004, Pan Feng et al. [50] analyzed the optical-intensity flicker characteristics of a transmission link using a multi-beam transmission technique for a ground-to-ground optical communication system. They found that the multi-beam transmission technique effectively suppressed the turbulence-induced optical-intensity flicker in the entire turbulence region from weak to strong.

In 2005, Trisno et al. [51] used a time-delay diversity technique to suppress the fading of optical signals, due to atmospheric turbulence, and obtained signal-to-noise gains of 3.8, 4.3, and 4.4 dB by increasing the time delay.

In 2006, Anguita et al. [52] used a multi-beam transmission technique, analyzed two space–time coding modes, and demonstrated that both modes had higher signal-to-noise gains than a single-beam transmission under the influence of atmospheric turbulence.

In 2008, Neda et al. [53] analyzed a ground-based optical-flux system using multi-beam emission and multiple avalanche-photodiode (APD) detection reception and discussed the gain-size selection for merging multiple APD output signals.

In 2011, Ke et al. [54] used a diversity technique at the receiver side. The light-intensity flicker effect caused by atmospheric turbulence was effectively suppressed by multi-aperture reception, verifying that the BER and aperture-smoothing factor of the system were further reduced with an increase in the number of reception apertures.

In 2012, Ke et al. [55] analyzed the light-intensity probability distribution of signals transmitted through multiple beams under different turbulent environments. The results showed that a multiple-beam transmission could reduce the BER of the system, decrease the light-intensity fluctuation effect, and improve the performance of the communication system under the same turbulent environment.

In 2014, Jiang [56] demonstrated that aperture diversity could reduce the optical-intensity scintillation index and that the system performance was better than that of a large-aperture receiving system by using a diversity technique and numerical simulations with a phase screen and comparing it with a large-aperture receiving system.

In 2015, Kaur et al. [57] investigated the combined effect of atmospheric-turbulence motion on optical-wireless multiple-input multiple-output (MIMO) communication systems under the influence of different weather conditions, and compared the BER and outage capacity using aperture-averaging techniques under the same conditions. The results showed that MIMO techniques under different weather conditions could effectively suppress atmospheric turbulence.

In 2016, Bhatnagar et al. [58] studied the effect of the alignment error on an optical MIMO communication system under a gamma–gamma turbulent channel model and analyzed the performance of a communication system under equal-gain merging and maximum-ratio merging. They concluded that the maximum-ratio merging criterion could suppress alignment errors.

In 2017, Meng [59] investigated optical-wireless communication systems using the split-set merging technique and showed through experimental simulations that the light-intensity flicker effect caused by atmospheric turbulence could be most effectively suppressed using the equal-gain merging technique.

In 2019, Kumar et al. [60] analyzed the performance of optical-wireless MIMO systems under gamma–gamma turbulent channels using different diversity techniques. The results showed that the optical-wireless MIMO system reduced the BER of the system and improved its reliability under a gamma–gamma turbulent channel.

In 2019, Xu [61] used a diversity technique to derive the probability distribution of light intensity under multi-beam emissions. Improved simulation experiments showed that, under the same turbulent conditions, the technique had the correct anti-turbulence ability, compared to single-aperture emission, which effectively mitigated the light-intensity scintillation effect.

In 2021, Ding [62] analyzed the variation in the receiving-system performance by using a spatial-diversity technique for a partially coherent optical transmission system. The experimental results showed that the signal-to-noise ratio of the communication system could be improved by using a diversity technique, and the performance was improved under the influence of turbulence.

In 2021, Dai [13] demonstrated that the scintillation effect could be suppressed by increasing the number of receiving apertures in a large-aperture Cassegrain receiving system with a reduced light-intensity scintillation index through a diversity technique.

In 2022, Liu [63] investigated optical-wireless MIMO communication technology. The atmospheric-transmission characteristics of a single-input single-output (SISO) system and MIMO system were comparatively studied using an LC-SLM to simulate atmospheric turbulence. It was demonstrated that the MIMO communication system had a lower link cost and improved communication performance compared to the SISO communication system.

5.1.3.2 Suppression Principle

The application of diversity technology to optical-wireless communication systems involves transmitting and receiving transmission signals by increasing the number of antennas at the transmitter and receiver ends. The block diagram of the system is shown in Fig. 5.3, where the transmitter end of the system transmits the laser beam using multiple transmitting antennas. When it reaches the receiver end, the transmitted laser beam is received using multiple receiving antennas. The atmospheric-channel attenuation experienced by each signal is unrelated, which effectively

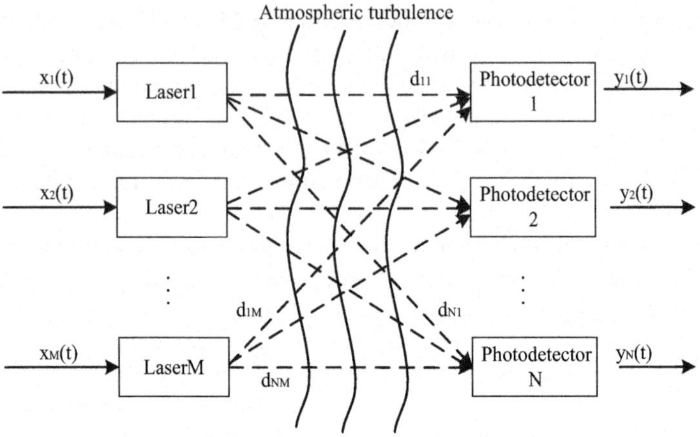

Fig. 5.3 Block diagram of a diversity system [63]

suppresses the atmospheric-turbulence effect and improves the communication quality of the entire communication system.

Diversity technology includes transmit and receive diversity. Among them, transmit diversity often uses multi-beam transmission technology, which is a free-space laser-transmission technology that can effectively suppress the scintillation effect by transmitting two or more independent and incoherent lasers through different paths to the receiving end for superposition. Consequently, the light-intensity scintillation on the far-field receiving plane is smoothed.

For receive diversity, antenna-array receiving technology is more commonly used, by setting multiple independent small-aperture receiver antenna arrays instead of large-aperture receivers. Under the same receiving-area conditions, the method can receive multipath signals to a greater extent, reduce the impact of turbulent channels, and improve the reliability of the communication system. Moreover, the structure is simple, flexible, and has a lower cost.

The receiver side combines multiple received fading signals to obtain a gain signal, which improves the quality of the received signal. Therefore, the merging technique is the most important aspect of the diversity-reception technique. The most commonly used diversity merging techniques are maximum-ratio, selection, and equal-gain combining [64].

Selection combining is a relatively simple merging technique, in which each branch is demodulated so that their signal-to-noise ratios are the same; then, a branch is selected at the receiving end for demodulation. This approach is difficult to implement in practice because the signal-to-noise ratio of the communication system is not easily measured.

Maximum-ratio combining uses the ratio of the voltage value of each branch signal and the noise power as the weight; then, the phase correction of each signal is carried out at the receiving end, and the sum of each signal is added according to the gain coefficient. In the detector receives the coherent detection, this can be in the case of received signal is not ideal, by the method of maximal ratio combining demodulation.

Equal-gain combining does not require the weighting of each signal. It directly assumes that each path has the same weight and directly superimposes the same signal. The results obtained using equal-gain combining and maximum-ratio combining are similar.

Under the same experimental conditions, the improvement in system performance was most evident for maximum-ratio merging, followed by equal-gain merging, and then, selective merging.

5.1.4 Adaptive-Optics Technology

Adaptive-optics technology can be used to adjust the shape and surface morphology of optical components in real-time, based on changes in the optical system. It is primarily used in fields such as astronomy and optical communication to improve the performance and effectiveness of optical systems. Adaptive-optics technology can effectively improve the performance of optical systems, particularly in environments with atmospheric turbulence. It can effectively reduce the aberration blurring caused by atmospheric turbulence and improve the transmission distance of astronomical observation and wireless laser communication.

5.1.4.1 Current Research Status

In 1996, Tyson et al. [65] concluded from a turbulence-theory analysis that conventional adaptive-optics (AO) techniques were effective in reducing the optical-signal attenuation and fluctuation in the uplinks and downlinks of a ground–space free-space optical communication system under weak turbulence.

In 1998, Roggemann et al. [66] used adaptive optics and proposed the joint use of two deformation mirrors to correct the light-intensity scintillation effect caused by atmospheric turbulence in laser atmospheric transmissions.

In 1998, Barbier et al. [67] used adaptive-optics techniques and simulated the performance of communication systems using the phase-screen method. The effects of the dynamic range and adaptive-optical aperture size of the wavefront sensor on the mean and variance of the optical intensity were analyzed.

In 2001, Li et al. [68] proposed a correction method using two adaptive-optics systems in series. The control performance and noise-transfer characteristics of this approach were analyzed. The simulation results showed that the tandem structure could substantially improve the utility of the adaptive-optics system in controlling the wavefront aberrations of atmospheric turbulence.

In 2004, Thomas et al. [69] used adaptive optics to compensate for the tilt aberration of a wavefront using a tilt mirror and the higher-order aberration using a micro-electromechanical system (MEMS) deformation mirror. This dual-deformation mirror correction method produced a two-fold increase in the received signal intensity, compared to when only the tilt aberration was corrected.

In 2010, Xia et al. [70] used an adaptive-optics technique to effectively correct the aberrated wavefront and experimentally demonstrated that this technique could improve the communication-system gain, reduce the system transmit power, and lower the BER under the same BER conditions.

In 2011, Han [71] used adaptive optics to compensate for free-space optical communication systems and provided the corresponding expressions for the BER and coupling efficiency, as well as the relationship between the coupling efficiency and Strehl ratio.

In 2013, Wu [72] verified through simulation experiments that adaptive optics could effectively correct the aberrated wavefront and improve the phase matching of local-oscillation light and signal light to reduce the system BER.

In 2014, Hashmi et al. [73] used adaptive-optical techniques to simulate an interstellar optical communication system under moderate optical background noise and strong turbulence. The results showed that the Strehl ratio of the system improved from 0.30 to 0.75, after reaching a steady state, with adaptive-optical closed-loop control.

In 2015, Cvijetic et al. [74] conducted experiments on coherent optical communication systems under the influence of atmospheric turbulence in oceans using adaptive optics. The results showed that adaptive optics could compensate for the phase distortion of the optical signal and reduce the BER of the communication system.

In 2015, Wright et al. [75] used adaptive optics to build an optical-communication telescope ground-observation communication system and corrected the aberrated wavefront. The Strehl ratio of the received beam improved from 0.03 to 0.66 after correction.

In 2017, Escarate et al. [76] designed a model-predictive controller to calibrate an adaptive-optics system with 140 units and two deformation mirrors spaced 3 m apart. The results showed that this method could effectively correct the amplitude and phase of the laser beam.

In 2018, Li [77] studied several methods to suppress atmospheric turbulence and conducted a wavefront detection-free adaptive-optical atmospheric-turbulence suppression experiment using a stochastic parallel gradient-descent (SPGD) algorithm. The results showed that the corrected fiber-coupling efficiency could be improved from 2.7 to 9 dB at different turbulence intensities.

In 2019, Lin et al. [78] predictively estimated state variables for adaptive-optical systems and implemented closed-loop correction using a linear quadratic Gaussian control technique based on state regulation. The results showed that the method better improved the system response speed and stability compared to the proportional-integral-derivative (PID) control algorithm.

In 2019, Chen et al. [79] used an adaptive-optics technique to apply a continuous deformation mirror with 137 cells to an adaptive-optics system. This yielded a BER of less than 10^{-6} for the communication system by correcting the aberrated wavefront, which was significantly reduced, compared to the uncorrected BER.

In 2020, Banet et al. [80] proposed a method for compensating for beacons using adaptive optics. Under the same turbulent conditions, the Shack–Hartmann wavefront sensor of an adaptive-optics system with compensated beacons exhibited a 17% performance improvement, compared with an uncompensated fiducial adaptive-optics system.

In 2022, Yang et al. [81] analyzed a Karhunen–Loève (K-L)-mode SPGD algorithm using adaptive optics, with an optical-wavefront aberration under the influence of atmospheric turbulence as the correction object, and verified that the convergence speed of the SPGD algorithm in this mode was improved by 47.5%, compared to the conventional SPGD algorithm.

In 2023, Li et al. [82] investigated free-space optical communication under the influence of atmospheric turbulence and proposed an advanced multifeedback SPGD (AMF-SPGD) algorithm to enhance the wavefront-aberration correction performance. They verified that the algorithm improved the Strehl ratio and reduced the BER of the communication system.

Xi'an University of Technology has investigated many aspects of adaptive optics [83]. In 2018, Wu et al. [84] addressed the fact that a beam was distorted under the influence of atmospheric turbulence and corrected the distorted wavefront using adaptive optics, without wavefront detection, based on the SPGD algorithm. The corrected Strehl ratio was improved from 0.15 to 0.81, which resulted in a larger received spot and mitigated the effect of atmospheric turbulence on the wavefront distortion.

In the same year, Wang et al. [85, 86] used adaptive optics without wavefront detection, based on the Gerchberg–Saxton (GS) and SPGD algorithms for wavefront correction of vortex beams; the simulation results showed that the GS algorithm outperformed the SPGD algorithm in terms of correction results.

In 2019, Ke et al. [87] used an adaptive-optics technique with wavefront detection based on double fuzzy control to correct an aberrated wavefront. The simulation results showed that this method could effectively achieve this correction. In the same year, Yang et al. [88, 89] used adaptive optics with wavefront detection and effectively corrected aberrated wavefronts using computational optical-field imaging.

Zhang et al. [90] proposed an improved, parallelized SPGD algorithm using wavefront-free detection adaptive optics. The experimental results showed that this method improved the algorithm convergence and wavefront-correction speeds. Han et al. [91] used two liquid–crystal spatial-light modulators for wavefront-aberration simulation and wavefront correction, according to the principle of adaptive optics. The corrected Strehl ratio was improved from 0.38 to 0.82, and this method reduced the experimental cost.

In 2020, Li [92] used adaptive optics without wavefront detection and investigated a wavefront-correction method using the deformation-mirror eigenmode. The Strehl ratio of the correction result improved from 0.56 to 0.88, verifying that the method could effectively correct wavefront aberrations caused by atmospheric turbulence and that the algorithm converged quickly. In the same year, Cui [93] used a phase-diversity (PD)-based wavefront detection-free adaptive-optical technique to correct vortex beams. The experimental results showed that this method improved the signal power, reduced inter-code crosstalk, and effectively mitigated the turbulence effect.

In 2021, Cheng [94] used a PID control algorithm and a large-amplitude aberration wavefront-residual correction algorithm to correct the problem of insufficient range for a deformation mirror in an adaptive-optics system. The experimental results showed that the large-amplitude aberration wavefront could be effectively corrected using the residual-correction algorithm.

5.1.4.2 Suppression Principle

Adaptive-optics technology can automatically improve the wavefront quality of beams. Therefore, the application of adaptive optics to optical-wireless communication systems has great potential for suppressing the effects of atmospheric turbulence. Adaptive-optics systems are divided into two categories: those with wavefront sensors [95] and those without wavefront sensors [96].

Adaptive-Optics Systems with Wavefront Sensors

An adaptive-optics system with a wavefront sensor is shown in Fig. 5.4 and consists of three main components: a wavefront sensor, a wavefront controller, and a wavefront corrector. The general correction process is as follows: The emitted beam is affected by atmospheric turbulence and the beam wavefront is distorted. The distorted beam first passes through the corrector for the first time to collect information, and then arrives at the beam splitter. Part of the beam is kept and part of the beam is transmitted to the wavefront sensor to measure the beam-wavefront information. This information is converted into electrical signals to the wavefront controller for wavefront reconstruction. A suitable control algorithm is selected to generate the control signal, and the wavefront corrector receives the control signal, allowing the corrector to correct the distorted wavefront, forming a closed-loop system.

Wavefront sensors are one of the main components in adaptive-optics systems. Their role is to measure aberrated wavefront information in real time and send it to the wavefront controller. Commonly used wavefront sensors are transverse-shear interferometers [97], Shack–Hartmann sensors [98], curvature-wavefront sensors [99], and point-diffraction interferometer sensors [100].

The wavefront controller converts the wavefront information from the wavefront sensor into a control signal and then performs two functions: wavefront reconstruction and control-signal generation. Wavefront correctors are essential components of adaptive-optics systems to correct wavefront errors in a timely manner. Common wavefront correctors include liquid–crystal spatial-light modulators (LC-SLM) [101] and deformable mirrors (DM) [102].

In adaptive-optics systems with wavefront sensors, the correction accuracy of the distorted wavefront varies between the different classic and iterative control

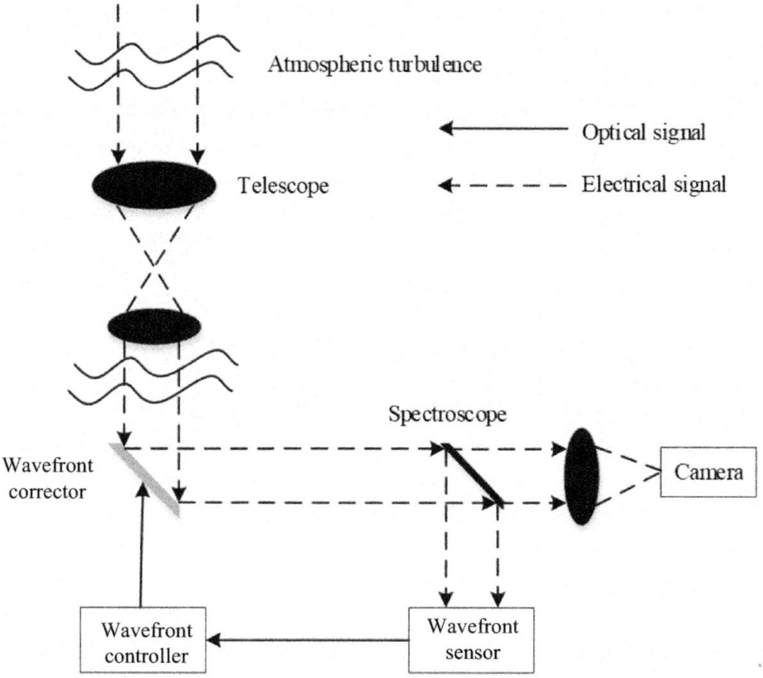

Fig. 5.4 Adaptive-optical system with wavefront sensor [83]

algorithms. Among the classic control algorithms, the PID control algorithm [103] is the most widely used in adaptive optics.

The main principle of the PID control algorithm is as follows: First, the command matrix of the system is determined using the push–pull method. Subsequently, the command matrix receives the error slope and converts it into an error voltage signal. Then, the obtained voltage is sent to the deformation mirror for wavefront correction by the PID control system. Finally, the wavefront sensor collects the wavefront slope again to achieve closed-loop control.

The Gauss–Seidel iterative algorithm is a common iterative control algorithm and its correction effect in the field of adaptive optics is related to the iteration accuracy error. The calibration process of the Gauss–Seidel iterative algorithm [103] is as follows: First, the response matrix is calibrated. Then, the wavefront sensor is used to detect the wavefront slope at that time, and the relevant parameter values are calculated. These values are substituted into an iterative formula to calculate the control voltage and iterative-judgment conditions.

If the iterative-judgment condition value is less than the set iteration-accuracy error, the algorithm converges, and the deformable-mirror control voltage is directly sent to stop the iteration; otherwise, the feedback is repeated until the conditions are met. At this point, a single wavefront correction is completed and the system can

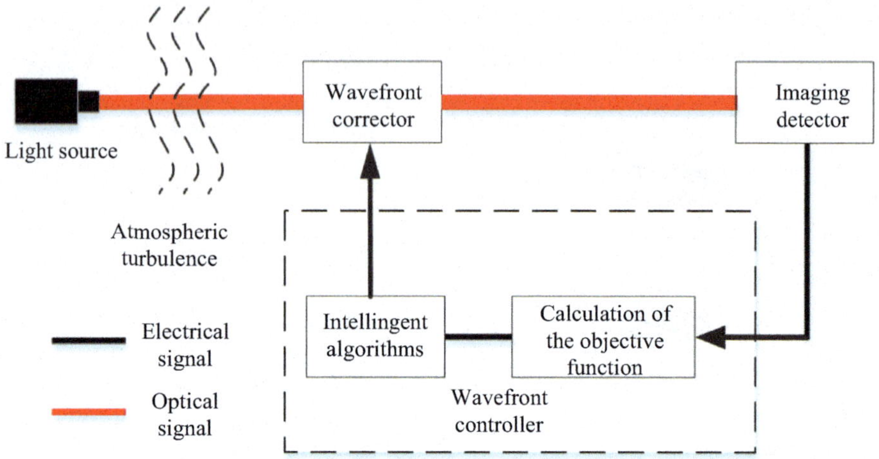

Fig. 5.5 Adaptive-optics system without wavefront detection [104]

proceed to the next slope-acquisition process and repeat the above steps to complete the closed loop.

Unfortunately, iterative algorithms have challenges with parameter ranges and other issues. Therefore, there are certain limitations to their practical applications. The principle of the PID control algorithm is simple, and its overall performance is better than that of iterative algorithms.

Adaptive-Optics Systems Without Wavefront Sensors

As shown in Fig. 5.5, an adaptive-optical system without wavefront detection does not require wavefront sensors to detect and reconstruct the wavefront. The wavefront-detection system consisted of three parts: an imaging detector, a wavefront corrector, and a wavefront controller.

The calibration process is as follows. First, the wavefront corrector corrects the initial phase of the received beam, and then the beam is incident on the imaging detector. The imaging detector collects the required system-performance parameter values and transmits them to the wavefront controller. The wavefront controller again drives the optimization algorithm to calculate the voltage signal required for the deformable mirror, iterating sequentially until the aberration is minimized.

Adaptive-optical intelligent-optimization algorithms without wavefront detection are divided into two main categories: patternless and patterned optimization methods.

The mode-optimization method involves selecting a suitable base mode to describe the wavefront error and correcting the wavefront distortion by optimizing the base-mode coefficients. Mode-free optimization involves directly controlling a deformable mirror to correct wavefront distortion and optimize the system performance-evaluation metrics. Currently, the SPGD algorithm [105], simulated-annealing (SA) algorithm [106], and genetic algorithm (GA) [107] are commonly used among pattern-free optimization algorithms. The SPGD algorithm can estimate the gradient, converges quickly, is more stable, has simple parameters, and is easy to implement.

The basic working principle of the SPGD algorithm [84] involves estimating the control parameters based on the change values of the performance-evaluation index and those of the control parameters, and to search for control parameters in the gradient-descent direction using an iterative approach. The main process is to generate the initial voltage signal of the deformed mirror and a random-disturbance voltage signal conforming to the Bernoulli distribution. Positive and negative disturbance voltages are added, respectively.

On the basis of the initial voltage signal, the change in the system objective function under the voltage is calculated. Finally, the voltage signal to be loaded in this iteration is calculated according to the iterative formula, and repeated until the objective function meets the initial conditions set by the system. This is repeated until the target function satisfies the initial conditions set by the system, so that the system-evaluation index reaches the optimal value, thus correcting the distorted wavefront.

Many researchers have proposed improved SPGD algorithms [82, 90], all of which can ultimately be very effective in improving the performance of communication systems and enhancing the algorithms' convergence speed and correction effect.

5.2 Atmospheric-Turbulence Suppression Experiment

5.2.1 Multi-beam Transmission Technology Experiment

See Fig. 5.6.

Multiple independent laser beams with the same wavelength are emitted from different transmitting antennas, and the spacing between each antenna is required to meet the constraint condition $S_2 \geq \lambda L$ (where λ represents the laser wavelength and L is the communication distance). The laser beams were noncoherently superposed at the communication-receiving end. This method can effectively suppress the

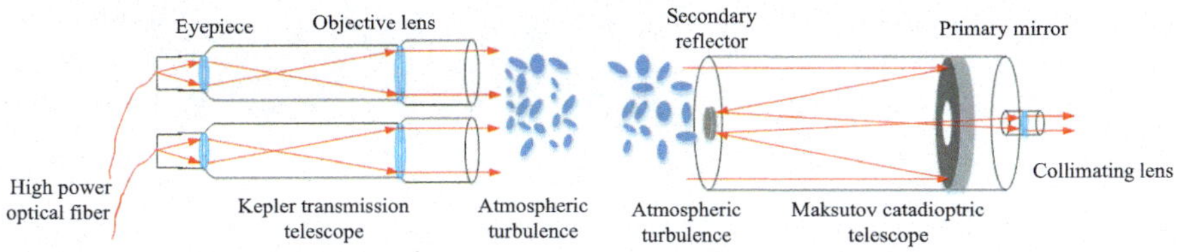

Fig. 5.6 Diagram of a 2 × 1 system

light-intensity flickering caused by atmospheric turbulence, and the optical power at the receiving end plays a smoothing role, effectively improving the communication quality [108].

In recent years, researchers have studied the improvement of the bit-error rate in intensity modulation/direct detection (IM/DD) optical communication systems using multi-beam transmission technology. We applied multi-beam transmission technology to coherent-optical communication field experiments at a communication distance of 100 km, as shown in Fig. 5.7. The improvement in the coherent optical communication system performance by multi-beam transmission technology was analyzed based on the experimental results.

We conducted a multi-beam transmission experiment with a communication distance of 100 km from the Erlangjian Scenic Area in Qinghai Lake to Quanji Township in Gangcha County, China. Figure 5.7 shows a map of the experimental sites. Figure 5.8 shows the fluctuation curve of the signal optical power at the receiving end when the system adopts single-beam and double-beam transmissions.

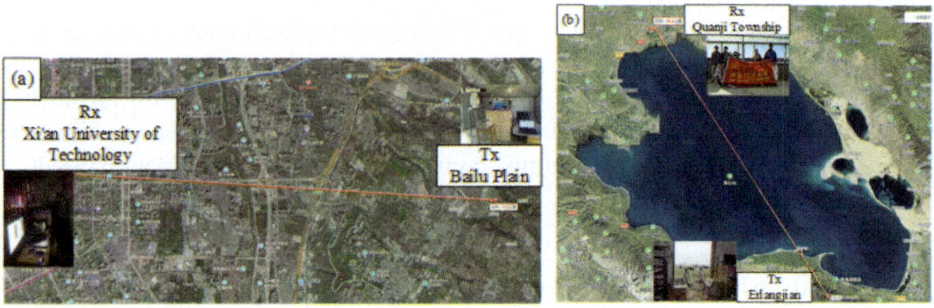

Fig. 5.7 Experimental communication link map. **a** 10-km link; **b** 100-km link

Fig. 5.8 Variation curve of the signal-light power under different transmission modes

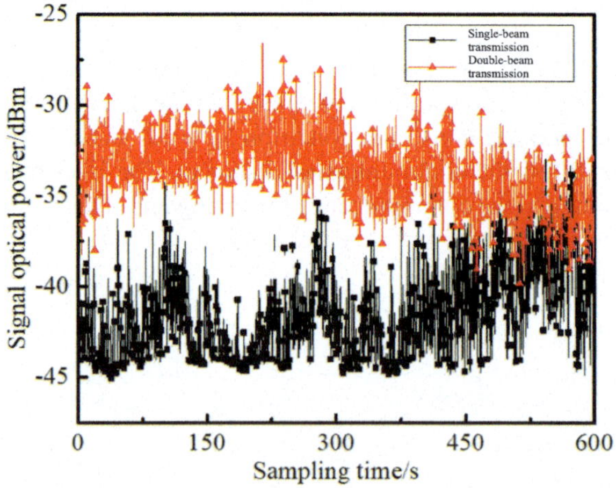

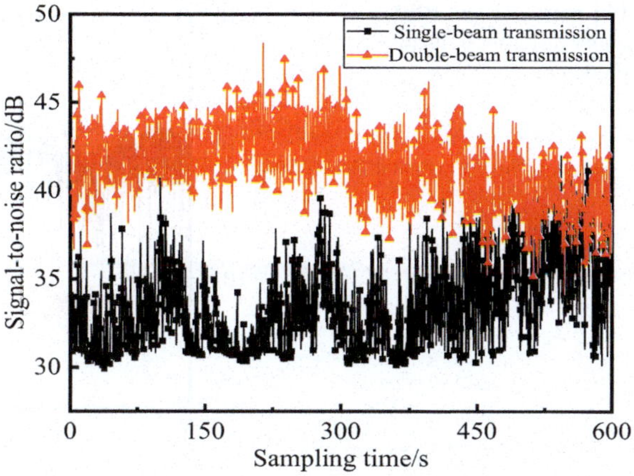

Fig. 5.9 System SNR under different transmission modes

From the figure, it can be seen that the average optical power at the receiving end for a single-beam transmission is – 41.53 dBm, with a variance of 6.01 dBm^2. The average optical power at the receiving end for double-beam transmission is – 33.47 dBm, with a variance of 3.79 dBm^2, indicating that the system can effectively reduce the jitter variance of the signal optical power at the receiving end when using double-beam transmissions.

In the system, the following parameters were set: electronic power $e = 1.6 \times 10^{-19}$ C, quantum efficiency $\eta = 0.8$, Planck's constant $h = 6.626 \times 10^{-34}$ J s, load impedance $R_L = 50\ \Omega$, background-radiation power $P_B = 10^{-9}$ W, detector bandwidth Δf IF = 100 MHz, Boltzmann constant $K = 1.38 \times 10^{-23}$ J/K, ambient temperature $T_b = 300$ K, and local-oscillator power is 1 mW. The signal-to-noise ratio variation curves of the heterodyne-detection system with different beam transmissions are shown in Fig. 5.9.

As shown in the Fig. 5.9, when a single laser beam is transmitted, the average signal-to-noise ratio of the system is 33.44, with a variance of 6.01. When double beams are transmitted, the average signal-to-noise ratio of the system is 41.5, with a variance of 3.8. This indicates that using multi-beam transmission technology can effectively improve the average signal-to-noise ratio of the communication system, while reducing its fluctuation range.

The error-rate curve of the optical-wireless communication system after transmissions with different numbers of beams is shown in Fig. 5.10. From Fig. 5.10, it can be seen that the average system error rate during a single-beam transmission is 6.5×10^{-9} with a variance of 3.14×10^{-7}. The average error rate of the system during a double-beam transmission is 9.98×10^{-11}, with a variance of 1.95×10^{-20}. Compared to a single-beam transmission, a double-beam transmission can reduce the mean and variance values of the system error rate by three orders of magnitude, improving the performance of the communication system.

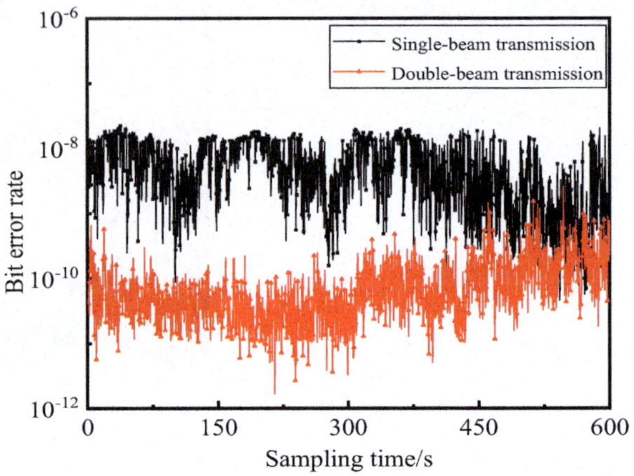

Fig. 5.10 System BER under different transmission modes

5.2.2 Adaptive-Optics Wavefront-Correction Technology Experiment

After a long-distance laser transmission, the wavefront is distorted, owing to the influence of atmospheric turbulence, which directly leads to a decrease in the coupling and mixing efficiencies in the coherent detection of wireless laser communication. Therefore, it is necessary to correct the distorted wavefront.

The distorted wavefront phase can usually be expanded based on Zernike polynomials; the tilted component of the wavefront is usually referred to as the low-frequency component, whereas the remaining components are referred to as the high-frequency component. The wavefront-tilt component accounts for approximately 86% of the total wavefront distortion. Therefore, it is necessary to use a combination of multiple correctors (such as a combination of a fast-steering mirror (FSM) and a deformable mirror (DM)) for wavefront correction.

Using a fast reflector to correct low-order aberrations and a deformable mirror to correct high-order aberrations, the orthogonality based on Zernike polynomials allows for a clear division of the correction-region space between the two. An adaptive-optical system composed of a piezoelectric fast reflector based on two-dimensional motion and a deformable mirror (DM69) with independent units is shown in Fig. 5.11.

After light waves are transmitted at different distances, such as 600 m, 1, 5, 10, and 100 km, the peak-to-valley (PV) values, proportion of Zernike-coefficient tilt components, and variance of the tilt-component changes are measured. After calculation and processing, the data shown in Table 3.4 are obtained.

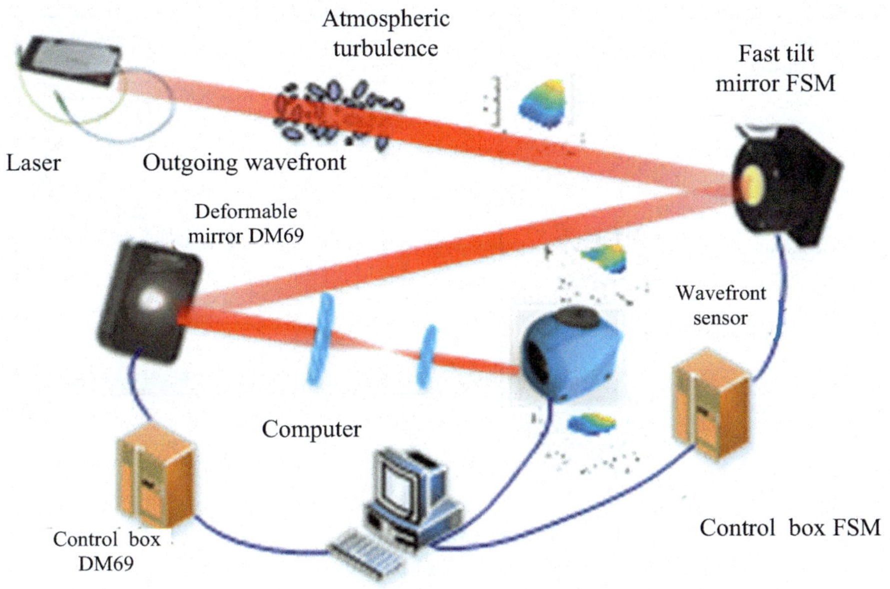

Fig. 5.11 Adaptive-optical wavefront-correction system based on the combination of a fast reflector and deformable mirror

From Table 5.1, it can be seen that the wavefront distortion increases with increasing transmission distance, and the proportion of tilted components accounts for approximately 80% of the overall distortion. The variance of the tilt distortion also increases with distance, indicating that the speed of change also increases with the communication distance.

Figure 5.12 shows the measured wavefront PV values at different distances. The degree of wavefront distortion and rate of wavefront change increase with an increase in the transmission distance. When the transmission distance reaches 100 km, the light-intensity flickering caused by strong turbulence completely shatters the collected wavefront, and the incomplete sampling points make reconstruction difficult, resulting in discontinuous data collection.

Figure 5.13 shows the wavefront phase diagram of the laser after a 100-km transmission, corrected only by the fast reflector, only by the deformable mirror, and

Table 5.1 Measured wavefront values at different distances

Distance	Indoor	600 m	1 km	5 km	10 km	100 km
PV value (μm)	4.99	2.18	5.67	11.25	14.55	200.7
Tilt percentage (%)	72.34	56.30	81.65	84.44	73	44.95
Variance Tilt-x	2.07×10^{-4}	0.221	0.136	0.38	4.23	2.58×10^{5}
Variance Tilt-y	2.07×10^{-4}	0.891	0.134	0.25	3.34	8.63×10^{6}

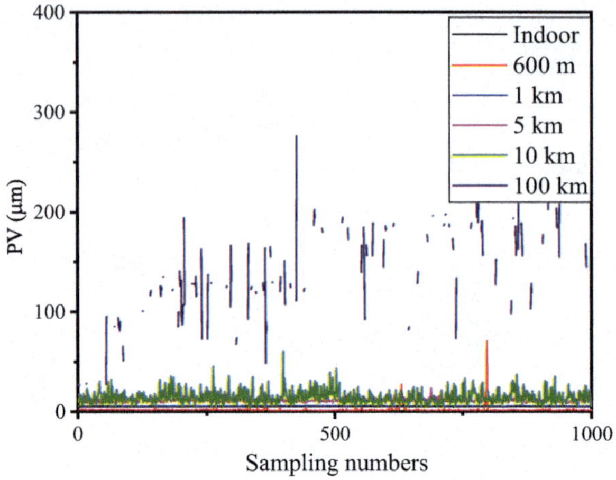

Fig. 5.12 Measured wavefront PV values at different distances

simultaneously corrected by both. From Fig. 5.13, it can be seen that the wavefront PV value decreased from 450 to 300 μm after only being corrected by the fast reflector. It decreased from 450 to 300 μm after only being corrected by the deformable mirror. When both the fast reflector and deformable mirror were used simultaneously, the wavefront PV value decreased from 450 to 200 μm. This indicates that the correction effect of an adaptive-optical system based on the combination of a fast reflector and a deformable mirror is better than that of a single fast reflector or deformable mirror.

For a 100-km communication link, the light-intensity flicker and beam drift caused by strong turbulence cause a discontinuous wavefront acquisition, and some sampling points fail to reconstruct, resulting in worse correction effects than those under medium to weak turbulence conditions. As the communication distance increases, the atmospheric-turbulence intensity increases and the effect of the wavefront correction deteriorates. This is because the effect of wavefront correction is related to the detection accuracy, body noise of the wavefront sensors, wavefront-reconstruction accuracy, closed-loop bandwidth, and atmospheric-measurement environmental conditions.

5.3 Development Trends

5.3.1 Development Trends

Optical-wireless communication systems are susceptible to atmospheric turbulence, which distorts the wavefront of the transmission beam and reduces its optical power, thereby increasing the bit-error rate of the communication system and reducing the communication quality. At present, researchers have used various technologies, such

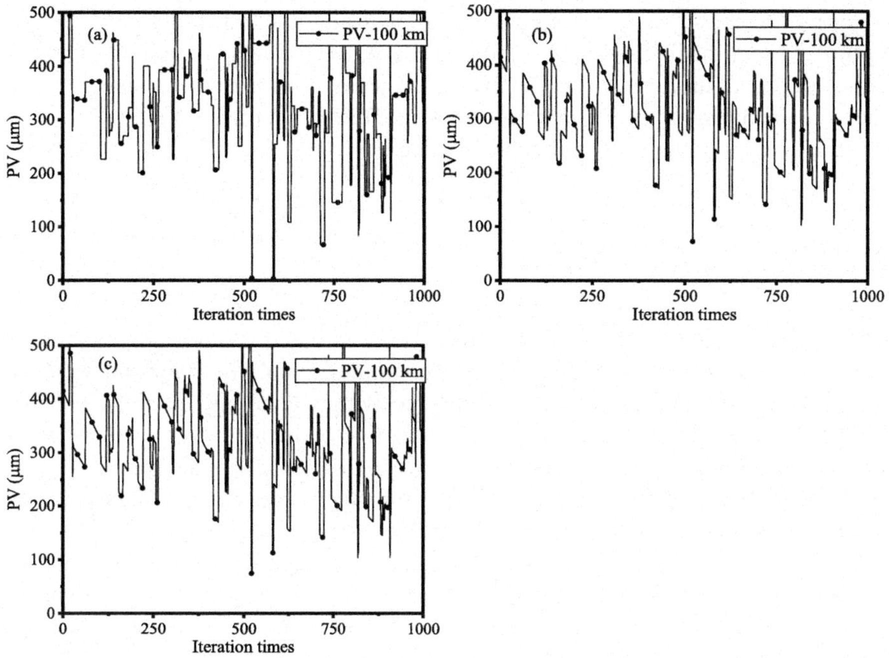

Fig. 5.13 Wavefront correction PV curves at 100 km. **a** FSM correction only; **b** DM correction only; **c** simultaneous FSM and DM correction

as large-aperture reception, partially coherent light, diversity, and adaptive optics, to suppress the impact of atmospheric turbulence on optical-wireless communication.

For large-aperture receiving technology, it is possible to appropriately increase the size of the receiving antenna to reduce the light-intensity flickering. Currently, many scholars use large-aperture receiving technology to suppress the flickering effect of turbulence and establish different smoothing-factor models, based on different turbulence intensities, to design the large-aperture receiving systems. They select as many appropriate apertures as possible to reduce the weight at the receiving end.

Partially coherent light technology also has a positive effect on atmospheric-turbulence suppression, and studying new types of partially coherent beams is highly useful in suppressing atmospheric turbulence.

Diversity technology is a fundamental anti-fading technology in wireless communication, which can effectively suppress the attenuation caused by atmospheric turbulence. This technology is low-cost, simple, easy to apply in engineering, and has broad development prospects in the future.

Adaptive optics is currently the most widely used technique for suppressing atmospheric turbulence because it can correct distorted wavefronts in real time and directly improve the quality of communication systems. Various methods have been proposed to improve the coupling efficiency of single-mode fibers and communication quality through adaptive-optics technology.

In recent years, many scholars have adopted multiple deformable mirrors and optimized them using appropriate control algorithms in adaptive-optical systems, which improves the correction effectiveness. The application of many new technologies and methods for improving the correction effect of adaptive-optical systems has enabled them to more fully realize their value.

5.3.2 Summary and Outlook

This chapter analyzed the domestic and foreign research on methods to suppress atmospheric turbulence in optical-wireless communication. The main suppressors of atmospheric turbulence are large-aperture receiving technology, partially coherent light technology, diversity technology, and adaptive-optics technology. The principle of each technology for suppressing atmospheric turbulence was analyzed.

To a certain extent, these technologies can effectively suppress atmospheric turbulence, compensate for wavefront distortion, and improve the performance of optical-wireless communication systems. Although these technologies can improve the performance of optical communication systems, they are not perfect in engineering practice; Table 5.2 shows the advantages and disadvantages of each key technology. Therefore, further research and discussion on methods to suppress atmospheric turbulence are of great importance.

1. Optical-wireless MIMO technology can fully use spatial spectrum resources and effectively suppress atmospheric turbulence effects; however, this technology is not yet mature in the optical-wireless field and further research is needed for optical communications.
2. In partially coherent optical technology, various new types of partially coherent beams have been studied. Partially coherent vortex beams have a better suppression effect on atmospheric turbulence, and further research on their generation methods is required.
3. For adaptive-optics technology, further study is needed on deformation mirrors and control algorithms, using multiple deformation-mirror combinations for correction, and algorithm convergence to improve the accuracy of deformation-mirror correction.

Table 5.2 Advantages and disadvantages of key technologies for suppressing atmospheric turbulence

Related technology	Advantages	Disadvantages
Large-aperture receiving technology	An appropriate increase in the size of the receiving aperture reduces the variance of the optical-power fluctuations and effectively suppresses the light-intensity flicker effect of atmospheric turbulence	The increased aperture makes the receiver larger and heavier, and therefore less portable
Partially coherent light technology	Partially coherent light with good turbulence resistance and effective suppression of light-intensity flicker effects	Large beam-divergence angle and severe beam extension at the receiving end
Diversity technology	Multi-beam emission with incoherent superposition at the receiving end smooths the optical power and effectively suppresses the light-intensity flicker caused by atmospheric turbulence	Multi-transmitter, multi-receiver diversity systems are complex and limited in their practicality
Adaptive optics with wavefront sensor	Automatically improves the optical wavefront quality, enabling the effective correction of distorted wavefronts	For long communication distances, it is difficult for the wavefront sensor to measure the wavefront-aberration information, which makes it impossible to reconstruct the wavefront and compensate for the wavefront aberration in real time
Adaptive optics without wavefront sensor	Correction of distorted wavefronts without wavefront reconstruction	Higher cost; requires additional equipment and increased system complexity

References

1. Ke X, Deng L (2016) Optical-wireless communication. Beijing: Science Press
2. Ke XZ, Ning C, Wang J (2018) Crosstalk analysis of orbital angular momentum-multiplexed state under atmospheric turbulence. Infrared Laser Eng 47(11):402–408
3. Kerr RJ, Dunphy RJ (1973) Experimental effects of finite transmitter apertures on scintillations. J Opt Soc Am 63(1):1–8
4. Heba Y, Davi CC (2006) Aperture averaging analysis and aperture shape invariance of received scintillation in free-space optical communication links. Free-Space Laser Commun 63(4):63041E.1–63041E.11
5. Chen J, Ai Y, Ye D et al (2008) Signal fading under 16 km horizontal laser link aperture reception conditions. Semicond Optoelectron 29(06):928–931
6. Prokes A (2009) Atmospheric effects on availability of free space optics systems. Opt Eng 48(6):635–644
7. Chunyi C, Huamin Y, Huilin J et al (2009) Analysis of large-aperture reception performance and aperture size selection in atmospheric optical communication. China Laser 36(11):2957–2961

8. Yan L, Huilin J, Chunyi C et al (2011) Experimental analysis of light intensity fluctuation and spot area for laser atmospheric turbulence transmission. Infrared Laser Eng 40(03):515–519
9. Lu H, Wang L, Li Y et al (2012) Computational simulation of aperture averaging effect in space laser communication. Appl Opt 33(03):619–623
10. Wang Z (2015) Research on large aperture and composite aperture receiving antennas in optical-wireless communication. Harbin Institute of Technology, Harbin
11. Li Z, Yang G, Bi M (2017) Study of spatial diversity to suppress atmospheric scintillation index of vortex beams. Opt Commun Technol 41(02):48–51
12. Soni GG, Tripathi A, Mandloi A et al (2019) Compensating rain induced impairments in terrestrial FSO links using aperture averaging and receiver diversity. Opt Quant Electron 51(7):1–11
13. Dai T (2021) Long-range atmospheric laser communication system and atmospheric turbulence suppression technology. Changchun University of Science and Technology, Changchun
14. Xi X, Wu H (2017) Design of near-surface UV communication transmitting optical system based on Cassegrain structure. Appl Opt 38(02):205–209
15. Wu J, Boardman A (1991) Coherence length of a Gaussian-Schell beam and atmospheric turbulence. J Mod Opt 38(7):1355–1363
16. Ricklin JC, Davidson FM (2003) Atmospheric optical communication with a Gaussian-Schell beam. J Opt Soc Am 20(5):856–866
17. Dogariu A, Amarande S (2003) Propagation of partially coherent beams: turbulence-induced degradation. Opt Lett 28(1):10–12
18. Borah DK, Voelz DG (2010) Spatially partially coherent beam parameter optimization for free space optical communications. Opt Express 18(20):20746–20758
19. Drexler K, Roggemann M, Voelz D (2011) Use of a partially coherent transmitter beam to improve the statistics of received power in a free-space optical communication system: theory and experimental results. Opt Eng 50(2):025002.1–025002.7
20. Zhang L, Chen Z, Xiong M et al (2014) Scintillation index of partially coherent light transmission in atmospheric turbulence. Intense Lasers Part Beams 26(09):77–81
21. Ke X, Ya Z, Wei C (2016) Effect of non-Kolmogorov atmospheric turbulence on the Rayleigh interval and turbulence distance of radial partially coherent array beams. J Photonics 45(11):23–30
22. Bo Z (2020) Laguerre-related beam modulation and its atmospheric turbulence transport characteristics. Nanjing University of Technology, Nanjing
23. Yu J, Zhu X, Wang F et al (2019) Experimental study of reducing beam wander by modulating the coherence structure of structured light beams. Opt Lett 44(17):4371–4374
24. Jie Z (2019) Partial coherent optical atmospheric turbulence transmission beam parameter simulation technique. Changchun University of Technology, Changchun
25. Yang N, Zhao L, Xu Y et al (2022) Transmission characteristics of partially coherent radially polarized distorted beams in non-uniform atmospheric turbulence. Laser Infrared 52(08):1167–1176
26. Ke X, Deng L (2016) Theory of partially coherent optical transmission in optical-wireless communication. Science Press, Beijing
27. Ke X, Han M, Wang M (2014) Spreading and drifting of partially coherent light on horizontal transmission paths in atmospheric turbulence. J Opt 34(11):78–82
28. Ke X, Han M, Wang M (2015) Spreading and drifting of partially coherent light on the oblique-range transport path in atmospheric turbulence. J Photon 44(03):149–153
29. Ke X, Yu Z (2015) Light intensity scintillation effect of partially coherent light in atmospheric turbulence. J Opt 35(01):56–62
30. Ke X, Wang W (2015) Light intensity and expansion of partially coherent light in oblique range and horizontal atmospheric turbulence. J Appl Sci 33(02):142–154
31. Ke X, Wang W (2015) Beam extension and angular expansion of partially coherent light in atmospheric turbulence. Infrared Laser Eng 44(09):2726–2733
32. Ke X, Wang J (2015) Comparison of upward and downward transmission polarization characteristics of partially coherent beams in atmospheric turbulence. J Phys 64(22):148–155

33. Ke X, Wang C (2016) Light intensity distribution of partially coherent vortex beams during transport in atmospheric turbulence. Adv Lasers Optoelectron 53(11):83–91
34. Ke X, Ya Z (2016) Study of transmission characteristics of partially coherent array beams in optical-wireless communication systems. Adv Lasers Optoelectron 53(10):38–49
35. Ke X, Zhang H (2016) Study on the spreading characteristics of partially coherent Gaussian pulses in atmospheric turbulence. Adv Lasers Optoelectron 53(08):110–116
36. Wang J, Ke X (2017) Scattering characteristics of partially coherent beams transmitted in atmospheric turbulence. Infrared Laser Program 46(07):135–142
37. Ke X, Xue Y (2017) Effect of atmospheric turbulence scale on partially coherent optical transmission properties. J Photon 46(01):35–42
38. Ke X, Yao X (2016) Arrival angle undulation of partially coherent light in atmospheric turbulence. J Photon 45(12):30–34
39. Ke X, Wang S (2017) Light intensity evolution of partially coherent Airy beams in atmospheric turbulence. J Photon 46(07):7–15
40. Ke X, Zhang L (2017) Beam extension and drift of partially coherent Airy light in atmospheric turbulence. J Photon 46(01):43–49
41. Ke X, Wang C (2017) Light intensity distribution of partially coherent off-axis vortex beams in atmospheric turbulence. J Opt 37(01):44–50
42. Wang J, Ke X, Wang M (2019) Influence of source parameters and atmospheric turbulence on the polarization properties of partially coherent electromagnetic vortex beams. Appl Opt 58(24): 6486–6494
43. Wang J, Ke X, Wang M (2019) Detecting the topological charge of partially coherent electromagnetic vortex beams through the orientation angle of polarization. J Quant Spectrosc Radiat Transfer 22(8):11–16
44. Gbur G, Wolf E (2002) Spreading of partially coherent beams in random media. J Opt Soc Am 19(8):1592–1598
45. Wolf E, Collett E (1978) Partially coherent sources which produce the same far-field intensity distribution as a laser. Opt Commun 25(3):293–296
46. Chen Y, Gu J, Wang F et al (2015) Self-splitting properties of a Hermite-Gaussian correlated Schell-model beam. Phys Rev A, 91(1):013823.1–013823.12
47. Mei Z, Korotkova O (2013) Cosine-Gaussian Schell-model sources. Opt Lett 38(14):2578–2580
48. Shu J, Xu H, Zhou ZL et al (2019) Beam wander of the multi-Gaussian Schell-model beam in anisotropic turbulence. Prog Electromag Res 78(4):185–192
49. Ibrahim MM, Ibrahim AM (1996) Performance analysis of optical receivers with space diversity reception. IEE Proc—Commun 143(6):369–372
50. Feng P, Jing M, Liying T et al (2004) Scintillation characterization of multiple transmitters for ground-to-satellite laser communication. Infrared Compon Appl 56(40):448–454
51. Trisno S, Smolyaninov II, Milner SD et al (2005) Characterization of time delayed diversity to mitigate fading in atmospheric turbulence channels. Univ Md/CollE Park (U S) 58(92):589215-1–589215.10
52. Anguita JA, Neifeld MA, Vasic BV (2006) Multi-beam space-time coded systems for optical atmospheric channels. Free-Space Laser Commun 63(04):63041B.1–63041B.9
53. Cvijetic N, Wilson SG, Brandt-Pearce M (2008) Performance bounds for free-space optical MIMO systems with APD receivers in atmospheric turbulence. IEEE J Sel Areas Commun 26(3):3–11
54. Ke X, Song P, Pei G (2011) Research on multi-aperture reception technology in wireless laser communication. J Opt 31(12):22–28
55. Ke X, Chen J, Pei G (2012) Research on multi-beam emission technology in wireless laser communication. Opt Eng 39(07):1–7
56. Long J (2014) Research on the performance and application of spatial diversity technology in free-space optical communication. Beijing University of Posts and Telecommunications, Beijing

57. Kaur P, Jain VK, Kar S (2015) Performance analysis of free space optical links using multi-input multi-output and aperture averaging in presence of turbulence and various weather conditions. IET Commun 9(8):1104–1109
58. Bhatnagar MR, Ghassemlooy Z (2016) Performance analysis of gamma–gamma fading FSO MIMO links with pointing errors. J Lightwave Technol 34(9):2158–2169
59. Meng D (2017) Performance study of outdoor visible light communication in weak turbulent channels. Chang'an University, Xi'an
60. Kumar N, Khandelwal V (2019) Simulation of MIMO-FSO system with Gamma-Gamma fading under different atmospheric turbulence conditions. In: IEEE.2019 International conference on signal processing and communication. IEEE, pp 117–124
61. Xu Y (2019) Research on field-of-view range expansion technology for space laser communication reception. Changchun University of Technology, Changchun
62. KeNan D (2021) Research on diversity technology in atmospheric partial coherent optical communication. Huazhong University of Science and Technology, Hubei
63. Liu Z (2022) Research on atmospheric transmission characteristics based on optical-wireless MIMO. Changchun University of Technology, Changchun
64. Ke X, Liu M (2015) Diversity reception technique in optical-wireless communication for turbulent channels. J Opt 35(01):88–95
65. Tyson RK (1996) Adaptive optics and ground-to-space laser communications. Appl Opt 35(19):3640–3646
66. Roggemann MC, Lee DJ (1998) Two deformable mirror concept for correcting scintillation effects in laser beam projection. Mich Technol Univ 33(81):66–75
67. Barbier PR, Rush DW, Plett ML et al (1998) Performance improvement of a laser communication link incorporating adaptive optics. Univ Md/CollE Park (U S) 34(32):93–102
68. Li X, Jiang W (2001) Control performance analysis of two adaptive optical systems with tandem correction. J Opt 21(09):1059–1064
69. Weyrauch T, Vorontsov MA (2004) Free-space laser communications with adaptive optics: atmospheric compensation experiments. J Opt Fiber Commun Rep 1(4):355–379
70. Xia L, Li X (2010) Wavefront correction experiments for atmospheric optical communication based on adaptive optics. Inf Electron Eng 8(03):331–335
71. Han L (2013) Performance and compensation method of spatial optical communication under atmospheric turbulence. Harbin Institute of Technology, Harbin
72. Wu Y (2013) Research on the application of adaptive optics in atmospheric optical communication. Graduate School of Chinese Academy of Sciences (Institute of Optoelectronics Technology), Beijing
73. Hashmi AJ, Eftekhar AA, Adibi A et al (2014) Analysis of adaptive optics-based telescope arrays in a deep-space inter-planetary optical communications link between Earth and Mars. Opt Commun 33(3):120–128
74. Cvijetic M, Li M (2015) Coherent free space optics communications over the maritime atmosphere with use of adaptive optics for beam wavefront correction. Appl Opt 54(6):1453–1462
75. Wright MW, Morris JF, Kovalik JM et al (2015) Adaptive optics correction into single mode fiber for a low earth orbiting space to ground optical communication link using the OPALS downlink. Opt Express 23(26):33705–33712
76. Escarate P, Aguero JC, Zuniga S et al (2017) Model predictive control for laser beam shaping. IEEE Lat Am Trans 15(4):626–631
77. Li B (2018) Research on turbulence suppression technology in spatial optical communication. Beijing University of Posts and Telecommunications, Beijing
78. Lin H (2019) Research on the control technology of adaptive optical system based on model recognition. University of Chinese Academy of Sciences (Institute of Optoelectronics Technology, Chinese Academy of Sciences), Chengdu
79. Mo C (2019) Research on key technologies of ground station for satellite-terrestrial laser communication based on large-aperture telescope. University of Chinese Academy of Sciences (Institute of Optoelectronics Technology, Chinese Academy of Sciences), Chengdu

80. Banet Matthias T, Spencer MF (2020) Compensated-beacon adaptive optics using least-squares phase reconstruction. Opt Express 28(24):36902–36914
81. Yang H, Su H, Zhang Z (2023) Wavefront correction of SPGD control algorithm based on K-L mode[J/OL]. China Laser, 1–14 [2023-05-10]
82. Zhaokun L, Tao S, Xiongchao L et al (2023) Advanced multi-feedback stochastic parallel gradient descent wavefront correction in free-space optical communication. Opt Commun 53(3):43–51
83. Ke X, Yang S, Wu J et al (2021) Research progress on adaptive optics for optical-wireless communication systems at Xi'an University of Technology. Intense Lasers Part Beams 33(08):30–52
84. Wu J, Ke X (2018) Adaptive optical correction of wavefront-free sensors. Adv Lasers Optoelectron 55(03):030103-1–030103-7
85. Wang XY (2018) Research on adaptive optical correction technology for vortex beams. Xi'an University of Technology, Xi'an
86. Ke X, Wang X (2018) Experimental study on wavefront aberration correction of vortex light. J Opt 38(03):197–203
87. Ke XZ, Zhang D (2019) Fuzzy control algorithm for adaptive optical system. Appl Opt 58(36):9967–9975
88. Ke X, Ke Y, Zhang Y (2019) Experimental study of wavefront distortion detection by CAPIS technique. Adv Lasers Optoelectron 56(12):30–37
89. Ke Y (2019) Research on wavefront sensing technology for computational optical field imaging. Xi'an University of Technology, Xi'an
90. Ke X, Zhang Y, Zhang Y et al (2019) Graphics processor acceleration for wavefront-free sensing adaptive wavefront correction systems. Adv Lasers Optoelectron 56(07):96–104
91. Ke X, Han KN (2019) Wavefront simulation and wavefront correction for liquid crystal spatial light modulators. Adv Lasers Optoelectron 56(05):170–175
92. Li M (2020) Experimental study on the correction of beam front aberrations by the Eigen mode method. Xi'an University of Technology, Xi'an
93. Cui N-M (2020) Experimental study on the correction of wavefront aberrations of vortex beams by phase difference method. Xi'an University of Technology, Xi'an
94. Cheng S (2021) Correction techniques for large distortion wavefront in free space coherent optical communication. Xi'an University of Technology
95. Zhou RZ (1996) Adaptive optics theory. Beijing Institute of Technology Press, Beijing
96. Wu J (2018) Experimental study of coherent optical communication system without wavefront detection. Xi'an University of Technology, Xi'an
97. Jian L (2009) Characterization of transverse shear interference and wavefront reconstruction. Dalian University of Technology, Liaoning
98. Liu C (2017) Application simulation study of 16QAM coherent optical communication transmission system. Beijing University of Posts and Telecommunications, Beijing
99. Xu Q (2020) Adaptive optical control technology for wavefront curvature detection. North China University of Technology, Hebei
100. Medecki H, Tejnil E, Goldberg KA et al (1996) Phase-shifting point diffraction interferometer. Opt Lett 21(19):1526–1528
101. Yao KN, Wang JL, Liu XY et al (2014) Closed-loop adaptive optics system with a single liquid crystal spatial light modulator. Opt Express 22(14):17216–17226
102. Liang JY, Wang HR, Zhang N et al (2023) Research progress on deformation mirror and its control algorithm [J/OL]. Opt Commun Res, 1–17 [2023-04-27]
103. Zhang D (2020) Adaptive optical wavefront distortion control and experimental research. Xi'an University of Technology, Xi'an
104. Wu J, Ke X (2018) Adaptive optical correction of wavefront-free sensors. Adv Lasers Optoelectron 55(3):030103-1–030103-7
105. Rui B (2011) Adaptive optics with wavefront-free sensors for FSO systems. Huazhong University of Science and Technology, Wuhan

106. Zommer S, Ribak EN, Lipson SG et al (2006) Simulated annealing in ocular adaptive optics. Opt Lett 31(7):939–941
107. Yang H, Li X, Jiang W (2008) Comparison of several stochastic parallel optimal control algorithms for adaptive optical systems. Intense Lasers Part Beams 20(01):11–16
108. Ke X (2014) Space-time coding theory in optical-wireless MIMO systems. Science Press, Beijing

Chapter 6
Visible-Light Communication Path-Loss Model

The objective of this chapter is to develop an analytical model of the intervehicle visible-light communication (VLC) channel, focusing on the channel path loss under different weather conditions, as this is an important metric for quantifying the communication channel. Finding mathematical expressions to describe the channel path loss is essential for establishing link budgets and predicting communication performance, based on the signal-to-noise ratio and bit-error rate.

6.1 Channel Theory for Visible-Light Communication

The basic visible-light communication (VLC) channel can be modeled as a baseband linear time-invariant (LTI) system with a non-negative impulse response $h(t) > 0$, as shown in Fig. 6.1.

The specific expression is as follows [1]:

$$Y(t) = RX(t) * h(t) + n(t), \tag{6.1}$$

where $Y(t)$ is the electrical receiving signal, R is the responsivity of the photodetector, $X(t)$ is the input optical power, which cannot be negative ($X(t) > 0$), $h(t)$ is the baseband channel impulse response (CIR), * is a convolution, and $n(t)$ is the receiver noise, which can be modeled as additive white Gaussian noise.

To ensure the reliable, safe, and efficient design and implementation of VLC systems, it is necessary to understand the main characteristics of VLC channels, which are described in this section.

X. Ke and C. Ke, *Noise Models in Optical-Wireless Communication Systems*,
Optical Wireless Communication Theory and Technology,
https://doi.org/10.1007/978-981-97-7550-7_6

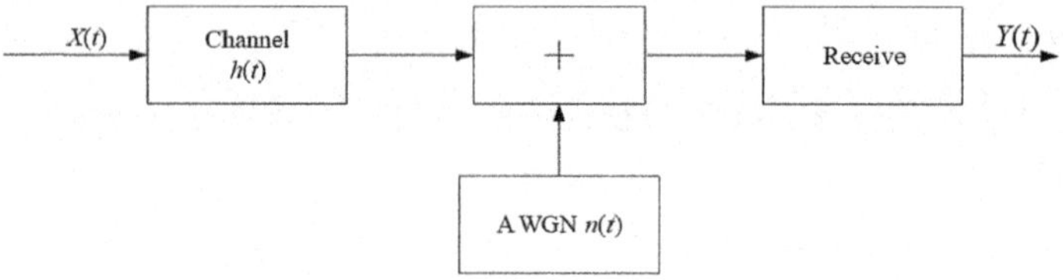

Fig. 6.1 Simplified channel model

6.1.1 Channel Impulse Response

The channel impulse response (CIR) is the time evolution of the received signal when the transmitter emits wireless light. The CIR provides all the statistical data that can describe the characteristics of the channel; it is fixed for some physical configuration of transmitters, receivers, and reflectors.

Mathematically, for a particular source S and receiver R, the total channel impulse response at time t, which is the sum of all reflections from the line-of-sight response and reflected links, can be expressed as [2]

$$h(t, S, R) = \sum_{k=0}^{\infty} h^{(k)}(t, S, R) \tag{6.2}$$

where $h(0)(t, S, R)$ is the direct-view response and $h(k)(t, S, R)$ is the response of the light after k reflections. The direct-view impulse response can be expressed as

$$h^{(0)}(t, S, R) = \frac{1}{(d)^2} S(\Phi, m, \lambda) A_{eff}(\psi) \delta\left(t - \frac{d}{c}\right) \tag{6.3}$$

where $S(\Phi, m, \lambda)$ represents the functional relationship between the radiated power of the transmitter and the wavelength, δ represents the Dirac function, m represents the light-wave modulus, d represents the distance between the transmitter and receiver, c represents the speed of light, $A_{eff}(\psi)$ represents the effective area of the receiver, and ψ represents the angle of incidence.

6.1.2 Channel Direct-Current Gain

One of the most critical characteristics of a VLC system is its channel direct-current (DC) gain. It can be expressed as [2]

$$H(0) = \int_{0}^{\infty} h(t) dt \tag{6.4}$$

The channel DC gain $H(0)$ is used to identify the losses in the channel that relate the average received optical power P_r to the transmitted optical power P_t, as follows:

$$P_r = H(0)P_t \tag{6.5}$$

In this chapter, the Lambertian model is used to describe the angular distribution of the radiation-intensity pattern of a light-emitting diode (LED) light source. The transmit power P_t of the LED transmitter can be calculated as [3]

$$P_t = \left[\frac{m+1}{2\pi}\right] P_{tt} \cos^m(\varphi) \tag{6.6}$$

where P_{tt} is the average transmitted optical power and the Lambertian order m is related to the half-power half-angle $\varphi_{1/2}$.

$$m = -\frac{\ln(2)}{\ln(\cos(\varphi_{1/2}))} \tag{6.7}$$

The detector is modeled as an active region that collects incident angles ψ smaller than the detector field of view (FOV). The following equation provides the effective receiving area of an optical power meter [3]:

$$A_{effect}(\psi) = \begin{cases} A_{PD} T_s(\psi) g(\psi) \cos(\psi), & 0 \le \psi \le \Psi_c \\ 0, & \psi > \Psi_c \end{cases} \tag{6.8}$$

where A_{PD} represents the size of the receiving aperture of the detector, $T_s(\psi)$ is the gain loss of the filter, $g(\psi)$ is the gain of the concentrator, and Ψ_c is the field-of-view angle of the receiver.

$$g(\psi) = \begin{cases} \frac{n^2}{\sin^2 \Psi_c}, & 0 \le \psi \le \Psi_c \\ 0, & \psi > \Psi_c \end{cases} \tag{6.9}$$

The received optical power can be obtained by Eq. (6.10):

$$P_r = P_t A_{effct}(\psi) \tag{6.10}$$

From Eqs. (6.6)–(6.10), we obtain

$$P_r = \frac{(m+1)A_{PD}}{2\pi d^2} \cos^m \varphi T_s(\psi) g(\psi) \cos(\psi) P_t \tag{6.11}$$

Therefore, in a line-of-sight transmission, $H(0)$ can be obtained using the following formula [4]:

$$H(0) = \begin{cases} \frac{(m+1)A_{PD}}{2\pi d^2} \cos^m \varphi T_s(\psi) g(\psi) \cos(\psi), & 0 \le \psi \le \Psi_c \\ 0, & 0 \ge \psi_c \end{cases} \tag{6.12}$$

where A_{PD} is the photodetector (PD) surface area, m is the Lambertian mode number, ψ_c is the receiver FOV, φ is the irradiance angle, and d is the distance between the LED and the receiving point. In addition, $T_s(\psi)$ is the optical-filter gain and $g(\psi)$ is the optical-concentrator gain. ψ denotes the angle of incidence at the PD.

6.1.3 Root-Mean-Square Delay Spread

The root-mean-square (RMS) delay spread is defined as the square root of the second central moment of the CIR. It can be used to quantify temporal dispersion and estimate intersymbol interference. The RMS delay spread is given by Eq. (6.13) [5]:

$$\tau_{RMS} = \sqrt{\frac{\int_{-\infty}^{\infty} (t - \tau_0)^2 h^2(t) dt}{\int_{-\infty}^{\infty} h^2(t) dt}} \tag{6.13}$$

where τ_0 is the average additional delay, t is the propagation delay time, and $h(t)$ is the channel impulse response. The RMS delay spread can be obtained after determining the impulse response of the system. The average extra delay is given by Eq. (6.14) [5]:

$$\tau_0 = \frac{\int_{-\infty}^{\infty} t \times h^2(t) dt}{\int_{-\infty}^{\infty} h^2(t) dt} \tag{6.14}$$

The RMS delay spread is a key performance metric for calculating the upper bound of reliable data-transfer rates. The maximum bit rate (R_b) transmitted over an optical-wireless channel can be approximated as

$$R_b \le \frac{1}{10\tau_{RMS}} \tag{6.15}$$

This allows the allocation of symbol-length constraints to avoid intersymbol interference (ISI). It should be noted that this equation is valid only if no equalization techniques are used.

The optical path loss (OPL) describes the attenuation of the received power of an optical signal during space propagation. The OPL depends on the transceiver and the environmental parameters. These include the reflectivity of touching objects, scattering by atmospheric particles, attenuation due to weather conditions, radiation patterns of light sources, and the orientation of the transmitter and receiver front ends. The OPL can be expressed as follows:

$$OPL = -10\log_{10} H(0) \tag{6.16}$$

where *OPL* represents the optical path loss, and *H(0)* represents the channel DC gain.

6.2 Analysis of a VLC System Model of the Internet of Vehicles

6.2.1 Receiver Half-Field Angle

Figures 6.2a, b show the composition of a typical outdoor VLC system, which consists of a transmitting vehicle Tx, based on vehicle taillights, and a receiving vehicle Rx, based on a PD. For illustration, only one taillight is used here; the second taillight has a nearly identical geometric configuration.

In an outdoor VLC system, the position of Tx relative to Rx, the beam profile of the headlight, and the field of view (FOV) of Rx are critical and affect the link performance. For line-of-sight transmissions, the channel DC gain, that is, the ratio of the received signal power at Rx to the transmitted power, can be expressed as [6]

$$H_{LOS} = \frac{Pr}{Pt} = \begin{cases} \frac{A_{PD}T_s(\varphi_h,\varphi_v)g(\varphi_h,\varphi_v)R_{TD}(\theta_h,\theta_v)}{L^2}, & 0 \le \varphi_h \le \phi_h \, and \, 0 \le \varphi_v \le \phi_v \\ 0, & \varphi_h > \phi_h \, or \, \varphi_v > \phi_v \end{cases} \tag{6.17}$$

where A_{PD} is the surface area of the photodetector PD; φ_h and φ_v are the horizontal and vertical incidence angles, respectively; and $R_{\mathrm{Tx}}(\theta_\mathrm{h},\theta_\mathrm{v})$ is Tx at the horizontal and vertical irradiance angles θ_h and θ_v. In the radiation pattern below, ϕ_h and ϕ_v respectively represent the horizontal and vertical half-angle FOVs of Rx, L represents the link span, $T_s(\varphi_\mathrm{h},\varphi_\mathrm{v})$ and $g(\varphi_\mathrm{h},\varphi_\mathrm{v})$ are the optical filter (OF) and optical concentrator (OC) gains, respectively.

Using imaging optics at Rx, an image of Tx can be captured on the PD-based sensor, if Tx is within the FOV of Rx. First, the horizontal semi-field of view (HS-FOV) required by Rx1 and Rx2 is analyzed according to the worst-case scenario (the vehicle is on the other side of the lane), as shown in Fig. 6.3.

For Rx1 and Rx2 to capture the beams emitted by a single or two Txes, the required HS-FOVs are φ_{12}, φ_{22}, φ_{11}, and φ_{21}, respectively. To obtain the required HS-FOV between two vehicles driving in the same lane, we need to consider the width of the lane and the vehicle, as well as the position of Rx and Tx. The required HS-FOV of Rx can be expressed as [7]

$$\varphi_{12} = \arctan\left(\frac{D - \left(\frac{L_{vz}-d_{Tx}}{2}\right) - \left(d_{Tx} - \frac{l_{Tx}}{2}\right) - \left(\frac{L_{rb}-d_{Rx}}{2}\right) - d_{Rx}}{L}\right) \tag{6.18}$$

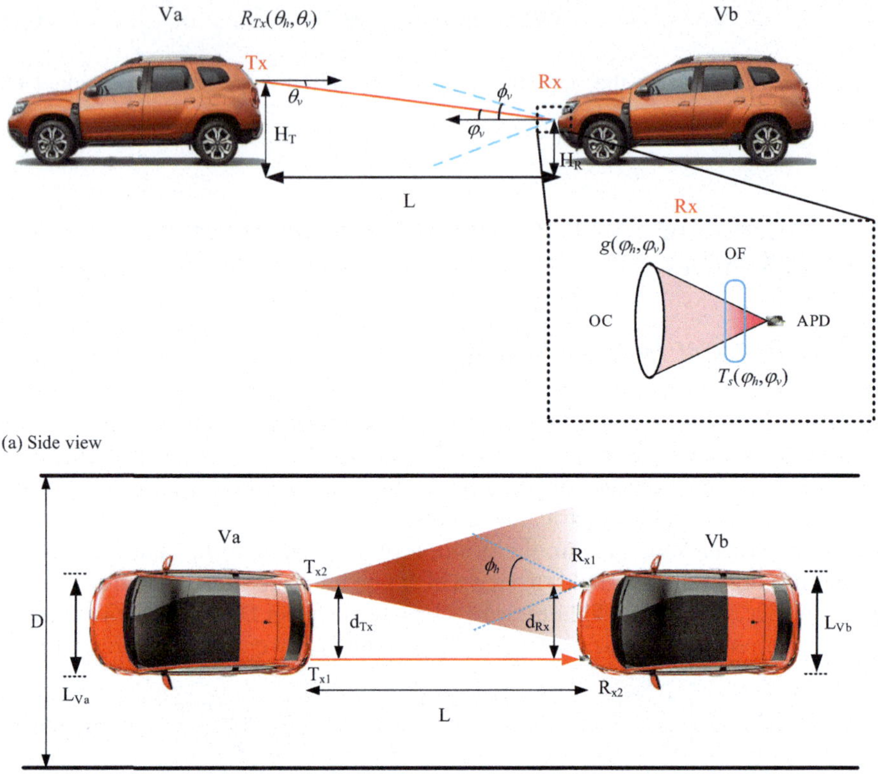

(a) Side view

(b) Top view

Fig. 6.2 Outdoor VLC-system configuration diagram

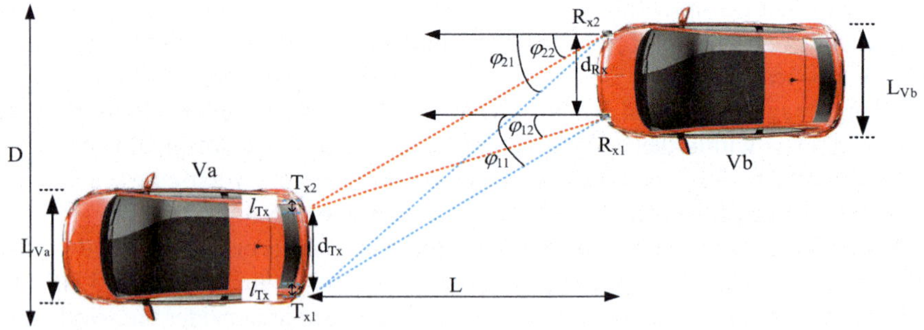

Fig. 6.3 Outdoor VLC configuration of the horizontal FOV

$$\varphi_{22} = \arctan\left(\frac{D - \left(\frac{L_{va} - d_{Tx}}{2}\right) - \left(d_{Tx} - \frac{l_{Tx}}{2}\right) - \left(\frac{L_{\eta b} - d_{Rx}}{2}\right)}{L}\right) \tag{6.19}$$

$$\varphi_{11} = \arctan\left(\frac{D - \left(\frac{L_{\gamma a} - d_{Tx}}{2}\right) + \frac{l_{Tx}}{2} - \left(\frac{L_{\gamma b} - d_{Bx}}{2}\right) - d_{Bx}}{L}\right) \tag{6.20}$$

$$\varphi_{21} = \arctan\left(\frac{D - \left(\frac{L_{\gamma u} - d_{Tx}}{2}\right) + \frac{l_{Tx}}{2} - \left(\frac{L_{\gamma b} - d_{Bx}}{2}\right)}{L}\right) \tag{6.21}$$

where D is the width of the road lane, L represents the horizontal distance between Tx and Rx, d_{Tx} and d_{Rx} are the intervals between Tx and Rx, respectively, L_{Va} and L_{Vb} are the vehicle widths, respectively, and l_{Tx} represents the length of Tx. Furthermore, Eqs. (6.18)–(6.21) can be applied to vehicles on different lanes, where D becomes $2D$ or $3D$ for scenarios where one vehicle is in the next lane or the next two lanes, respectively.

The vertical semi-FOV (VS-FOV) required in the case shown in Fig. 6.2a is as follows [8]:

$$\varphi_v = \arctan\left(\frac{\Delta H}{L}\right) \tag{6.22}$$

where $\Delta H = |H_{\text{T}} - H_{\text{R}}|$ represents the offset between the Tx and Rx vertical heights. The required HS-FOV and VS-FOV as functions of the system link span L are shown in Fig. 6.4.

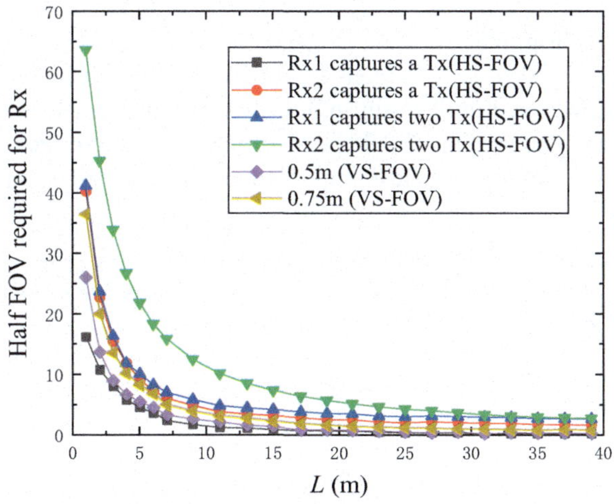

Fig. 6.4 Relationship between the HS-FOV and VS-FOV required by the receiving end and L

Table 6.1 Key geometric parameters and simulated values

Parameter	Symbol	Value
Lane width	D	2.5 m
Tx vehicle width	L_{Va}	1.7 m
Rx vehicle width	L_{Vb}	1.7 m
Distance between Rx1 and Rx2	d_{Rx}	1.2 m
Distance between Tx1 and Tx2	d_{Tx}	1.2 m
Tx length	l_{Tx}	0.1 m
Horizontal link span	L	1–50 m
Height difference between Tx and Rx	ΔH	0.25–0.75 m
Lens focal length	f	25.4–50.0 mm
PD surface area	A_{PD}	0.2–2 cm^2

Figure 6.4 shows the required HS-FOV and VS-FOV for the Rx of two vehicles in the same lane as a function of link span L, which was generated using Eqs. (6.18)–(6.22) and the parameters listed in Table 6.1. From Fig. 6.4, it can be seen that the HS-FOV decreases exponentially with the link span. For example, for Rx2, when the link span is reduced from 10 to 3 m, the required HS-FOV increases by approximately seven times. For Rx1, the required HS-FOV is smaller than that for Rx2 because it is closer to Tx. Furthermore, for VS-FOV, a higher exponential distribution is shown for a larger height difference ΔH below a link span of 20 m.

6.2.2 *Relationship Between the PD, Lens, and Half Field of View*

See Fig. 6.5.

Using a lens to focus light onto the PD can increase the received optical power; however, it will result in a decrease in the half FOV at Rx. The different PD sizes and f of the lens should be investigated to determine whether they meet the half FOV required for vehicle communication. The half FOV represented by f and the size D_{PD}

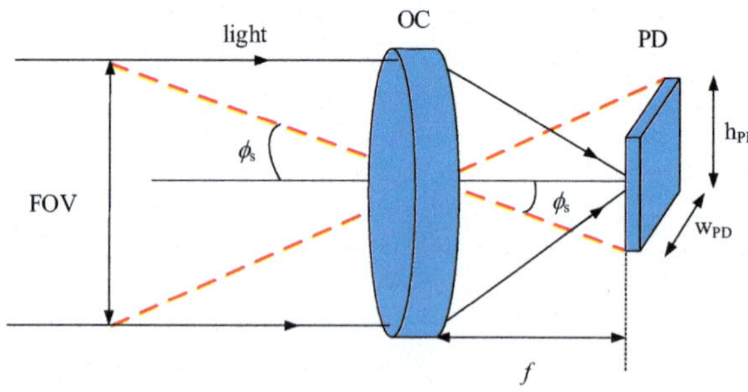

Fig. 6.5 Relationship between f, semi-FOV, and the PD size

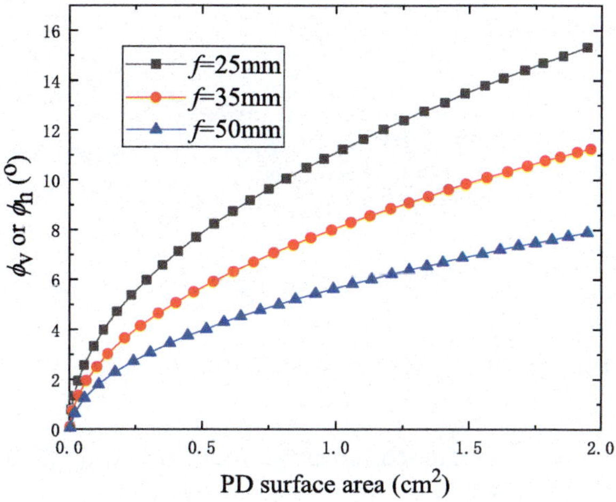

Fig. 6.6 Relationship between ϕ_h or ϕ_v and the PD surface area under different f conditions

of the PD are given by Eqs. (6.23) and (6.24) [9]:

$$\phi_s = \arctan\left(\frac{D_{PD}}{2f}\right) \tag{6.23}$$

$$\phi_s = \begin{cases} \phi_h, \ D_{FD} = w_{FD} \\ \phi_v, \ D_{FD} = h_{FD} \\ \phi_d, \ D_{FD} = \sqrt{h_{FD}^2 + w_{FD}^2} \end{cases} \tag{6.24}$$

where h_{PD} and w_{PD} represent the height and width of PD, respectively, and ϕ_h, ϕ_v, and ϕ_d are the horizontal, vertical, and half-width FOV of Rx, respectively.

Figure 6.6 shows the relationship between ϕ_h or ϕ_v and the PD surface area for different f values. It can be seen from the figure that when using a converging lens at the PD, ϕ_h or ϕ_v decreases with increasing f and shows a logarithmic increase with the increasing surface area of the PD.

6.2.3 Transmitter Field of View

Figure 6.7 shows the beam coverage of Tx. L_1 depicts the horizontal distance from Tx to the start of the overlap of the Tx1 and Tx2 beams, which can be expressed as [9]

$$L_1 = d_{Tx}/(\tan\theta_{in} + \tan\theta_{out}) \tag{6.25}$$

where in and out are the inner and outer half-power angles of the Tx beam, respectively, as shown in Fig. 6.7.

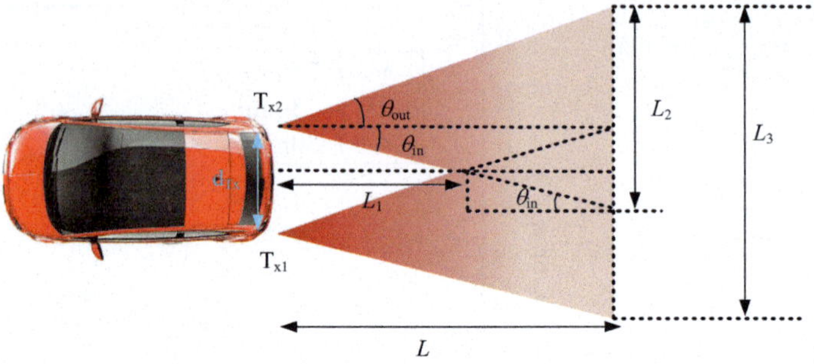

Fig. 6.7 Vehicle transmitter field of view

L_2 is the beam-end coverage width of a single Tx and L_3 is the beam-end coverage width of two Txes, which can be expressed as

$$L_2 = L(\tan\theta_{in} + \tan\theta_{out}) \tag{6.26}$$

$$L_3 = d_{Tx} + L(\tan\theta_{in} + \tan\theta_{out}) \tag{6.27}$$

The beams from Tx1 and Tx2 may not overlap; therefore, the coverage length is $2L_2 = 2L(\tan\theta_{\rm in} + \tan\theta_{\rm out})$. Therefore, it can be seen from Eq. (6.27) that the relationship between the half-power angle (assuming $\theta_{\rm in} = \theta_{\rm out}$) and L_3 under a certain range of link spans is as shown in Fig. 6.8.

Figure 6.8 shows the relationship between the half-power angle of Tx and L_3. It can be seen from the figure that the half-power angle of Tx increases logarithmically with the increase of L_3, and the half-power angle of Tx increases with the increase

Fig. 6.8 Relationship between the half-power angle of Tx and L_3

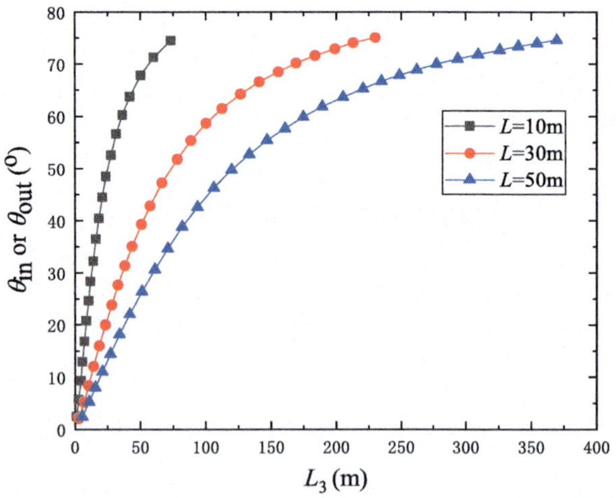

of the link span. L_3 reached saturation points at approximately 65 m, 220 m, and 370 m, corresponding to $L = 10$ m, 30 m, and 50 m, respectively.

For a given transmit power PT, the received power can be obtained as

$$P_r(dBm) = P_T(dBm) + H_{LOS}(dB) - L_{sm}(dB) \tag{6.28}$$

where L_{sm} is the link safety margin, and H_{LOS} is the channel DC gain of the line of sight (LOS), given by

$$H_{los}(dB) = 10\log_{10}(A_{PD}T_s(\varphi_h, \varphi_v)g(\varphi_h, \varphi_v)R_{Tx}(\theta_h, \theta_v)/D_v^2 \tag{6.29}$$

where A_{PD} is the surface area of the photodetector PD, $R_{\text{Tx}}(\theta_{\text{h}}, \theta_{\text{v}})$ is the radiation pattern of Tx at the horizontal and vertical irradiance angles θ_{h} and θ_{v}, respectively, D_v is the distance between vehicles, $T_s(\varphi_{\text{h}}, \varphi_{\text{v}})$ denotes the optical filter, and $g(\varphi_{\text{h}}, \varphi_{\text{v}})$ denotes the gain of the optical concentrator.

6.3 Visible-Light Path-Loss Model of the Internet of Vehicles

The overall system model is illustrated in Fig. 6.9. Vehicle Vb is located in front of vehicle Va, and the two vehicles are located in the same lane. The headlight of vehicle Va is the signal transmitter, vehicle Vb is the receiver of visible-light signals, and vehicle Vc is the vehicle in the adjacent lane. Car Vc passes by the lanes adjacent to cars Va and Vb. The headlight beam of the car is reflected on the surface of Car Vc and the road surface. The reflected power of the beam at all points through which the vehicle passes can be described using a spatial rectangular model.

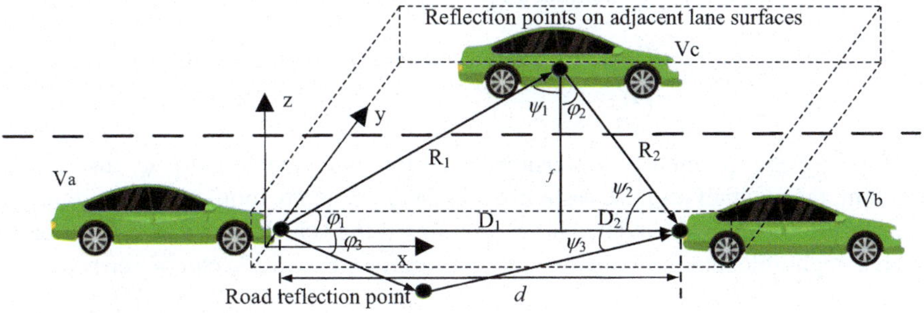

Fig. 6.9 Model of a dual-vehicle road-lamp system

6.3.1 Analysis of a Direct-Look Transmission

The incident ray creates an angle ψ normal to the receiver surface; the channel impulse response for the line-of-sight link is as follows [10]:

$$h(t) = \frac{A(m+1)}{2\pi d^{\gamma}} \cos^{m}(\varphi) T_s(\psi) g(\psi) \cos(\psi) \pi(d) \delta(t - \frac{d}{c}) \lim_{x \to \infty} \tag{6.30}$$

where γ is the path-loss index, t is the propagation time, A is the effective area of the optical power meter, d is the intervehicle distance, ϕ is the irradiance angle, ψ is the incident angle at the detector, c is the speed of light, $T_s(\psi)$ is the filter gain, $g(\psi)$ is the concentrator gain, $\tau(d)$ is the attenuation coefficient under different weather conditions, d/c is the propagation delay of the signal, m = – 0.6931/ln(cos($\psi_{1/2}$)) is the Lambertian order, and $\psi_{1/2}$ is the half-power angle of the radiation.

The absorption and scattering of light in the atmosphere are calculated using the Beers–Lambert law:

$$\tau(d) = I_t / I_0 = \mathrm{e}^{(-\gamma(\lambda) d)} \tag{6.31}$$

where I_0 is the transmitted light intensity, I_t is the received light intensity, d is the propagation distance, $\tau(d)$ is the transmittance at distance d from the transmitter, and $\gamma(\lambda)$ is the extinction coefficient per unit length.

The direct current (DC) gain $H(0)$ is given by the following formula [10]:

$$H(0) = \int_{\infty}^{-\infty} h(t) dt \tag{6.32}$$

Assuming that the filter gain $T_s(\psi)$ in Eq. (6.30) and the concentrator gain $g(\psi)$ have a single gain, the DC channel gain $H(0)$ is [10]

$$H(0) = \begin{cases} \frac{(m+1)A}{2\pi d^{\gamma}} \cos^{m}(\varphi) \cos(\psi) \tau(d), & 0 \leq \psi \leq \Psi_c \\ 0, & \psi > \Psi_c \end{cases} \tag{6.33}$$

Because the geometric dimensions of communication links d, φ, and ψ αρε random, the channel impulse response and DC gain at any moment are random.

$\varphi = \psi = 0$ is considered a valid assumption if the vehicles are in a stable platoon with a negligible lateral-angle deviation. Therefore, Eq. (6.33) can be rewritten as

$$H(0) = \frac{A(m+1)}{2\pi d^{r}} \tau(d) \tag{6.34}$$

The channel gain or path loss in decibels as a function of the distance d between the vehicles is as follows:

$$P_L = 10\log_{10}(A(m+1)) + 10\log_{10}(\tau(d)) - 10\log_{10}(2\pi) - 10\gamma\log_{10}(d). \tag{6.35}$$

The first two terms, $10\log_{10}(A(\mathrm{m}+1)) - 10\log_{10}(2\pi)$, are constants. The distance d between the vehicles is a random variable.

When the traffic flow exceeds 1000 vehicles/h, the lognormal distribution fits the vehicle spacing well. It is defined as follows [11]:

$$f_d = \frac{1}{d\sqrt{2\pi}\sigma_s}\exp\left(-\frac{(\ln d - \mu_s)^2}{2\sigma_s^2}\right) \tag{6.36}$$

where f_d is the vehicle distance function, d is the intervehicle distance, and μ_s and σ_s are the mean and variance, respectively.

A lognormal distribution can be used to account for the safety of the distance between two consecutive vehicles and the speed limit of the road. When the distance between vehicles satisfies Eq. (6.36), the path loss in Eq. (6.35) has a normal distribution, as follows:

$$P_L(x;\mu_m,\sigma_m) = \frac{1}{\sigma_m\sqrt{2\pi}}\exp\left(-\frac{(x-\mu_m)^2}{2\sigma_m^2}\right) \tag{6.37}$$

where the standard deviation is $\sigma_m = 10\gamma\sigma_s/\ln(10) + 10\log_{10} e^{-\gamma(\lambda)}\sigma_s$ and the mean is $\mu_m = 10\frac{A(m+1)}{2\pi} - 10\gamma\frac{\mu_s}{\ln(10)} + 10\log_{10} e^{-\gamma(\lambda)}\mu_s$.

However, when the traffic flow exceeds 1000 vehicles/h, the influence of the angular deviation between vehicles becomes significant. Assuming that alignment is maintained between vehicle runs, the angle $\varphi = \psi$ in Eq. (6.33). Therefore, the path loss is given by

$$\begin{aligned} P_L = {} & 10\log_{10}(A(m+1)) + 10\log_{10}(\tau(d)) - 10\log_{10}(2\pi) \\ & - 10\gamma\log_{10}(d) + 10(m+1)\log_{10}(\cos(\varphi)) \end{aligned} \tag{6.38}$$

Assuming that the angle φ is uniformly distributed, $\varphi \sim U(0,\varphi_0)$; then, the cosine distribution is given by Eq. (6.39) [11]:

$$f_z(z) = \frac{1}{\varphi_0\sqrt{1-z^2}} \quad \cos(\varphi_0) < z < 1 \tag{6.39}$$

Using the principle of random variable transformation, we can prove that the last term of Eq. (6.38) has the following distribution:

$$f_Y(y) = \frac{1}{g\varphi_0\sqrt{e^{-2y/g}-1}} \quad 0 < y < 10(m+1)\log_{10}\cos(\varphi_0), \tag{6.40}$$

where $g = 10(m + 1)e$ and e is a Napierian number. Thus, the path loss in Eq. (6.38) is given by the convolution of the normal distribution in Eq. (6.37) and the log-cosine term distribution in Eq. (6.40). The convolution is expressed by Eq. (6.41):

$$P_L(x; \mu_m, \sigma_m) = \frac{1}{g\varphi_o\sigma_m\sqrt{2\pi}} \int_{\alpha}^{0} \frac{\exp\left(-\frac{(x-y-\mu_m)^2}{\sigma_m^2}\right)}{\sqrt{e^{-2y/g} - 1}} dy, \tag{6.41}$$

where $\alpha = 10(m + 1)\log(\cos(\varphi_o))$.

6.3.2 *Reflection Component Analysis*

In this chapter, the reflection from the vehicle surface is considered to be diffuse. The intervehicle distance ($D = D_1 + D_2$) in Fig. 6.9 depends on the relative speeds of vehicles Vb and Vc. The irradiance angle φ_1 and the incidence angle ψ_2 depend on the relative positions of Vb and Vc. φ_2 and ψ_1 are the angle of irradiance and angle of incidence relative to the normal of the reflecting surface, respectively. Therefore, the channel-reflection component of the ith reflector is as follows [12]:

$$h_{rgf}(t) = \sum_{i=1}^{N} \frac{\zeta A(m+1)}{(2\pi)^2 R_1^{\gamma} R_2^2} \cos^m\left(\varphi_1^i\right)\cos\left(\varphi_2^i\right)\cos\left(\psi_1^i\right)\cos(\psi_2^i)\tau(R^i)\delta\left(\frac{t-(R_1+R_2)}{c}\right), \tag{6.42}$$

where N is the total number of reflectors, R_1 is the distance from the transmitter to the reflector, R_2 is the distance from the reflector to the receiver, and ζ is the reflection index of the reflecting surface. Based on the triangular relationship (see Fig. 6.9), a single reflector can be represented as follows:

$$h_{ref}(t) = \frac{\zeta A(m+1)f^2 D_1^m D_2}{(2\pi)^2 R_1^{(m+\gamma+1)} R_2^4} \tau(R_1)\tau(R_2), \tag{6.43}$$

where D_1 and D_2 are the longitudinal intervehicle spacings and f is the transverse intervehicle spacing. For a single reflection, the path loss of the channel-reflection component is given by

$$\begin{aligned} P_L = {} & 10\log_{10}(\zeta A(m+1)) + 20\log_{10}(f) + 10m\log_{10}(D_1) + 10\log_{10}(D_2) \\ & + 10\log_{10}(\tau(R_1)) + 10\log_{10}(\tau(R_2)) - 20\log_{10}(2\pi) \\ & - 10(m+\gamma+1)\log_{10}(R_1) - 40\log_{10}(R_2). \end{aligned} \tag{6.44}$$

In this chapter, we only consider the reflection during the peak traffic period because the reflection at a low traffic density is negligible when the average distance between vehicles is 250 m. Therefore, we assume that D_1 and D_2 have a lognormal distribution with mean μ_s and standard deviation σ_s, and f is a constant. Therefore, the lognormal distributions of R1 and R2 are μ_r and σ_r [12]:

$$\mu_r = \frac{1}{2}\ln\left(\frac{E^2}{\sqrt{V+E^2}}\right) \tag{6.45}$$

$$\sigma_r = \frac{1}{2}\sqrt{\ln\left(\frac{V+E^2}{E^2}\right)}, \tag{6.46}$$

where

$$E = \exp\left(2\left(\mu_s + \sigma_s^2\right)\right) + f^2 \tag{6.47}$$

$$V = \exp\left(4\left(\mu_s + 2\sigma_s^2\right)\right) - \exp\left(4\left(\mu_s + \sigma_s^2\right)\right). \tag{6.48}$$

Assume that the path loss in Eq. (6.44) has a normal distribution, as follows:

$$PL_{dB}(x;\mu_k,\sigma_k) = \frac{1}{\sigma_k\sqrt{2\pi}}\exp\left(-\frac{(x-\mu_k)^2}{2\sigma_k}\right), \tag{6.49}$$

where

$$\begin{aligned}\sigma_k^2 = &\left(10m\frac{\sigma_s}{\ln(10)}\right)^2 + \left(10\frac{\sigma_s}{\ln(10)}\right)^2 + \left(10(m+\gamma+1)\frac{\sigma_r}{\ln(10)}\right)^2 \\ &+\left(40\frac{\sigma_r}{\ln(10)}\right)^2 - Cov(d_1, r_1) - Cov(d_2, r_2)n\end{aligned}, \tag{6.50}$$

where $Cov(d_1, r_1)$ and $Cov(d_2, r_2)$ are covariances. The average is as follows:

$$\begin{aligned}\mu_k = &10\frac{\rho A_r(m+1)f^2}{(2\pi)^2} + \left(10m\frac{\mu_s}{\ln(10)}\right) + \left(10\frac{\mu_s}{\ln(10)}\right) \\ &+\left(10(m+\gamma+1)\frac{\mu_r}{\ln(10)}\right) + \left(40\frac{\mu_r}{\ln(10)}\right)\end{aligned}. \tag{6.51}$$

The expected received power in the optical power meter is

$$P_{rx}(t) = P_{tr}(t)P_L. \tag{6.52}$$

6.3.3 Path-Loss Simulation

When the distance D between vehicles is log-normally distributed, $\mu_s = 2.95$ and $\sigma_s = 0.30$. When $\psi_{1/2} = 30°$, the probability-density function (PDF) of the path-loss

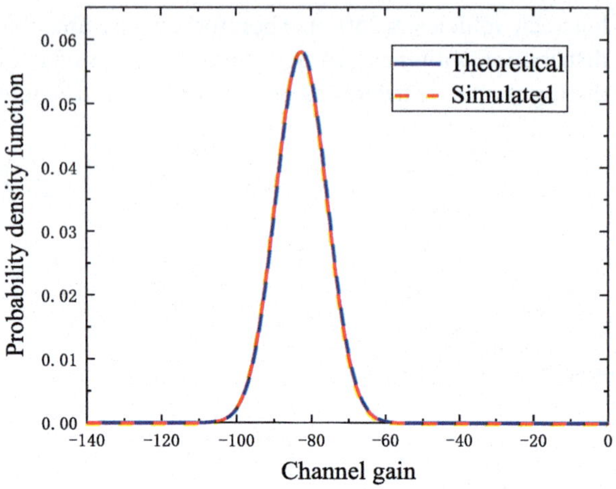

Fig. 6.10 PDF of VLC path loss

distribution is as shown in Fig. 6.10, when the rainfall rate is 15 mm/h. The average path-loss value under this weather condition is − 82.46 dB and the standard deviation value is 13.71 dB.

6.4 Visible-Light Communication Attenuation in Different Weather Conditions

6.4.1 Rain Attenuation

Rainfall attenuates infrared signals and light waves, and in severe cases, it interrupts wireless communication links. Therefore, it is necessary to investigate outdoor visible-light communication in rainy weather conditions. Based on the Mie theory, the influence of raindrop scattering on optical-signal propagation was studied. Using the Mie theory and raindrop spectral distribution, the light-scattering intensity distribution of single particles was studied. In addition to the influence of different wavelengths on the extinction-efficiency factor, the method for calculating visible-light attenuation in rain was analyzed in detail, and the relationship between attenuation and the rainfall rate was calculated.

The raindrop-size distribution, also known as the raindrop spectrum, is an important parameter for studying rainfall characteristics, atmospheric laser communication, radar meteorology, and radio-wave propagation, and has been extensively measured and modeled. The Laws–Parsons raindrop-size distribution is still considered the most typical and widely used.

Recent raindrop-spectrum measurements mostly use advanced measurement equipment, such as raindrop spectrometers. The raindrop spectrum has been widely measured worldwide, and a large number of raindrop-size distribution models have been obtained, based on measurement data from different regions, such as the Laws–Parsons distribution, Marshall–Palmer distribution, Joss distribution, and Gamma distribution. Among them, the Laws–Parsons, Marshall–Palmer, and Joss distributions are widely used.

The Laws–Parsons raindrop-size distribution (L-P distribution) is discrete, and the volume distribution of the raindrops is [13]

$$f(D, R) \approx (33.44 \pm 1.8)R^{-1.28}D^{5.93} \cdot \exp(-0.538R^{-0.186}D), \tag{6.53}$$

where D is the diameter of the raindrop and R is the rainfall rate.

The distribution model of raindrops is [13]

$$N(D) = 6Vf(D) \cdot (\pi D^3)^{-1}, \tag{6.54}$$

where $V = \frac{R}{3.6\times10^6 U}$, $U = \int_0^\infty f(D)V(D)dD$, V(D) represents the terminal velocity of the raindrop when the diameter of the raindrop is D, and R represents the rainfall rate.

The terminal velocity of two commonly used raindrops with diameter D can be expressed as [14]

$$V(D) = 9.65 - 10.3e^{-0.6D}\ \text{m/s} \tag{6.55}$$

$$V = \left(\frac{0.787}{r^2} + \frac{503}{\sqrt{r}}\right)^{-1} \cdot 10^{-6} = 10/\left(\frac{0.3148}{D^2} + \frac{2.2495}{\sqrt{D}}\right)\ \text{m/s}. \tag{6.56}$$

If the raindrop-volume distribution follows the gamma distribution, its distribution form is [14]

$$f(D) = aD^b \exp(-cD). \tag{6.57}$$

At this time, the empirical formula for the speed of raindrops is [14]

$$R(D) = p - q\exp(-sD), \tag{6.58}$$

$$U = \int_0^\infty f(D)R(D)dD. \tag{6.59}$$

Substituting Eqs. (6.57) and (6.58) into Eq. (6.59), we obtain

$$U = \Gamma(b+1)\frac{a}{c^{b+1}}\left[p - \frac{q}{\left(1+\frac{q}{c}\right)^{b+1}}\right]. \tag{6.60}$$

Then, by substituting the relevant coefficients in Eqs. (6.53) and (6.55) into Eq. (6.60), we obtain

$$U = \frac{\Gamma(6.93)}{1000}\frac{33.44}{5.38^{6.93}}[9.506 - \frac{10.15562}{(1 + 0.1115R^{0.186})^{6.93}}]R^{1.161}, \tag{6.61}$$

where R represents the rainfall rate per hour in millimeters. Thus, the above formula can be approximated as

$$U \approx 0.000888R^{1.256}. \tag{6.62}$$

By substituting Eq. (6.52) into Eq. (6.54), the expression for the L-P raindrop-size distribution can be obtained as follows:

$$n(D)_{LP} = 2.0 \times 10^7 R^{-0.384} D^{293} \cdot \exp(-5.38R^{-0.186} D), \tag{6.63}$$

where D represents the diameter of the raindrops and R represents the rainfall rate.

The Mie theory was used to calculate the light scattering by single spherical rain particles. The single spherical rain-particle scattering of 650-nm-wavelength light was calculated and simulated. The corresponding negative refractive index of water is $m = 1.331 - 1.64i \times 10^{-8}$ [15]. When the incident light intensity is $I_0 = 1$ W, distance between scatterer and light source is r = 0.1 m. The polar-coordinate distribution of scattered light intensity (the logarithm is taken) and scattering angle is shown in Fig. 6.11.

Single spherical rain particles with radii of 100 and 500 μm, 1 and 2 mm were used, and a scattered light-intensity simulation was conducted under the above conditions. As shown in Fig. 6.11, the scattering is mainly concentrated in the forward

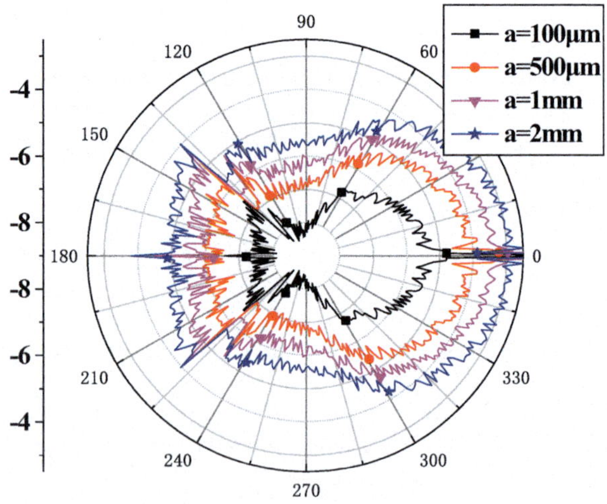

Fig. 6.11 Scattering of rain particles with different diameters

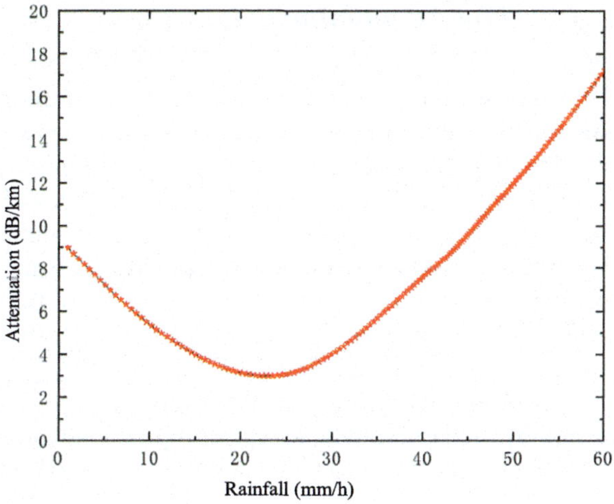

Fig. 6.12 Relationship between the rainfall rate and scattering attenuation

and backward directions. In light rain, that is, when the raindrop size is small, the backscattering is larger. As the size of the raindrops increases, the forward scattering increases.

The light-intensity scattering coefficient under rainy conditions is [16]

$$\gamma_{rain} = \frac{I_i}{I_0} = \exp(-\rho\sigma_a d) - \exp(-\rho\sigma_t d), \tag{6.64}$$

where I_0 is the emitted light intensity, I_i is the scattered light intensity, d is the propagation distance, ρ is the particle number density, σ_a is the absorption cross section of the medium, and σ_t is the total cross section.

The relationship between the rainfall rate and the scattering attenuation of a 650-nm-wavelength visible light in rainy weather conditions is shown in Fig. 6.12.

It can be seen from Fig. 6.12 that the scattering attenuation has an extreme value in the rainfall rate of 0.1–45 mm/h, and the influence of rain on the received signal is the smallest at the extreme point. This is because, in light rain, the backscattering of rain particles is larger and the attenuation is larger; with the increase in rain force, the radius of the rain particles increases, the number remains almost unchanged, and the forward scattering increases.

With the increase in rain particles, the attenuation is small, and when the rain continues to increase, the radius of the rain particles will remain at a certain size, and the total optical thickness of the rain particles increases, resulting in larger attenuation. Therefore, visible-light attenuation is greater during light rain than during moderate or heavy rain. When the rainfall rate exceeds 45 mm/h, the scattering attenuation is greater than that of light rain, and the attenuation will continue to increase as the rainfall increases.

6.4.2 Fog Attenuation

Fog is a colloidal system composed of slowly settling water droplets or ice crystals suspended in the air near a surface. Fog is divided into different types according to its visibility and water content. Fog with visibility lower than 50 m is dense fog, fog with visibility lower than 200 m is thick fog, and fog with visibility lower than 500 m is moderate fog. Generally, the particle number density of mist is 50–100/cm^3, while that of dense fog is 500–600/cm^3. The mass of air droplets or ice crystals per unit volume is called the water content of the fog and is expressed in W (g/m^3). The water content of the fog varies with the intensity of the fog; the greater the fog concentration, the greater the water content [17].

In addition to describing the characteristics of fog based on water content, visibility is usually used to express the type of fog. The relationship between the fog visibility and water content can be expressed as in Eq. (6.65) [17]:

$$V = C\frac{r_{gf}}{W}, \tag{6.65}$$

where r_{ef} is the effective radius of the fog (usually the root-mean-square radius in microns), W is the water content (g/m^3), C is a constant with a value of 2.5, and V is the visibility.

The expression of the water content of advective fog is

$$W = (18.35V)^{-1.43} = 0.0156V^{-1.43}. \tag{6.66}$$

The expression of the water content of radiation fog is

$$W = (40.0V)^{-1.54} = 0.00316V^{-1.54}. \tag{6.67}$$

The volume of a fog droplet is very small, and it is regarded as a small water droplet at an average temperature. Therefore, the complex dielectric constant of fog can be replaced with that of water. The refractive index and absorptivity of water for commonly used wavelengths of light are listed in Table 6.2 for reference.

6.4.2.1 Fog-Size Distribution

The gamma droplet-size distribution model is widely used [18]:

$$n(r) = ar^2 \exp(-br), \tag{6.68}$$

where r is the radius of the fog droplet, n is the number of fog droplets per unit volume, and a and b are functions of the visibility V and water content W of the fog, respectively. Below, we use this model as the basis for deriving the relationship between the droplet spectral distribution and V and W.

Table 6.2 Optical refractive index and absorption rate of water [15]

λ (nm)	n (λ)	k (λ)	λ (nm)	n (λ)	k (λ)
300	1.349	1.60×10^{-8}	325	1.346	1.08×10^{-8}
350	1.343	6.50×10^{-9}	375	1.341	3.50×10^{-9}
400	1.339	1.86×10^{-9}	425	1.338	1.30×10^{-9}
450	1.337	1.02×10^{-9}	475	1.336	9.35×10^{-10}
500	1.335	1.00×10^{-9}	525	1.334	1.32×10^{-9}
550	1.333	1.96×10^{-9}	575	1.333	3.60×10^{-9}
600	1.332	1.09×10^{-8}	625	1.332	1.39×10^{-8}
650	1.331	1.64×10^{-8}	675	1.331	2.23×10^{-8}
700	1.331	3.35×10^{-8}	725	1.330	9.15×10^{-8}
750	1.330	1.56×10^{-7}	800	1.329	1.25×10^{-7}

6.4.2.2 Determination of Droplet-Size Distribution Parameters

In the gamma droplet-size distribution, σ is expressed as [18]

$$\sigma = \int_0^\infty 2\pi r^2 n(r)dr = 2\pi \int_0^\infty ar^4 e^{-br} dr = \frac{2\pi a 4!}{b^5} \tag{6.69}$$

$$V = \frac{3.912}{\sigma} = \frac{3.912b^5}{2\pi a 4!} \tag{6.70}$$

$$W = 10^6 \int_0^\infty \frac{4\pi}{3} \int_0^\infty ar^5 e^{-b\pi} dr = \frac{4\pi 5! a}{3b^6} \times 10^6. \tag{6.71}$$

It is simplified to obtain

$$a = \frac{9.781}{V^6 W^5} \times 10^{15} \tag{6.72}$$

$$b = \frac{1.304 \times 10^4}{VW}, \tag{6.73}$$

where V represents the visibility of fog and W represents the water content.

Substituting a and b into the Khrgian–Mazin distribution model and using Eqs. (6.66) and (6.67), the empirical formulas of water content and fog visibility, the droplet-size distribution in micron radius can be obtained [18]:

Advective fog:

$$n(r) = 1.059 \times 10^7 V^{1.15} r^2 \exp(-0.8359 V^{0.43} r)$$
$$= 3.37 \times 10^5 W^{-0.804} r^2 \exp\left(-0.2392 V^{-0.301} r\right) \mathrm{m}^{-3}\,\mathrm{m}^{-1} \quad (6.74)$$

Radiation fog:

$$n(r) = 3.104 \times 10^{10} V^{1.7} r^2 \exp(-4.122 V^{0.54} r)$$
$$= 5.400 \times 10^7 W^{-1.104} r^2 \exp\left(-0.5477 W^{-0.351} r\right) \mathrm{m}^{-3}\,\mathrm{m}^{-1}. \quad (6.75)$$

The Mie theory was used to calculate the scattering of light by single-sphere fog particles, and the single-sphere fog-particle scattering for 650-nm-wavelength light was calculated and simulated. The corresponding negative refractive index of water $m = 1.331 - 1.64i \times 10^{-8}$ [15]. When the incident light intensity is $I_0 = 1$ W, distance between scatterer and light source is r = 0.1 m. The polar-coordinate distribution of scattered light intensity (the logarithm is taken) and scattering angle is shown in Fig. 6.13:

When the radius of the fog particles is within the range of 1–60 μm, single-sphere fog particles with radii of 1, 10, 30, and 60 μm were selected, and the scattered light intensity was measured. From Fig. 6.13, it can be seen that the light-intensity distribution of the fog particles is very complex and that there are complex lobes in the side direction, which are obviously smaller than the forward scattering. The forward-scattering light intensity of the fog particles increases with the particle size.

The attenuation coefficient of an LED under foggy conditions is [19]

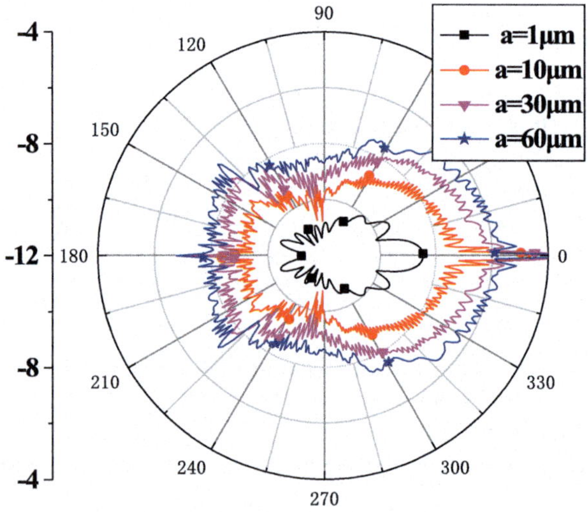

Fig. 6.13 Scattering diagram of fog particles with different diameters

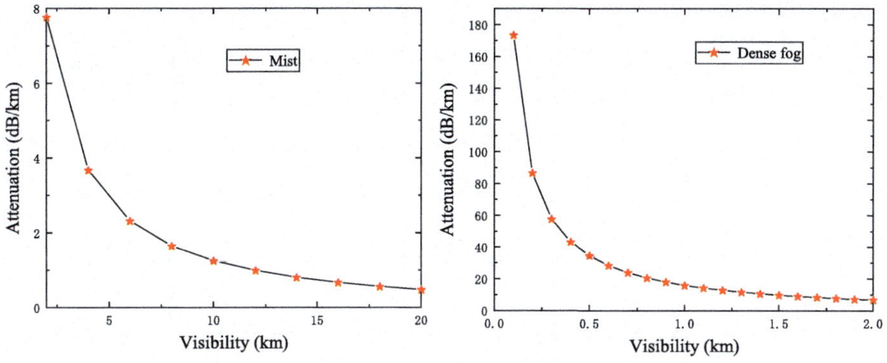

(a) Mist (b) Dense fog

Fig. 6.14 Relationship between the attenuation coefficient and visibility in foggy weather

$$\gamma_{fog}(\lambda) = \frac{17.35}{V}\left(\frac{\lambda}{550}\right)^{-\varphi}, \tag{6.76}$$

where γ_{fog} (λ) is the attenuation coefficient (dB/km), V is the visibility range, λ is the wavelength of the LED light, and φ depends on the visibility-distance range [19]:

$$\varphi = \begin{cases} 1.6, & V > 50\,\text{km} \\ 1.3, & 6\,\text{km} < V < 50\,\text{km} \\ 0.16V + 0.34, & 1\,\text{km} < V < 6\,\text{km} \\ V - 0.5, & 0.5\,\text{km} < V < 1\,\text{km} \\ 0, & V < 0.5\,\text{km} \end{cases}. \tag{6.77}$$

This chapter considers the relationship between the fog concentration and the attenuation of visible light at a 650-nm wavelength under two weather conditions: mist and dense fog, as shown in Fig. 6.14.

Figure 6.14a, b show the relationship between the attenuation coefficient and the visibility of mist and dense fog, respectively. It can be seen from the figure that the attenuation coefficient curves of dense fog and mist are similar, but on dense fog days, the concentration of fog particles is high, the visibility is low, and the scattering of light by fog particles is serious. Although the attenuation trends of dense fog and mist are similar, the attenuation is more severe for dense fog.

In a foggy channel environment with a visibility of 800 m, Fig. 6.15 shows the influence of different wavelengths of visible light on the foggy-weather attenuation coefficient. It can be seen from the figure that when the wavelength increases from 380 to 780 nm, the attenuation coefficient shows a decreasing trend, indicating that in the visible-light range, the shorter the wavelength, and the greater the attenuation in a foggy channel environment.

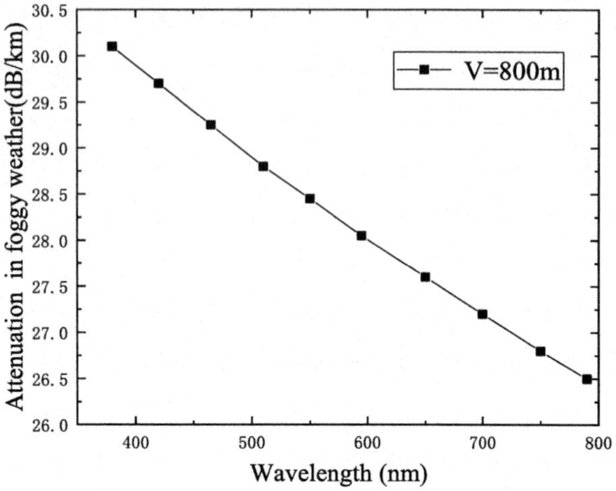

Fig. 6.15 Relationship between the attenuation coefficient in foggy weather and the wavelength

6.4.3 Snow Attenuation

The empirical relationship between the velocities of wet and dry snow was obtained by linear fitting, based on the measured data of wet and dry snow, and the relationship between the velocity of wet snow and the equivalent diameter is [20]

$$v = 2.5D_{eq}^{0.2}, \tag{6.78}$$

where v represents the speed of snow particles and D_{eq} represents their equivalent radius. The relationship between the speed of dry snow and the equivalent diameter is

$$v = 1.6D_{eq}^{0.14}. \tag{6.79}$$

Generally, the horizontal speed of snow particles is between ± 15 m/s, most of which falls in the range of ± 2.5 m/s. It should be noted that this conclusion is drawn for snowfall accompanied by wind. When the ambient wind speed is different, the characteristics of the horizontal velocity of snow particles may differ.

The gamma spectral-distribution function was used to fit the measured spectral distribution, and the fitted gamma spectral-distribution function is [20]

$$N(D) = 1.864 \times 10^3 D^{1.24} \exp(-2.0D). \tag{6.80}$$

The gamma function can accurately describe the spectral-distribution characteristics. Among them, the number density of snow particles with a diameter of 0.9 mm, measured using a precipitation microphysical-characteristics sensor (PMCS), is

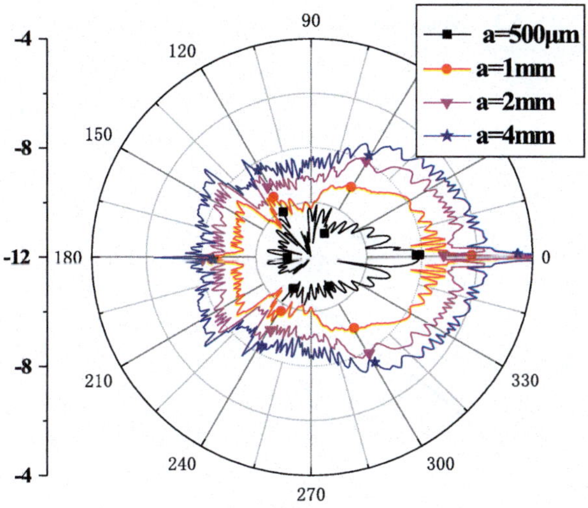

Fig. 6.16 Scattering diagram of snow particles with different diameters

the largest, which is 283.7 m^{-3} mm^{-1}. The maximum diameter of snow particles measured is 6.9 mm, and the minimum density of snow particles is 5.3×10^2 m^{-3} mm^{-1}.

The Mie theory was used to calculate the scattering of light by single-sphere snow particles, and single-sphere snow-particle scattering for 650-nm-wavelength light was calculated and simulated. The corresponding negative refractive index of water $m = 1.331 - 1.64i \times 10^{-8}$ [15]. When the incident light intensity is $I_0 = 1$ W, distance between scatterer and light source is r = 0.1 m. The polar-coordinate distribution of scattered light intensity (the logarithm is taken) and scattering angle is shown in Fig. 6.16:

Considering single-sphere snow particles with radii of 500 μm, 1, 2, and 4 mm, the scattering is mainly concentrated in the forward and backward directions, and the forward scattering increases as the size of the snow particles increases.

Because the cross-sectional area of snow particles is larger than that of raindrops, the attenuation of snow is greater than that of rain at the same precipitation rate. The attenuation γ_{snow} (dB/km), as a function of the snow-precipitation intensity S (mm/h), is given by Eq. (6.81) [21].

$$\gamma_{snow} = a \cdot S^b \tag{6.81}$$

Parameters a and b are functions of the operating wavelength λ (nm). The estimated values for wet and dry snow are listed in Table 6.3.

This chapter considers the relationship between the snowfall rate and attenuation of visible light with a 650-nm wavelength under dry and wet snow conditions, as shown in Fig. 6.17.

Figure 6.17a, b show the relationships between the wet and dry-snow attenuation coefficient and visibility, respectively. It can be seen from the figure that although

Table 6.3 Attenuation parameters caused by snowfall [21]

Snow form	a	b
Wet	0.000102 * λ + 3.79	0.72
Dry	0.0000542 * λ + 5.50	1.38

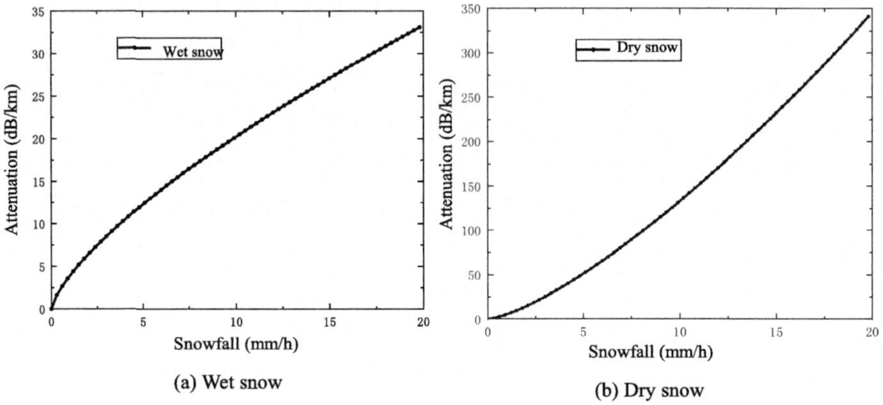

Fig. 6.17 Relationship between the attenuation coefficient and snowfall rate

the attenuation coefficients of both wet and dry snow increase with an increase in snowfall rate, the attenuation trends of the two curves are different. When the snowfall rate increases by the same amount, the attenuation coefficient of dry snow is larger, and the attenuation of dry snow is more severe than that of wet snow under the same snowfall rate.

In a channel environment with a snowfall of 2 mm/12 h, Fig. 6.18 presents the influence of the wavelength on the snow attenuation coefficient. It can be seen from the figure that the attenuation coefficient shows an increasing trend when the wavelength increases from 380 to 780 nm, indicating that in the visible-light band range, in a snowy channel environment, the longer the wavelength, the greater the attenuation of the visible-light signal.

6.4.4 *Attenuation of Dusty Weather*

The radius of dust particles in the atmosphere in dusty weather is generally in the range of 0.001–0.1 mm, and the number of particles between particles can be expressed by the scale-distribution function. The lognormal distribution accurately represents the actual dust particle-size distribution [22]:

$$n(r) = \frac{N}{2r\sigma\sqrt{2\pi}} \exp\left\{-\frac{1}{2}\left[\frac{In(2r) - \mu}{\sigma}\right]^2\right\}, \tag{6.82}$$

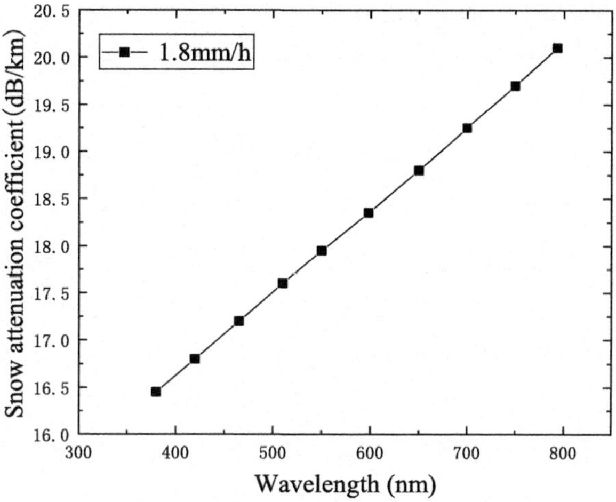

Fig. 6.18 Relationship between the snow attenuation coefficient and the wavelength

where N is the average number of particles per unit volume; μ and σ are the mean and standard deviation of In($2r$), respectively; and r is the particle radius. Particle-scale parameters are $\mu = -9.718$ and $\sigma = 0.405$ in a natural dust-storm atmosphere. The complex permittivity of dry sand is fixed and does not change with the wavelength. In this study, the negative dielectric constant of dry sand was $m = 1.55 - i0.005$ [15].

Using the Mie theory to calculate the light scattering by single spherical dust particles, the refractive index of dry sand does not change with the wavelength of light, and the negative refractive index corresponding to the wavelength of 650 nm is $m = 1.55 - 0.005i$.

When the incident light intensity is $I_0 = 1$ W, distance between scatterer and light source is r = 0.1 m. The polar-coordinate distribution of scattered light intensity (the logarithm is taken) and scattering angle is shown in Fig. 6.19:

Single-sphere dry sand particles with radii of 10, 20, 50, and 100 μm are shown in Fig. 6.19. The forward and backward scatters are much smaller than the forward scatter, and the scattered energy is mainly concentrated in the forward direction. The intensity of the forward-scattered light increases with an increase in particle radius.

If only single scattering is considered, the attenuation coefficient of the spherical dust particles follows a logarithmic-normal distribution [23]:

$$\beta = N_0 \int_0^{\infty} \pi r^2 Q_{cu}(r) n(r) dr, \tag{6.83}$$

where N_0 is the number of suspended particles per unit volume and $n(r)$ is the particle-size distribution probability-density function, as shown in Eq. (6.82).

The figure below shows the change curve of the attenuation of visible light by sand and dust with visibility.

Fig. 6.19 Scattering of dust particles with different diameters

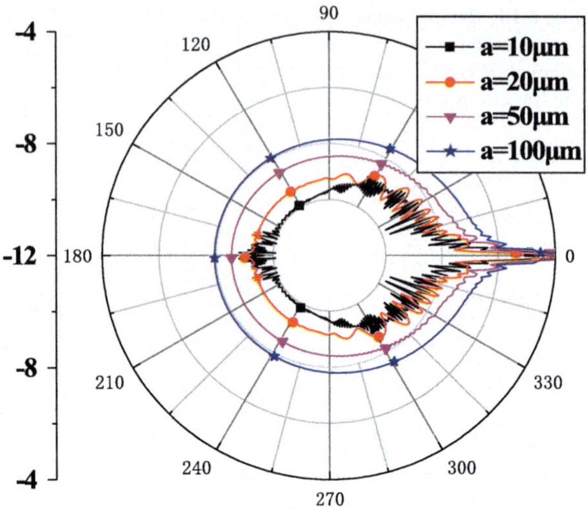

Figure 6.20 shows that the attenuation decreases with an increase in visibility under dusty weather conditions. When the visibility is lower than 200, the attenuation decreases faster as the visibility increases, whereas when the visibility is higher than 400, the attenuation decreases slowly as the visibility increases.

In the visible-light range, the complex refractive index of dust particles is $m = 1.55 - i0.005$. Figure 6.21 analyzes the influence of the wavelength on the attenuation coefficient and transmission power at a certain concentration of dust particles. It can be seen from the figure that the number concentration of dust particles is $N_0 = 1.3596 \times 10^9$ /m^3, the wavelength increases from 380 to 780 nm, and the attenuation coefficient of sandy and dusty weather shows an increasing trend. This indicates that in the visible-light range, the longer the wavelength, the greater the dust particles, and the greater the attenuation of the optical signal.

Fig. 6.20 Relationship between dusty-weather attenuation and visibility

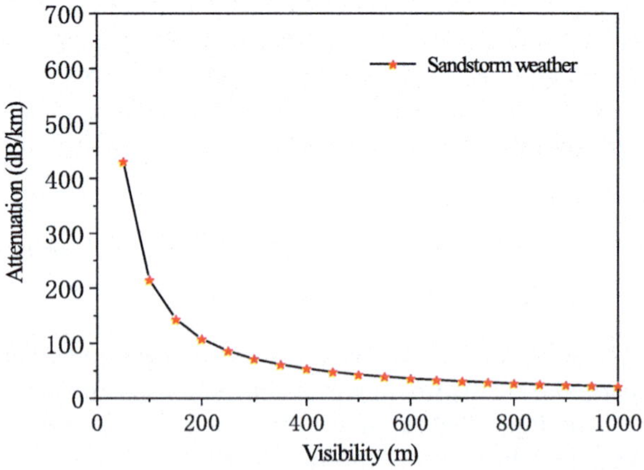

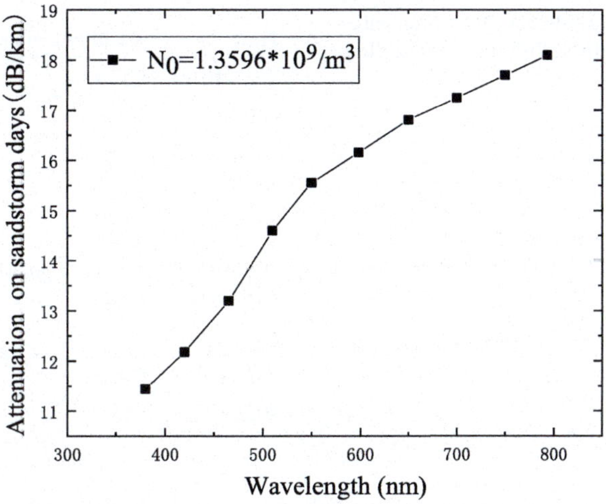

Fig. 6.21 Attenuation coefficient of sand and dusty weather varying with wavelength

6.5 Experimental Measurements from Different Road Sections and Weather Conditions

6.5.1 Analysis of Experimental Measurement Data from Different Road Sections

Figure 6.22 shows actual measurement data measured at different time periods on Jinhua Road and Xianning Road in Xi'an City, China. The probability-density function was fitted to the measured optical power.

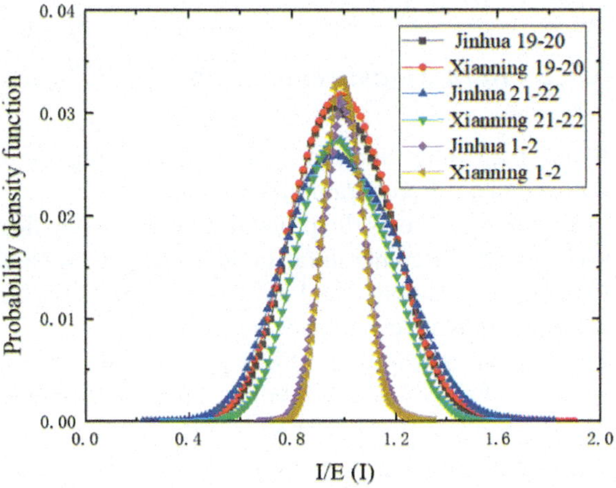

Fig. 6.22 PDF of received optical power at different time periods

Table 6.4 Measurements from different road sections

Time	Weather conditions	Measured section	Groups
2021-10-21	Clear	Xianning Road	6
2021-10-22	Clear	Xianning Road	8
2022-05-07	Clear	Xianning Road	9
2022-05-12	Clear	Xianning Road	8
2022-06-26	Clear	Xianning Road	5
2021-10-25	Clear	Jinhua Road	9
2021-10-27	Clear	Jinhua Road	7
2021-12-05	Clear	Jinhua Road	9
2022-03-28	Clear	Jinhua Road	10
2022-04-15	Clear	Jinhua Road	8

As shown in Fig. 6.22, the received optical-power probability-density function conforms to a Gaussian model. After normalizing the received optical power, the variance of the probability-density function of the received optical power is different in different time periods; however, the path loss of different roads is the same in the same time period.

From 19:00 to 20:00 in the evening, the traffic flow was large, and the variance in the probability-density function of the received optical power was relatively large. From 1:00 to 2:00 in the morning, there were few vehicles and the variance in the probability-density function of the received optical power was small. The main reason for this difference is the traffic flow. When the traffic flow is large, the received signal is affected by more factors, such as the light reflected from vehicles in the side lane, which increases the channel path loss. Table 6.4 lists the sampling data for certain road sections under sunny conditions for different time periods.

6.5.2 *Rain Measurement Data and Analysis*

Figure 6.23 shows the probability-density function fitting of the measured optical power under different rainfall conditions.

Figure 6.23 shows the probability-density function fitted to the measured optical power under different rainfall conditions. The probability-density function conforms to the Gaussian model, and its variance differs for different rainfall conditions after normalizing the received power values.

After the transmitted optical power at the transmitter side passes through the rainy channel, different rainfall events have different effects on its attenuation, and the magnitude of the path loss is different. The variance of the probability-density function of the received optical power is greater for light rain and rainstorms, indicating that the received optical power is more unstable in light rain and rainstorm

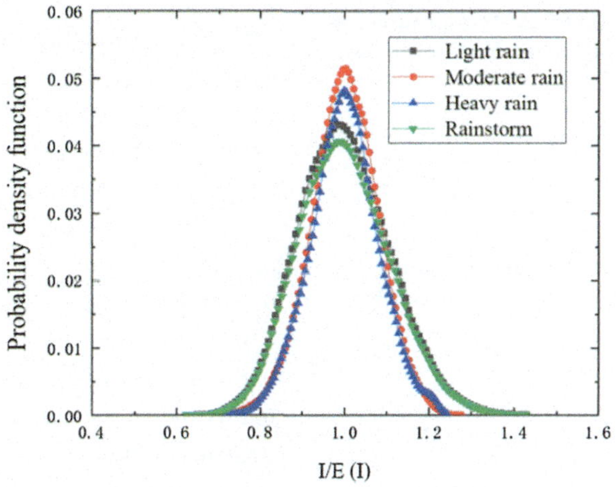

Fig. 6.23 PDF of received light power under different rainfall conditions

days, which will also have a greater impact on the communication quality in the actual communication process.

Figure 6.24 shows the mean value of the received optical power under different rainfall conditions. It can be seen that the mean value of the received optical power under light rain and rainstorm conditions is smaller than that received under moderate and heavy rain. That is, the mean value of the received optical power is smaller when the rain attenuation is larger, which corresponds to Fig. 6.12, proving the correctness of the relationship between the rainfall rate and scattering attenuation. The two graphs corroborate each other.

Table 6.5 lists the sampling data from different rainfall conditions; the rainfall was recorded based on the weather forecast from the weather station that day.

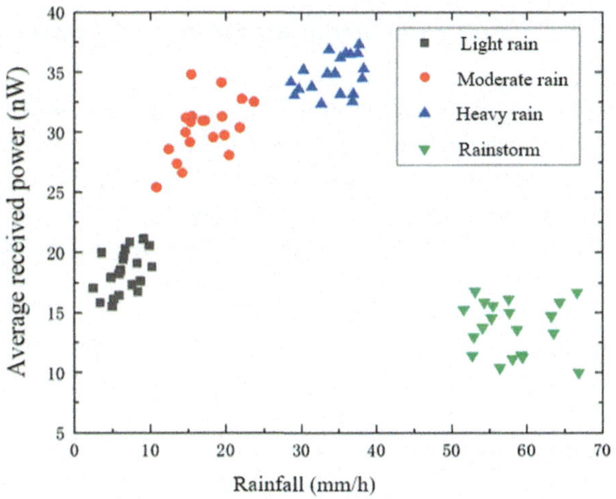

Fig. 6.24 Mean received light power under different rainfall forms

Table 6.5 Measurements of different rainfall conditions

Time	Weather condition	Rainfall (mm)	Groups
2021-08-31	Rainstorm	66	10
2021-09-27	Light rain	6	5
2021-09-27	Heavy rain	29	6
2021-09-28	Moderate rain	18	8
2021-10-03	Light rain	5	6
2021-10-05	Light rain	7	11
2021-10-19	Moderate rain	15	13
2021-11-29	Light rain	4	8
2021-12-10	Moderate rain	16	12
2022-03-16	Light rain	6	9
2022-04-11	Heavy rain	32	11
2022-04-27	Heavy rain	33	7
2022-04-28	Light rain	5	9
2022-05-13	Rainstorm	56	8
2022-07-18	Rainstorm	57	9
2022-08-05	Rainstorm	62	7
2022-08-09	Heavy rain	32	10
2022-08-28	Moderate rain	18	8
2022-09-19	Heavy rain	31	11
2022-11-11	Moderate rain	13	10
2022-11-27	Light rain	6	9
2023-03-24	Moderate rain	15	6
2023-04-02	Heavy rain	28	12

6.5.3 Fog Measurement Data and Analysis

Figure 6.25 shows the probability-density function fitting of the measured optical power under different fog concentrations.

As shown in Fig. 6.25, the probability-density function of the received optical power conforms to a Gaussian model. After normalizing the received power value, the variance of the probability-density function differs under different fog concentrations.

When the transmitted light of the transmitter passes through the channel in foggy weather, different fog concentrations have different effects on its attenuation. The variance of the probability-density function of the optical power of dense fog is larger, indicating that the received optical power is more unstable on dense-fog days. The communication process will also have a greater impact on the communication quality.

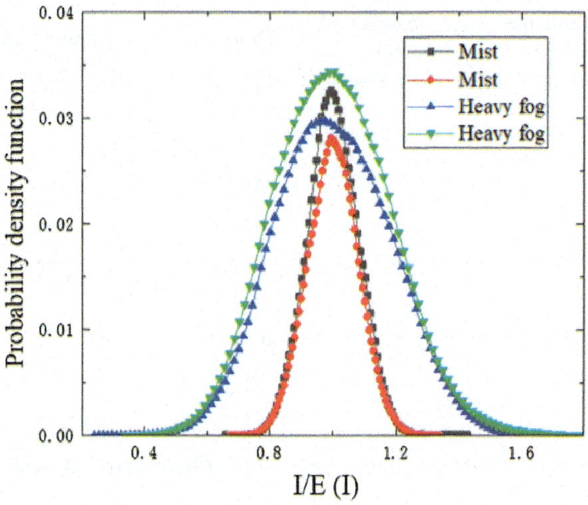

Fig. 6.25 PDF of received light power under mist and heavy fog

Figure 6.26 shows the average value of the received optical power under different fog concentrations. It can be observed that the average value of the received optical power under dense fog is smaller than that under misty weather conditions. That is, when the fog attenuation is large, the average value of the received optical power is small, which proves the correctness of the relationship between the fog visibility and scattering attenuation, as shown in Fig. 6.14. The two figures are mutually confirmed.

Table 6.6 lists the sampling data under different fog concentrations. The fog concentration is recorded according to the weather forecast of the weather station that day.

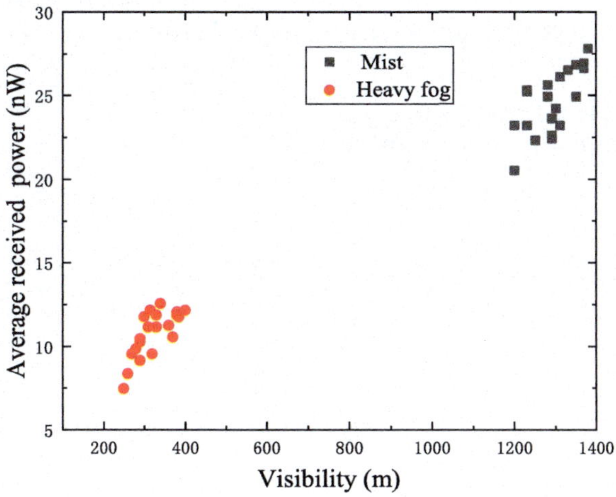

Fig. 6.26 Mean received light power under different fog forms

Table 6.6 Measurements of weather conditions with different fog concentrations

Time	Weather condition	Visibility (m)	Groups
2021-09-25	Heavy fog	400	11
2021-10-27	Heavy fog	330	13
2021-11-27	Mist	1600	10
2022-11-10	Heavy fog	350	12
2022-11-24	Heavy fog	210	9
2022-11-25	Heavy fog	330	11
2022-12-26	Mist	1200	8
2023-03-16	Mist	1300	11
2023-03-17	Mist	1500	10

6.5.4 Snow Measurement Data and Analysis

Figure 6.27 shows the probability-density function fitting of the measured optical power under different humidity and snow conditions.

As shown in Fig. 6.27, the probability-density function of the received optical power conforms to a Gaussian model. After normalizing the received power value, the variance in the probability-density function differs under different rainfall and snow conditions. After the transmitted light from the transmitter passes through the snowy channel, different forms of snowfall have different effects on its attenuation. The variance of the probability-density function of the optical power on dry snow days was larger, indicating that the received optical power was more unstable on dry snow days. This will greatly affect the communication quality in the actual communication process.

Figure 6.28 shows the average value of the received optical power under different snowfall conditions. It can be seen that the average value of the received optical power

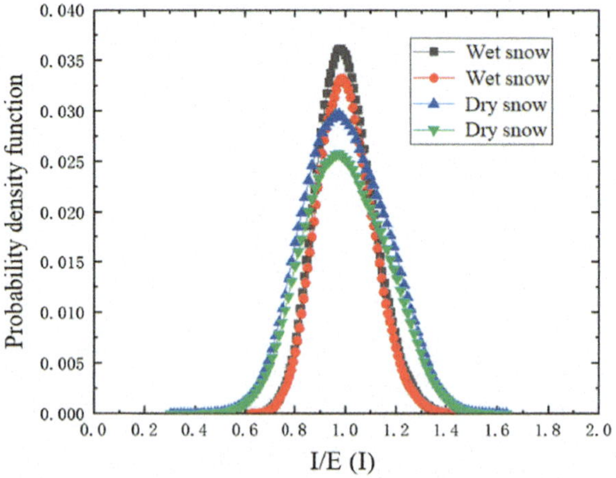

Fig. 6.27 PDF of received optical power under wet and dry snow conditions

under wet snow conditions is smaller than that received under dry snow conditions. That is, when the snow attenuation is large, the average value of the received optical power is small, which proves the correctness of the relationship between the snowfall rate and attenuation, as shown in Fig. 6.17. The two figures are mutually confirmed.

Table 6.7 lists the sampling data under snowy weather conditions, and the snowfall is recorded according to the weather forecast of the weather station that day.

Gaussian fitting was performed on the optical-power probability-density function measured under different weather conditions, the parameters of which are listed in Table 6.8. After mean normalization, the mean remained the same under different weather conditions, and the variance increased with the path loss.

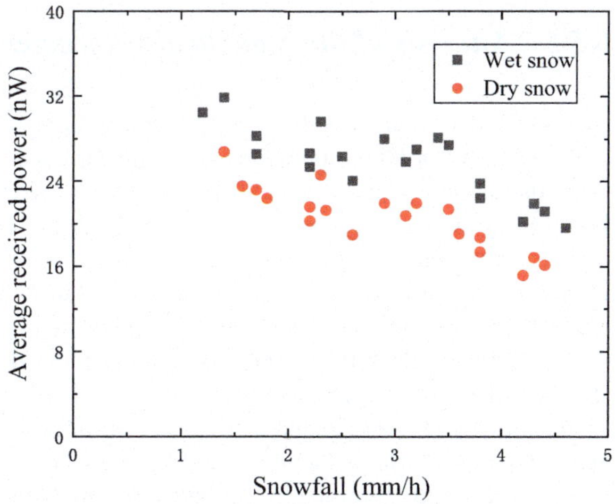

Fig. 6.28 Mean received optical power under different snowfall forms

Table 6.7 Measurements under different snowfall conditions

Time	Weather condition	Snowfall (mm)	Groups
2021-11-07	Dry snow	1.5	15
2021-12-25	Dry snow	1.8	15
2022-01-27	Wet snow	2.5	15
2022-01-28	Wet snow	2.3	15
2022-02-18	Wet snow	1.5	14
2022-11-24	Wet snow	2.5	16
2022-11-25	Dry snow	2.2	13
2022-11-30	Dry snow	1.6	15
2022-12-25	Dry snow	3.0	12

Table 6.8 Gaussian-model parameters under different weather conditions

Weather condition	μ	σ
Light rain	0.99	0.092
Moderate rain	0.99	0.053
Heavy rain	0.99	0.055
Rainstorm	0.99	0.098
Wet snow	0.99	0.114
Dry snow	0.99	0.146
Mist	0.99	0.133
Heavy fog	0.99	1.191

6.5.5 Analysis of the Statistical Characteristics of the Data

The first-order moment of the signal sequence represents the mean value of the signal and the second-order moment represents the variance. The third-order moment represents the slope and the degree of symmetry of the probability-density distribution function of the received signal about the midline. The fourth-order moment represents the kurtosis, as well as the degree of concentration of the received signal around its mathematical expectation. The measured data for the samples are presented in Tables 6.5, 6.6, and 6.7. The abscissa represents the variance. The skewness and kurtosis of the light-intensity signal samples under different weather conditions are shown in Figs. 6.29 and 6.30.

As shown in Figs. 6.29 and 6.30, as the variance of the data samples increases, the absolute values of both skewness and kurtosis increase. The skewness and kurtosis of foggy days are significantly higher than those of other weather conditions, the skewness and kurtosis of snowy days are higher than those of rainy days, and the skewness and kurtosis of light and heavy rain are higher than those of heavy and

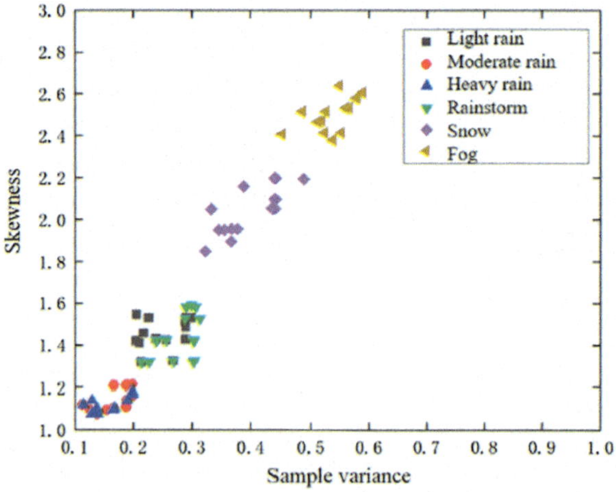

Fig. 6.29 Sample skewness

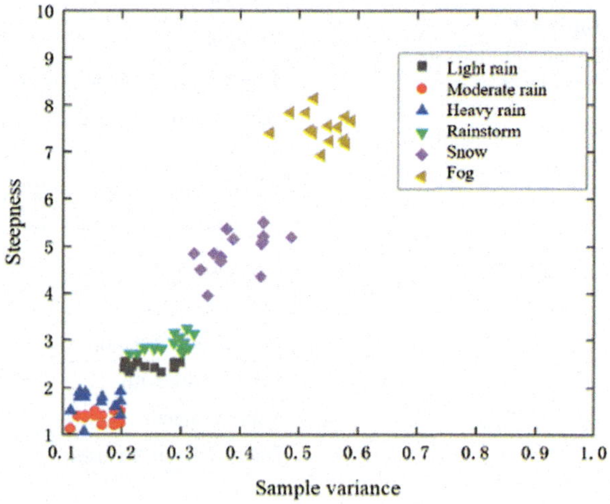

Fig. 6.30 Sample kurtosis

moderate rain. The greater the effect of weather conditions on the signal link, the greater the variance of the signal, and the greater the slope and steepness of the signal data samples.

6.6 Analysis of Communication Performance Under Different Weather Conditions

6.6.1 Received Optical Power at Different Vehicle Distances

The intervehicle visible-light communication channel is a special dynamic medium in which the spacing of the vehicles changes at any time with the road traffic. This subsection describes the simulations of the received optical power at different vehicle distances. Table 6.9 lists the numerical parameters.

Figure 6.31 analyzes the received optical power under different rainfall patterns. Clearly, the received optical power decreases with increasing distance for all rainfall forms. Under the same transmission distance, the order of the received optical power from largest to smallest is moderate rain, heavy rain, light rain, and rainstorm.

Figure 6.32 analyzes the received optical power under different fog-visibility conditions. The received optical power under different visibilities decreases with an increase in distance. At the same transmission distance, the received optical power decreases as the visibility decreases.

Figure 6.33 analyzes the received optical power for different snowfall patterns. Clearly, the received optical power decreases with increasing distance for all snowfall forms. A comparison of the influences of wet and dry snow under the same snowfall rate shows that the influence of dry snow on the received optical power is greater

Table 6.9 Numerical-analysis parameters

Parameter	Symbol	Value
Transmission optical power	P_t	314 [mW]
Detector physical area	A	0.79 [cm^2]
Filter gain	$T_s(\psi)$	1.0
Concentrator refractive index	n	1.7
Photodetector responsivity	r	0.35 [A/W]
Open-loop voltage gain	G	10
Field-effect transistor channel noise	Γ	1.5
Bandwidth of bandpass optical filters	λ	10 [nm]
LED wavelength	l	550 [nm]
Vehicle speed	V_s	60 [km/h]
Data-transmission rate	T_d	20 M [bit/s]

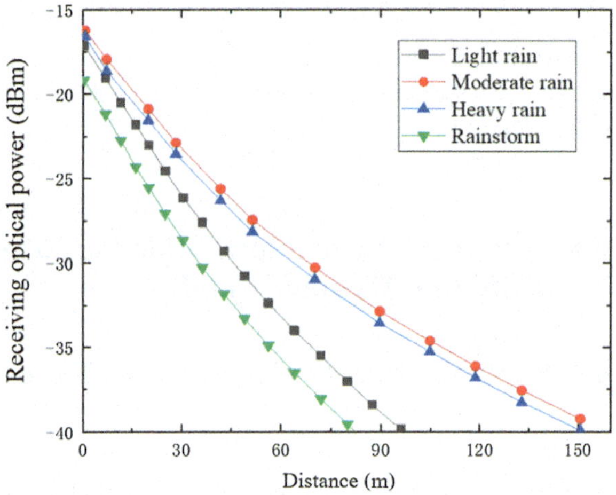

Fig. 6.31 Relationship between the received light power and distance on rainy days

than that of wet snow. At the same transmission distance, the received optical power decreases with an increase in the snowfall rate.

Figure 6.34 analyzes the received optical power for different dust-particle concentrations. The received optical power decreases with increasing distance in all dusty weather conditions. At the same transmission distance, the received optical power decreases with an increase in dust-particle concentration.

Figure 6.35 analyzes the received optical power for different transmission distances and road widths. From the figure, it can be seen that the received power shows a decreasing trend from the position of the light source, along the direction of the road width and the horizontal transmission direction. Under the same transmission distance, the wider the road width, the smaller the received optical power, and under the same road width, the farther the transmission distance, the smaller the received optical power.

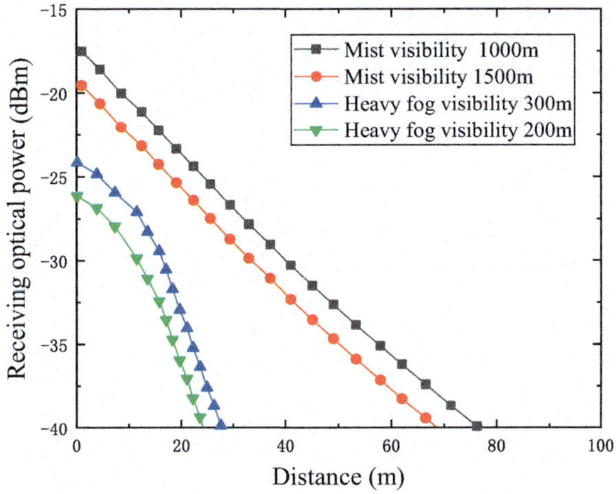

Fig. 6.32 Relationship between received light power and distance in foggy weather

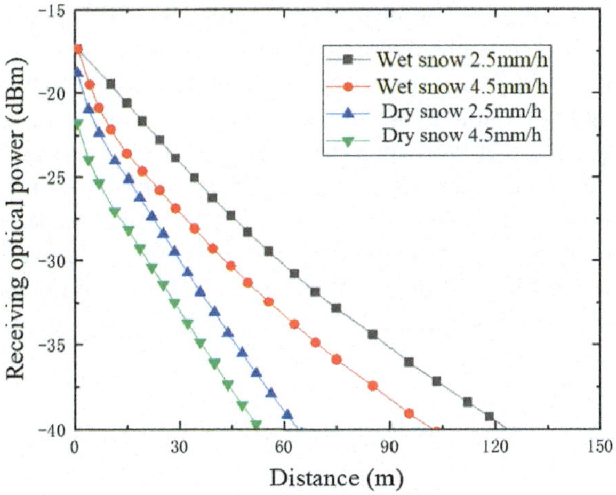

Fig. 6.33 Relationship between the received light power and distance on snowy days

6.6.2 *Bit-Error Rate of Received Information at Different Vehicle Distances*

Figure 6.36 compares and analyzes the bit-error rate (BER) performance of the outdoor visible-light communication system under different rainfall patterns. It can be observed from the figure that the bit-error rate of the system increases with an increase in the transmission distance under different weather conditions. Under rainy weather conditions, when the system bit-error rate is 10^{-6}, the effective communication distances of the outdoor visible-light communication system are as follows, from small to large: rainstorm, light rain, heavy rain, and moderate rain.

Fig. 6.34 Relationship between the received light power and distance on a dusty day

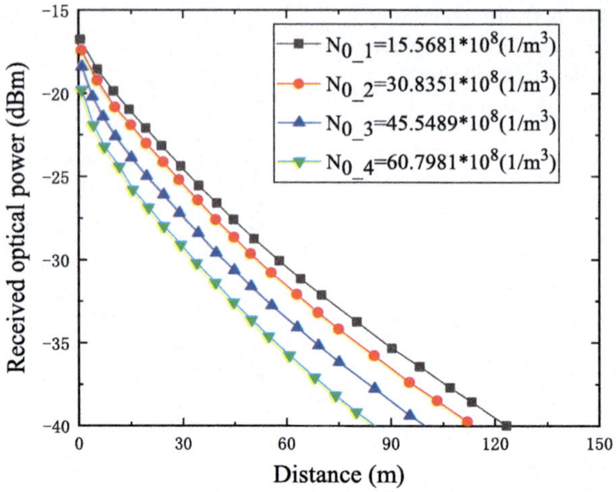

Figure 6.37 compares and analyzes the BER performance of the outdoor visible-light communication system with different snowfall rates on wet and dry snow days. It can be seen that the system BER increases with an increase in transmission distance, and with an increase in snowfall rate, the effective transmission distance gradually becomes shorter when the BER is 10^{-6}. Comparing the changes in the system BER under different snowfall rates of wet and dry snow, it can be seen that under the same BER and snowfall rate, the information-transmission distance of the dry-snow outdoor VLC system compared with that of the wet-snow outdoor VLC system is reduced by 3.6 m, 5.3 m, 13.9 m, 13.9 m, 21.4 m, and 34.5 m, respectively, for the different visibilities.

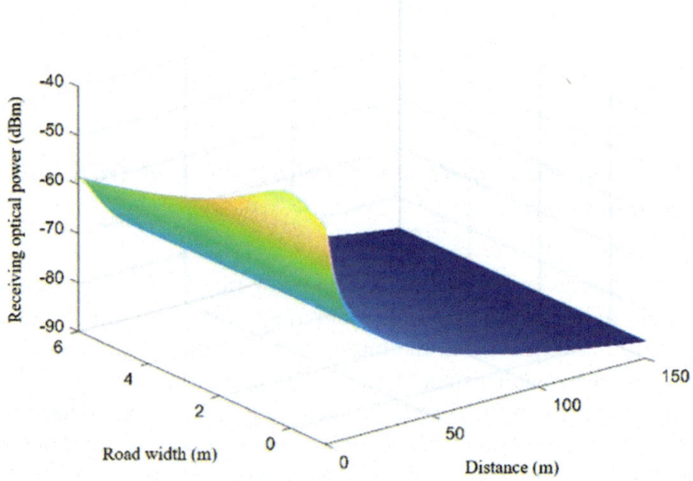

Fig. 6.35 Distribution of received light power for automotive headlights

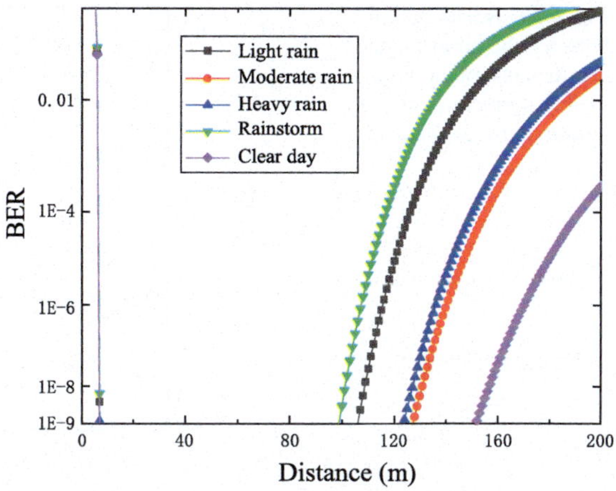

Fig. 6.36 Relationship between the error rate and distance of an outdoor visible-light communication system in rainy weather

Figure 6.38 compares and analyzes the bit-error rate performance of the outdoor VLC system under different fog visibilities. It can be seen from the figure that the bit-error rate of the system increases with the increase of the transmission distance under different fog visibilities. As the visibility decreases, when the bit-error rate is 10^{-6}, the effective transmission distance gradually reduces. The working ranges of the outdoor VLC system, for different visibilities from large to small, are as follows: sunny day, 2000, 1000, 500, 300, and 50 m.

Figure 6.39 analyzes the bit-error rate performance of the outdoor VLC system for four different dust particle-number concentrations: $N_{01} = 15.5681 \times 108$ (1/m^3), $N_{02} = 30.8351 \times 108$ (1/m^3), $N_{03} = 45.5489 \times 108$ (1/m^3), and $N_{04} = 60.7981 \times 108$ (1/m^3). It can be seen from the figure that the bit-error rate of the system increases with the increase of the transmission distance under different dust particle-number concentrations. Moreover, with the increase of the particle number concentration

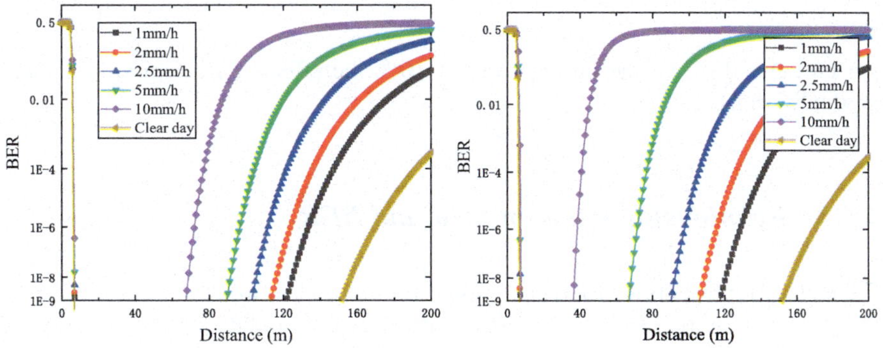

(a) Wet snow (b) Dry snow

Fig. 6.37 Relationship between the error rate and distance of an outdoor visible-light communication system in snowy weather

Fig. 6.38 Relationship between the bit-error rate and distance of an outdoor visible-light communication system in foggy weather

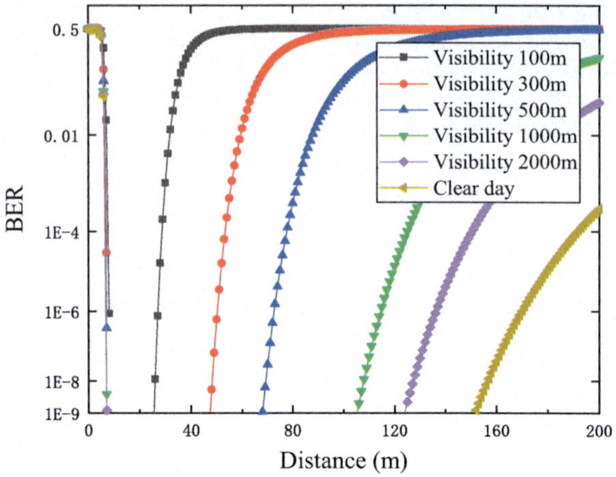

Fig. 6.39 Relationship between the bit-error rate and distance of an outdoor VLC system in dusty weather

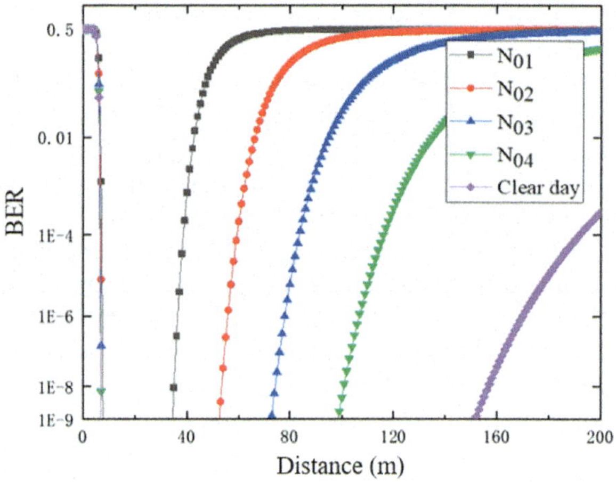

per unit cubed, when the bit-error rate is 10^{-6}, the effective transmission distance gradually shortens.

6.6.3 Relationship Between SNR and BER

Figure 6.40 represents the BER analyses performed under rainy conditions. When the signal-to-noise ratio (SNR) is the same under different rainfall conditions, the BERs for moderate and heavy rain are close to each other, and the BERs for rainstorms and light rain are larger than those for moderate and heavy rain.

From Fig. 6.40, it can be seen that at both ends of the rain-attenuation extremes, although the attenuation is approximately the same under different rainfall amounts,

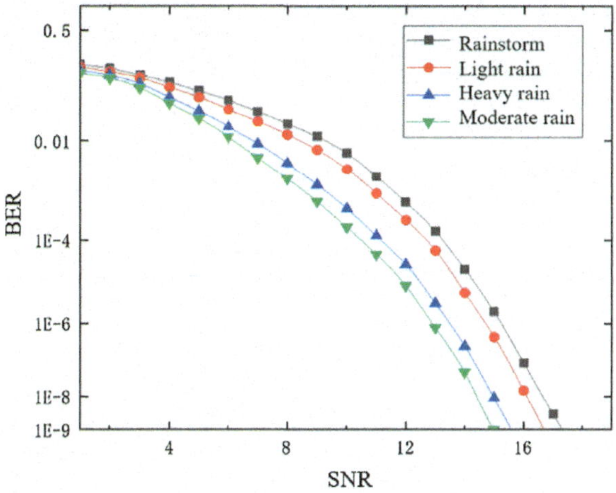

Fig. 6.40 Relationship between the signal-to-noise ratio and bit-error rate on rainy days

the BER of rainstorms is larger than that of light rain under the same signal-to-noise ratio, and the forward scattering of the raindrops is most serious under rainstorm weather conditions; therefore, the BER of rainstorms is larger under the same signal-to-noise ratio.

Figure 6.41 shows the BER analysis performed under snowy conditions. As shown in Fig. 6.41a, under the same SNR, the receiving bit-error rate increases gradually with an increase in the snowfall rate. From Fig. 6.41b, when the snowfall rate is 2.5 mm/h, the bit-error rate of dry snow is higher than that of wet snow under the same signal-to-noise ratio.

Figure 6.42 shows the BER analyses performed under foggy conditions. It can be seen from the figure that under the same signal-to-noise ratio, as the fog concentration increases, the receiving bit-error rate also increases, and the receiving bit-error rate of dense fog is much higher than that of thin fog.

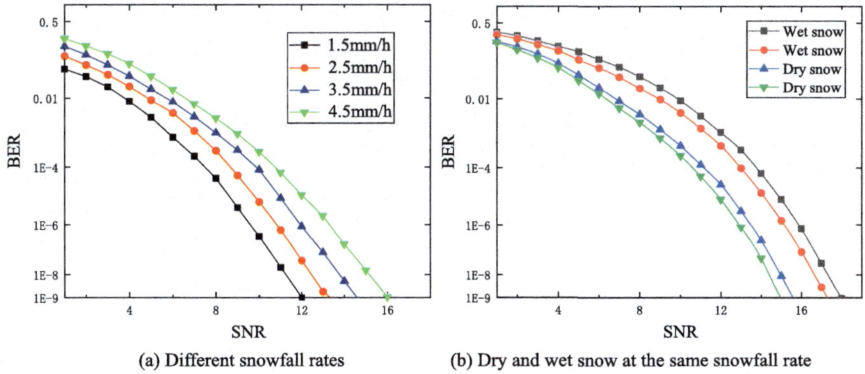

Fig. 6.41 Relationship between the signal-to-noise ratio and bit-error rate on snowy days

Fig. 6.42 Relationship between the signal-to-noise ratio and bit-error rate in foggy weather

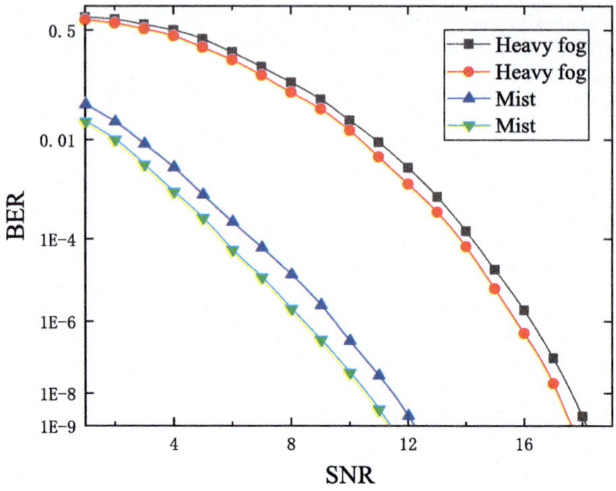

Figure 6.43 shows the BER analyses performed under sandy weather conditions. It can be seen from the figure that under the same signal-to-noise ratio, as the dust particle concentration increases, the receiving bit-error rate also increases gradually.

Fig. 6.43 Relationship between the signal-to-noise ratio and bit-error rate in sandstorms

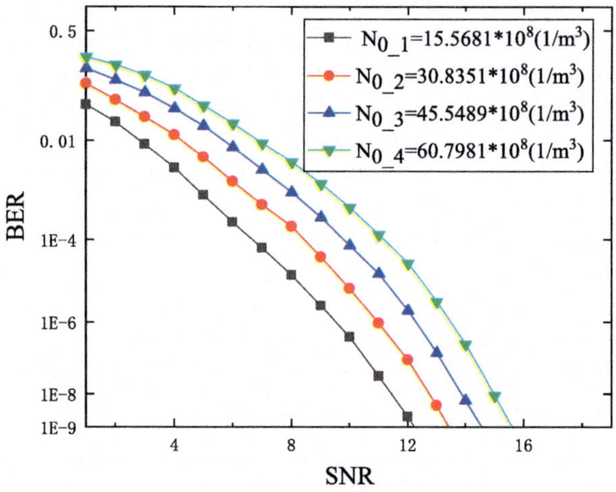

6.7 Summary and Outlook

6.7.1 Summary

In high traffic flow, the current mainstream radio-frequency communication signals between vehicles are interfered with, and the shortage of spectrum resources leads to a decline in communication quality, which cannot meet the requirements of effective communication between vehicles. In this chapter, through simulation analysis and experimental measurements, a model describing the path loss of outdoor VLC channels under different weather conditions was established, and its system performance was studied. The specific research objectives were as follows:

1. The outdoor VLC system model was analyzed, and the outdoor visible-light path-loss model was studied through formula derivation and theoretical simulations. The path-loss model conformed to the Gaussian model. The line-of-sight transmission and reflection components were studied in detail, and the path-loss model was simulated when the rainfall was 15 mm/h.
2. The simulations revealed the attenuation of visible light on rainy, foggy, and snowy days. The attenuation of visible light on rainy days from large to small was rainstorms, light rain, heavy rain, and moderate rain; the attenuation of visible light on snowy days increased with the amount of snowfall. When the snowfall was the same, the attenuation of dry snow was greater than that of wet snow. The attenuation of visible light on foggy days increased with a decrease in visibility, and the attenuation of dense fog was greater than that of thin fog. The attenuation of sand in dusty weather increased gradually with an increase in particle number concentration.
3. The performance of outdoor visible-light communication was analyzed from three aspects: received optical power, received signal BER, and relationship between the signal-to-noise ratio and BER. Under the same transmission distance, the received optical powers on a rainy day were medium rain, heavy rain, light rain, and rainstorms, in descending order.

   The optical power received on foggy days decreased with decreasing visibility, that received on snowy days decreased with increasing snowfall, and that received on dry snow was lower than that received on wet snow under the same amount of snowfall. The optical power received on dusty days decreased with increasing dust-particle concentration.

   When the system BER was 10^{-6}, the effective communication distances of the outdoor visible-light communication system on rainy days were in the following order, from small to large: rainstorm, light rain, heavy rain, and medium rain. The effective communication distance on foggy days decreased with a decrease in visibility, and that on snowy days decreased with an increase in snowfall rate. The effective communication distance on dusty days decreased with increasing dust-particle concentration.

4. The results of the road experimental measurements showed that the attenuation of the transmitted light intensity at the transmitter end is affected differently by different rainfall amounts after passing through a rainy-day channel. Moreover, the variance of the received optical power probability-density function was larger on light rain and rainstorm days, indicating that the received optical power value was more unstable.

   Different fog concentrations had different effects on the attenuation, and the received optical power values were more unstable on dense fog days. Different forms of snowfall had different effects on the attenuation, and the received optical power was more unstable on dry snow days than on wet snow days. The more severe the channel path loss, the greater the variance of the received optical power probability-density function, and the greater the skewness and kurtosis of the noisy data samples. When the attenuation was small, the mean value of the received optical power was large for each group of data. When the attenuation was large, the mean value of the received optical power was small for each group of data.
5. Single- and double-layer constellation diagrams of the measured data under different weather conditions were analyzed. The distribution of constellation maps on rainy, snowy, and foggy days gradually increased, the dispersion of constellation points gradually increased, and the impact on the effective data transmission gradually increased. When the fog visibility was 500 m, the dispersion of some of the constellation points was connected. In the actual communication process, it will be necessary to perform algorithm processing on the received constellation diagram to achieve ideal communication performance.

6.7.2 *Outlook*

As the channel characteristics of an outdoor visible-light communication system are complex, this chapter established a path-loss model for it and investigated the impact of different weather conditions on it by analyzing the measured data. The aspects that need to be studied further in the research and development of outdoor visible-light communication technology are as follows.

1. Formulation of technical standards. Because outdoor visible-light communication technology is still in the initial stages of development, relevant standards and specifications are lacking. Therefore, relevant technical standards must be formulated to ensure that outdoor visible-light communication technologies are widely used.
2. Fast communication connections. On an actual road, the distance and orientation changes between vehicles are very rapid. Therefore, a rapid communication connection is required to ensure the stability of the communication system.
3. Sunlight noise. This study was mainly aimed at researching and measuring the road environment at night. Research is lacking on the impact of solar radiation during the day.

4. Research and development of headlights. Vehicle-mounted visible-light communication technology requires the use of vehicle-mounted lamps; therefore, the research and development of lamps are also key. In the future, it will be necessary to develop more-intelligent vehicle lamps that meet vehicle-safety standards to ensure vehicle safety and communication quality.

Overall, outdoor visible-light communication technology has broad application prospects; however, it will face many challenges in future research and applications. If the advantages of VLC technology can be fully utilized and the related technology problems can be solved, this technology is expected to contribute to intelligent, efficient, and safe vehicle communication.

References

1. Béchadergue B, Chassagne L, Guan H (2019) Simultaneous visible light communication and distance measurement based on the automotive lighting. IEEE Trans Intell Veh 4(4):532–547
2. Kruse PW et al (1962) Elements of infrared technology. John Wiley Sons. Inc., New Yoke, London
3. Jiang L, Chan P, Ng D et al (2012) Hybrid visible light communications in intelligent transportation systems with position based services. In: IEEE Globecom workshops, Anaheim, CA, USA 2012, pp 1254–1259
4. Gfeller FR, Bapst UH (1980) Wireless in-house data communication via diffuse infrared radiation. Inf Syst 5(11):1474–1486
5. Mohamed M, Hranilovic S (2009) Optical impulse modulation for indoor diffuse wireless communications. IEEE Press 57(2):499–508
6. Memedi A, Tebruegge C, Jahneke J et al (2018) Impact of vehicle type and headlight characteristics on vehicular VLC performance. In: 2018 IEEE vehicular networking conference, 2018, pp 1–8
7. Tseng HY, Wei YL, Chen AL et al (2015) Characterizing link asymmetry in vehicle-to-vehicle visible light communications. In: IEEE vehicular networking conference, pp 88–95
8. Memedi A, Tsai HM, Dressler F (2017) Impact of realistic light radiation pattern on vehicular visible light communication. In: Global communications conference, 2017, pp 1–6
9. Komine T, Nakagawa M (2004) Fundamental analysis for visible-light communication system using LED lights. IEEE Trans Consum Electron 50(1):100–107
10. Cheng L, Tsai HM, Viriyasitavat W et al (2018) Comparison of radio frequency and visible light propagation channels for vehicular communications. IEEE Access 6(1):2634–2644
11. Yan GJ, Stephan O (2011) A probabilistic analysis of link duration in vehicular ad hoc networks. IEEE Trans Intell Transp Syst 12(4):1227–1236
12. Huang HQ, Yang AY, Feng LH et al (2017) Artificial neural-network-based visible light positioning algorithm with a diffuse optical channel. Chin Opt Lett 15(05):20–24
13. Dong QF (2006) Study of transmission effects of pulses at millimeterwavelengths through rain and fog medium. Xi'an University of Electronic Science and Technology, Xi'an
14. Pu JP, Zhang W, Jiang AJ et al (2010) Characteristics of gamma raindrop size distribution based on the precipitation particle spectrometer. J Meteorol Sci 30(05):701–707
15. Hale GM, Querry MR (1973) Optical constants of water in the 200 nm to 200 μm wavelength region. Appl Opt 12(3):555–563
16. Ke XZ, Yang LH, Ma DD (2008) Transmitted attenuation of laser signal in rain. Infrared Laser Eng 37(06):1021–1024
17. Zou JS (1982) Fundamentals of atmospheric physics. Meteorological Publishing House, Beijing

18. Wu ZS, Wang YP (1993) A new algorithm for electromagnetic scattering from multilayer spheres. J Electron Inf Technol 15(2):174–180
19. Guo YH, Chang YC, Zhen QC (2007) Free-space optical communication using visible light. J Zhejiang Univ A Sci 8(2):186–191
20. Liu XC, Gao TC, Liu L et al (2014) Research on microphysical property of snowfall based on particle imaging velocimetry technology. Acta Phys Sin 63(19):454–462
21. Zaki RW, Fayed HA, Abd EA et al (2019) Outdoor visible light communication in intelligent transportation systems: impact of snow and rain. Appl Sci 9(24):1–20
22. Wu ZS, You JG, Yang RK (2004) Study on laser attenuation character in sand and dust storms. Chin J Lasers 31(09):1075–1080
23. Ebrahim K, Al-Omary A (2021) Effects of sandstorms on vehicular-to-road visible-light communication. J Opt Commun 42(1):165–175

Chapter 7
Underwater Optical-Wireless Communication Channel Model

Underwater optical-wireless communication provides important technological support in underwater environmental monitoring, deep-sea exploration, and national-defense military fields. The study of optical channel transmission characteristics in seawater is the most difficult problem in underwater optical-wireless communication, which requires a comprehensive understanding and analysis of the complex physical, chemical and biological systems in seawater environment.

This chapter introduces the characteristics of seawater channels, basic theory of marine turbulence, attenuation characteristics of underwater suspended particles, and transmission of lasers underwater. It also includes a discussion on an experiment conducted to measure the channel characteristics under different conditions in different waters and during transmission.

7.1 Seawater Channel Characteristics

Seawater is a complex, irregular hybrid medium. The following elements affect the transmission characteristics of laser beams in seawater: turbulence caused by the movement of water molecules and the production of chlorophyll, the partial yellow substance produced by various fish, algae, etc. (such as fulvic acid and humic acid), it also includes all kinds of dissolved salts and mineral particles. When a laser beam is transmitted underwater, the random undulating marine turbulence, absorption, and scattering characteristics of the underwater suspended particles cause the laser-beam energy to decay, expand, drift, flicker, etc. This renders underwater communication difficult and affects the system communication performance.

X. Ke and C. Ke, *Noise Models in Optical-Wireless Communication Systems*,
Optical Wireless Communication Theory and Technology,
https://doi.org/10.1007/978-981-97-7550-7_7

7.1.1 Marine Turbulence Spectrum Model

Since the beginning of the 1990s, different ocean turbulence spectrum models have been proposed [1]. However, the ocean turbulence spectrum model proposed by Nikishov is the most widely used. Temperature and salinity factors are considered in this model, and it is suitable for isotropic ocean turbulence. The ocean turbulence spectrum is expressed as [1]

$$\begin{aligned} E_n(k) &= C_0 \chi_n \varepsilon^{-1/3} \kappa^{-11/3} \left[1 + C_1 (\kappa\eta)^{2/3}\right] \\ &\quad \times \frac{\omega^2 \theta \exp(-A_T \delta) + \exp(-A_S \delta) - \omega(1+\theta) \exp(-A_{TS} \delta)}{\omega^2 \theta + 1 - \omega(1+\theta)}, \end{aligned} \tag{7.1}$$

where $C_0 = 0.07$, $C_1 = 2.3$, $\chi_n = \alpha^2 \chi_T + \alpha^2/\omega^2 \chi_T - 2\alpha^2/\omega^2 \chi_T, \theta = K_T/K_S$, K_T is the turbulent heat dissipation coefficient, K_S indicates the turbulent salinity diffusion coefficient, and substitute χ_n into Eq. 7.1:

$$\begin{aligned} E_n(\kappa) &= C_0 \left(\alpha^2 \chi_T + \frac{\alpha^2}{\omega^2} \chi_T - \frac{2\alpha^2}{\omega} \right) \varepsilon^{-1/3} \kappa_{-11/3} \left[1 + C_1 (\kappa\eta)^{2/3}\right] \\ &\quad \times \frac{\omega^2 \theta \exp(-A_T \delta) + \exp(-A_S \delta) - \omega(1+\theta) \exp(-A_{TS} \delta)}{\omega^2 \theta + 1 - \omega(1+\theta)} \end{aligned} \tag{7.2}$$

For uniform oceanic turbulence with the same orientation, when $K_T = K_S$ and $\theta = 1$, the marine turbulence in Eq. (7.2) can be simplified to [1]

$$\begin{aligned} E_n(k) &= C_0 \alpha^2 \varepsilon^{-1/3} \kappa^{-11/3} \left[1 + C_1 (\kappa\eta)^{2/3}\right] \chi_T / \omega^2 \\ &\quad \times \omega^2 \exp(-A_T \delta) + \exp(-A_S \delta) - 2\omega \exp(-A_{TS} \delta). \end{aligned} \tag{7.3}$$

According to the relationship between the SCAPEL spectrum and the power spectrum function, the power spectrum function in ocean turbulence $\phi_n(\kappa)$ can be written as

$$\begin{aligned} \phi_n(\kappa) &= (4\pi)^{-1} \kappa^{-2} E_n(\kappa) \\ &= 0.388 \times 10^{-8} \varepsilon^{-1/3} \kappa^{-11/3} \left[1 + 2.35 (\kappa\eta)^{2/3}\right] \frac{\chi_T}{\omega^2} \\ &\quad \times \left[\omega^2 \exp(-A_T \delta) + \exp(-A_S \delta) - 2\omega \exp(-A_{TS} \delta)\right], \end{aligned} \tag{7.4}$$

where ε is the turbulent energy dissipation rate, and the value range is $(10^{-1}\text{–}10^{-10})$ m^2/s^3. η is the Kolmogorov scale and the value is $(6\times10^{-5} - 0.01)$ m. χ_T represents the temperature variance dissipation rate, and the value range from the sea to deep-sea is $(10^{-4}\text{–}10^{-10})$ K^2/s. $\omega = [-5, 0]$ is the ratio of the temperature and salinity inductions. $\omega = 0$ means the salinity changes play a main role, whereas $\omega = -5$ indicates that temperature changes play a main role. Other parameters include $A_T =$

1.863×10^{-2}, $A_S = 1.9 \times 10^{-4}$, $A_{TS} = 9.41 \times 10^{-3}$, and $\delta = 8.284(\kappa\eta)^{4/3} + 12.978(\kappa\eta)^2$.

Equation (7.5) represents the power spectrum $\phi_n(\kappa_x, \kappa_y)$ of the turbulence of the ocean and is expressed as follows:

$$\phi_n(\kappa_x, \kappa_y) = 0.388 \times 10^{-8} \mu_x \mu_y \varepsilon^{-1/3} \left[(\mu_x\kappa_x)^2 + (\mu_y\kappa_y)^2\right]^{-11/6}$$
$$\left\{1 + 2.35\left[(\mu_x\kappa_x)^2 + (\mu_y\kappa_y)^2\right]^{1/3} \eta^{2/3}\right\}(\chi_T/\omega^2)$$
$$\times \left[\omega^2 \exp(-A_T\delta) + \exp(-A_S\delta) - 2\omega \exp(-A_{TS}\delta)\right], \quad (7.5)$$

where μ_x and μ_y are the anisotropy factors in the x and y directions, respectively. The meanings of parameters ε, η, χ_T and ω are as explained above, and the expression of δ is

$$\delta = 8.284\eta^{4/3}\left[(\mu_x\kappa_x)^2 + (\mu_y\kappa_y)^2\right]^{2/3} + 12.978\eta^2\left[(\mu_x\kappa_x)^2 + (\mu_y\kappa_y)^2\right]. \quad (7.6)$$

7.1.2 Decay Characteristics of Underwater Particles

Seawater is classified according to the turbidity of the water body to facilitate the study of information transmission in seawater. It can be divided into the following four categories:

Class I is pure sea water, that is, seawater in the depths of the ocean; the attenuation coefficient is generally 0.056 m^{-1}.

Class II is clear seawater, that is, seawater on the ocean surface. Owing to the impact of rainfall and marine-biological activity, there are numerous particles containing chlorophyll (floating plants) and decomposing organic matter; the attenuation coefficient is generally 0.151 m^{-1}.

Class III is coastal seawater that contains a large number of organic solvents, floating plants, and large suspended particles. The attenuation coefficient is generally 0.398 m^{-1}.

Class IV is turbid port seawater. Owing to the influence of artificial operations, the suspended particle and organic solvent content are extremely high. This causes serious light attenuation; the attenuation coefficient is generally 2.17 m^{-1}.

7.1.2.1 Absorption Characteristics

During light transmission in seawater, the light is absorbed by water molecules, plankton, and algae in the seawater. This phenomenon is known as the light absorption characteristic of seawater. Factors affecting the seawater absorption characteristics can be divided into four categories: seawater bodies, particles containing chlorophyll (floating plants), suspended particles in seawater, and colored dissolved organic matter. Therefore, the absorption attenuation coefficient of seawater can be expressed as [2]

$$\alpha(\lambda) = \alpha_{\mathrm{w}}(\lambda) + \alpha_{\mathrm{chl}}(\lambda) + \alpha_{\mathrm{nap}}(\lambda) + \alpha_{\mathrm{y}}(\lambda), \tag{7.7}$$

where $\alpha_{\mathrm{w}}(\lambda)$ represents the light absorption attenuation coefficient of pure water, $\alpha_{\mathrm{chl}}(\lambda)$ represents the light absorption attenuation coefficient of particles containing chlorophyll (phytoplankton), $\alpha_{\mathrm{nap}}(\lambda)$ indicates the light absorption attenuation coefficient of suspended particles, and $\alpha_{\mathrm{y}}(\lambda)$ indicates the light absorption attenuation coefficient of yellow substances. In the same sea area, the absorption characteristics of seawater vary with changes in the salinity, depth, and measurement time.

Light Absorption Characteristics of Pure Water

The light in seawater is mainly absorbed by water molecules and dissolved salts. The concentration of inorganic salts in seawater is approximately 3.8%; the salts include NaCl, KCl, $CaCl_2$, etc. Numerous experiments indicate that, compared with water molecules, the effect of dissolved salt in seawater can be ignored because of its small absorption effect on light. Therefore, the light in seawater is primarily absorbed by water molecules.

Morel et al. measured various parameters of various samples of pure seawater using an optical method. First, the light attenuation coefficient of pure water was measured, and then the absorption coefficient of pure water for light was calculated by subtracting the scattering coefficient from the data. The relationship between the absorption coefficient of different wavelengths $\alpha_{\mathrm{w}}(\lambda)$ and the light wavelength λ [3] is shown in Fig. 7.1. When the wavelength of the light waves is between 350 and 550 nm, the light is a blue-green light band. With a gradual increase in the wavelength, the light absorption coefficient of pure water also increases. There is a small shoulder peak at 510 and 600 nm, and the absorption coefficient increases rapidly after this.

Light Absorption Characteristics of Floating Plants

Similar to plants on land, phytoplankton in seawater use chlorophyll for photosynthesis; therefore, their ability to absorb light waves depends on the amount of chlorophyll they contain. Different types of chlorophyll have different light absorption capacities at different wavelengths. The ratio of chlorophyll a in phytoplankton is

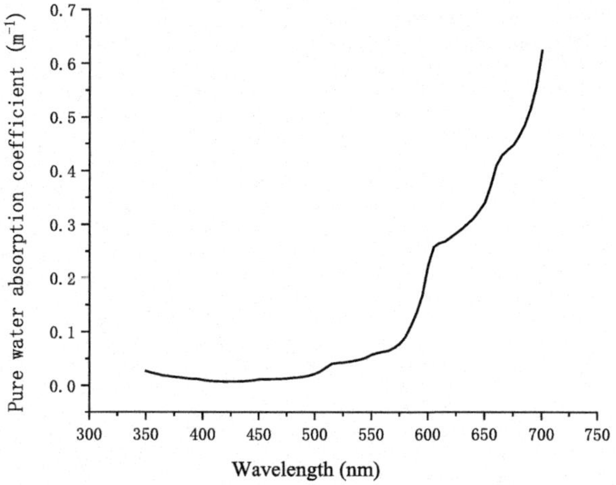

Fig. 7.1 Absorption coefficient of pure seawater for different wavelengths [4]

the highest. Therefore, the light absorption coefficient of phytoplankton can be determined by the concentration of chlorophyll a; the absorption coefficient of chlorophyll a to light waves is [4]

$$\alpha_{chl}(\lambda) = a_c^0(\lambda)\left(\frac{C_c}{C_c^0}\right)^{0.602}, \tag{7.8}$$

where C_c is the concentration of chlorophyll, $C_c{}^0 = 1$ mg/m^3, and $a_c{}^0(\lambda)$ is the absorption coefficient of the chlorophyll ratio in plants.

The absorption coefficient $a_c{}^0(\lambda)$ of chlorophyll ratio of light waves of different wavelengths in literature [5] is substituted into Eq. (7.8). When the chlorophyll concentration is 1 mg/m^3, chlorophyll's absorption coefficient of different light wavelengths is $\alpha_{chl}(\lambda)$. According to the calculation results, the absorption coefficient of floating plants $\alpha_{chl}(\lambda)$ and the long wavelength λ are shown in Fig. 7.2.

When the wavelength of light is between 550 and 650 nm, the absorption coefficient of chlorophyll α of phytoplankton is the smallest, with two peaks at wavelengths of 440 and 670 nm. This suggests that the chlorophyll of phytoplankton has the greatest absorption coefficient for blue-violet and orange-red light.

Absorption Characteristics of Light by Yellow Matter

The yellow substance in seawater refers to the yellow-brown color-soluble organic matter generated by rotten plants. In 1977, Prieur et al. observed through experimental measurements that the attenuation of light during transmission through seawater resulting from the absorption by yellow substances is exponential [5]. In 2002, Wu et al. provided a mathematical model for the absorption by yellow substances when light is transmitted in seawater through experimental measurements [6]:

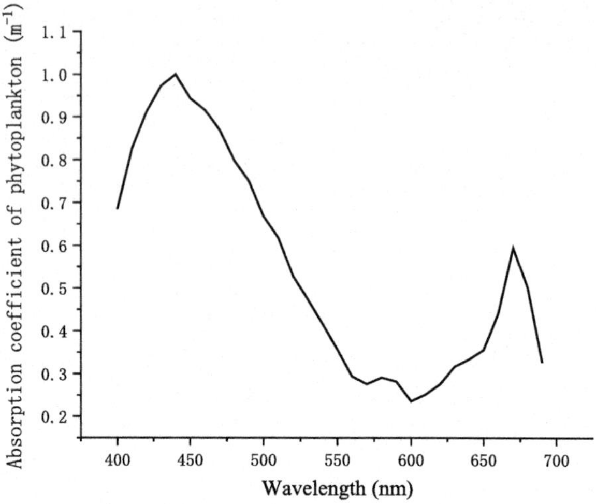

Fig. 7.2 Absorption coefficient of floating plants for different wavelengths [4]

$$\alpha_y(\lambda) = \alpha_y(\lambda_0) \times \exp\left[-S_y \times (\lambda - \lambda_0)\right], \tag{7.9}$$

where $\alpha_y(\lambda_0)$ is a normal number for the reference wavelength when the wavelength is λ_0. S_y is the exponential slope of the light absorption spectrum curve of the yellow material, and changes with the changes in water.

In 1977, Morel et al. determined the index slope parameter of the absorption coefficient curve S_y to be 0.0140 nm^{-1} [5]. In 1981, Bricaud et al. collected 10^5 water samples from different waters to obtain the S_y value range of 0.0100–0.0200 nm^{-1} [7]. In 1995, Pegau et al. measured the S_y of 26 lake-water samples and arrived at a mean value of 0.0170 nm^{-1} [8]. In 2002, Wu et al. determined that samples of waters from areas such as Jiaozhou Bay had an S_y value of 0.0131–0.0180 nm^{-1} [6].

The measured data of Wu et al. [6] were substituted into Eq. (7.9) to calculate the absorption coefficient of the yellow substance to light, where $\lambda_0 = 400$ nm, $\alpha_y(400) = 0.650\ m^{-1}$, and the exponential slope $S_y = 0.0141$. The light-absorption coefficient of the yellow substance with a change in wavelength was calculated through simulation. The relationship between the absorption coefficient $\alpha_y(\lambda)$ of the yellow substance and the optical wavelength λ was plotted, as shown in Fig. 7.3.

As can be seen from Fig. 7.3, in the visible-light band of 350–700 nm, the spectral diagram of the light absorption by yellow substances in seawater causes an exponential change and shows a monotonically decreasing trend with an increase in wavelength.

Light Absorption Properties of Suspended Particles

Suspended particles in seawater refer to suspended substances with particle sizes between several microns and several hundred microns; these are mainly biological excreta, debris, and suspended sediment particles, which also absorb light waves.

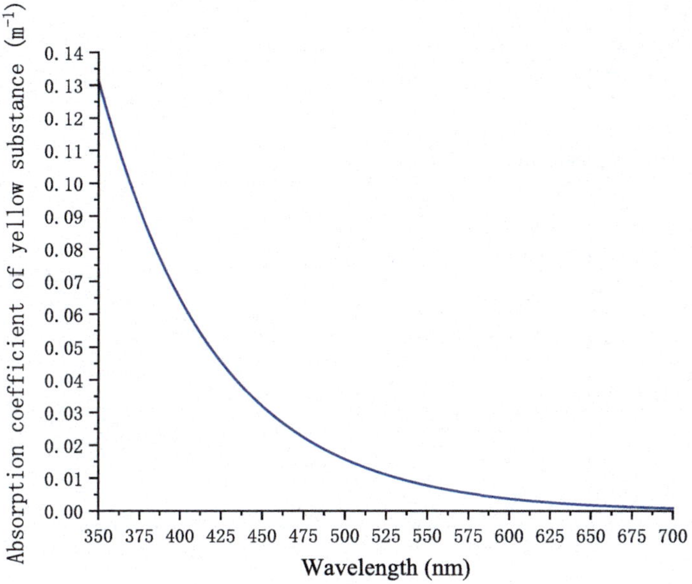

Fig. 7.3 Plot of the absorption coefficient of the yellow substance versus wavelength [6]

Suspended particles, insoluble in nature, are mainly found in coastal areas affected by river discharge and coastal erosion. The total amount of suspended particles in deep-sea areas farther away from the coast is lower because of particle deposition and anthropogenic influences.

Gilerson et al. determined the absorption spectra of suspended particles in seawater using marine optical spectrometry through in situ measurements at 42 stations along the Maryland coast of the Chesapeake Bay and adjacent rivers and harbors. The absorption spectra of suspended particles in seawater were identical to those of yellow matter, both being modeled as exponentially decreasing in magnitude with wavelength. The functional relationship of the transmission of light through seawater affected by the absorption properties of the suspended particles is expressed as [9]

$$\alpha_{nap}(\lambda) = \alpha_{nap}(\lambda_0) \times \exp\left(-S_{nap}(\lambda - \lambda_0)\right) \tag{7.10}$$

$$\alpha_{nap}(\lambda_0) = C_{nap} \times \alpha^*_{nap}(\lambda_0), \tag{7.11}$$

where $\alpha^*_{nap}(\lambda_0)$ is the specific absorption rate of suspended particles at the reference wavelength λ_0, C_{nap} is the concentration of non-algal suspended particles, and S_{nap} is the exponential slope of the spectral curve of light absorption by suspended particles.

Referring to the research data of Gilerson et al. in 2007 [9], the light absorption coefficient of suspended particles was calculated using Eq. (7.10), taking the reference wavelength $\lambda_0 = 400$ nm, $\alpha^*_{nap}(400) = 0.0375$ m^2/g, $C_{nap} = 10$ g/m^3, and the

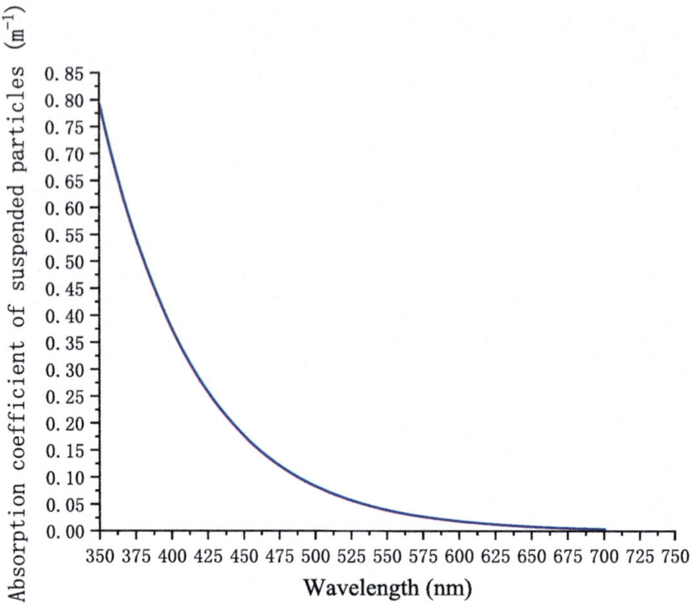

Fig. 7.4 Plot of the absorption coefficients of suspended particles for different wavelengths [6]

exponential slope $S_{nap} = 0.015$. The light absorption coefficient of suspended particles was calculated by simulation. The relationship between the absorption coefficient $\alpha_{nap}(\lambda_0)$ of suspended particles and the wavelength of light λ is plotted as shown in Fig. 7.4.

As can be seen from Fig. 7.4, the spectrogram of light absorption by suspended particles in seawater shows an exponential change in the form of a monotonically decreasing trend with an increase in the wavelength between 350 and 700 nm in the visible-wavelength band. When the wavelength of light is greater than 650 nm, the light absorption of suspended particles almost converges to 0 [9].

7.1.2.2 Scattering Characteristics

The absorption effect of seawater on light near land was investigated. In addition to the loss caused by the absorption properties, the water contained a large number of particles and organic matter, and the seawater attenuation was dominated by scattering. The minimum attenuation window of seawater moved from the blue band (470 nm) to the green band (550 nm).

The scattering of light by water molecules, chlorophyll-containing particles (phytoplankton), and suspended particles of different sizes in seawater also causes energy to be lost during the transmission process. The superposition of the scattering effects of these three types of substances constitutes the scattering effect of seawater on light. Therefore, the attenuation coefficient $\beta(\lambda)$ caused by the scattering of light by the above three types of substances can be expressed as [10]

$$\beta(\lambda) = \beta_w(\lambda) + \beta_{chl}(\lambda) + \beta_{\text{nap}}(\lambda), \tag{7.12}$$

where $\beta_w(\lambda)$ denotes the attenuation coefficient due to the scattering of light by pure seawater, $\beta_{chl}(\lambda)$ denotes the attenuation coefficient due to the scattering of light by plankton, and $\beta_{nap}(\lambda)$ denotes the attenuation coefficient due to the scattering of light by suspended particles. Each of these three types of effects is described below.

Scattering Characteristics of Pure Water

Water molecules in seawater not only have light absorption characteristics but also scattering characteristics that follow the Rayleigh distribution. In 1977, Morel experimentally determined the light scattering model of pure water. In the equation [11], the Rayleigh scattering coefficient $\beta_w(\lambda)$ of pure seawater is

$$\beta_w(\lambda) = 0.005826 \times \left(\frac{400}{\lambda}\right)^{4.322}. \tag{7.13}$$

Substituting the wavelength λ (350–700 nm) of the visible light band into Eq. (7.13), a plot of the scattering coefficient $\beta_w(\lambda)$ of pure seawater and the visible-light band λ (350–700 nm) can be derived, as shown in Fig. 7.5. The scattering coefficient of pure seawater for light, $\beta_w(\lambda)$, shows an exponentially decreasing trend with increase in the wavelength of the light wave.

Light Scattering Properties of Phytoplankton

The light scattering characteristics of phytoplankton are related to their chlorophyll concentration. To simulate the scattering of phytoplankton, Roesler et al. derived the equation for the light scattering coefficient $\beta_{chl}(\lambda)$ of phytoplankton from Mie theoretical calculations and in situ measurements as [11]

$$\beta_{chl}(\lambda) = 0.45\left(\frac{550}{\lambda}\right)C^{0.62}, \tag{7.14}$$

where C is the concentration of phytoplankton chlorophyll. Carina et al. optically modeled the variation in carbon concentration within the core of the particles by varying the real part of their refractive index and obtained the backward-scattering ratio of chlorophyll as [12]

$$\beta_{pc} = 2\pi\,\chi_p(124^\circ)\beta_p(124^\circ), \tag{7.15}$$

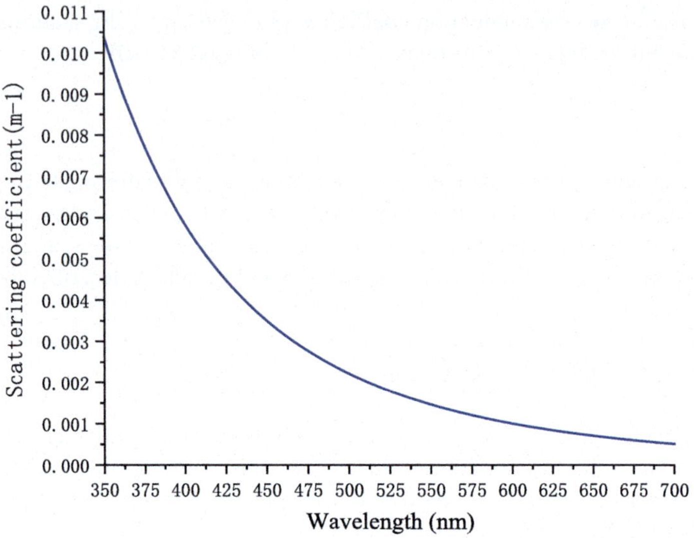

Fig. 7.5 Relationship between the scattering coefficients of pure seawater for different wavelengths [11]

where $\chi_p(124°)$ and $\beta_p(124°)$ are proportionality constants. $\chi_p(124°) = 1.076$ and $\beta_p(124°) = 0.00076$; thus, the backward-scattering coefficient of chlorophyll to light is

$$\beta_{cb}(\lambda) = \beta_{chl}(\lambda) \times \beta_{pc}. \tag{7.16}$$

The forward-scattering coefficient of chlorophyll is

$$\beta_{cf}(\lambda) = \beta_{chl}(\lambda) - \beta_{cb}(\lambda). \tag{7.17}$$

The backward-scattering coefficient $\beta_{cb}(\lambda)$ and forward-scattering coefficient $\beta_{cf}(\lambda)$ of phytoplankton to light were obtained by taking different values of phytoplankton chlorophyll concentration, $C = 0.01$, 0.05, 0.1, 0.5, and 1 mg/m^3, and substituting them into Eqs. (7.16) and (7.17). The backward and forward-scattering coefficients of phytoplankton to light versus the light wavelength λ are plotted as shown in Fig. 7.6.

From Fig. 7.6, it can be seen that the light-scattering coefficient of phytoplankton, $\beta_{chl}(\lambda)$, decreases with increasing wavelength at a constant concentration of chlorophyll. When the light wavelength is kept constant, the scattering coefficient of phytoplankton increases as the concentration of chlorophyll contained in the phytoplankton increases.

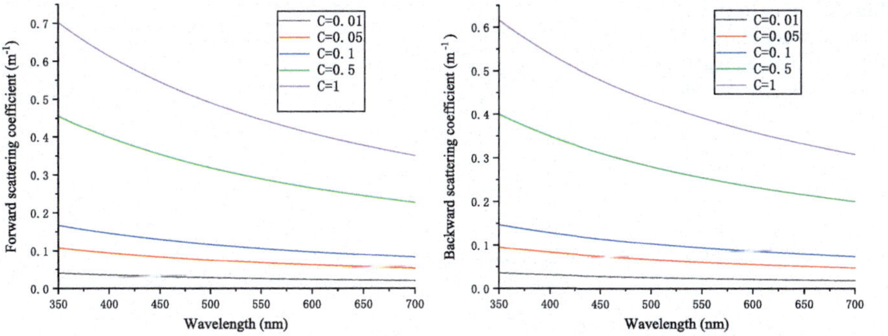

(a) Relationship between the forward-scattering coefficient and light wavelength

(b) Relationship between the backward-scattering coefficient and light wavelength.

Fig. 7.6 Scattering coefficient of phytoplankton as a function of the light wavelength [11–13]

Light Scattering Characteristics of Suspended Particles

Suspended particles in seawater are mainly composed of biological debris, excreta, decomposing matter, and suspended sediment particles. However, because the shape and size of suspended particles in seawater are difficult to determine, to facilitate the calculation of scattering coefficients, the scattering model of suspended particles is simplified by the Mie theory and Rayleigh scattering model. The suspended particles are divided into small and large particles according to their diameters; the equation of the scattering coefficients of suspended particles is [14]

$$\beta_{nap}(\lambda) = \beta^0_{snap}(\lambda)C_s + \beta^0_{lnap}(\lambda)C_l, \tag{7.18}$$

where $\beta^0{}_{\mathrm{snap}}(\lambda)$ is the specific scattering coefficient for small particles in the suspended particles, $\beta^0{}_{\mathrm{lnap}}(\lambda)$ is the specific scattering coefficient for large particles, C_s is the concentration of small particles, and C_l is the concentration of large particles.

Specific Scattering Properties of Light by Small Particles

If the diameter of suspended particles in seawater is less than 5 μm, they are called small particles. The density of small particles is 2 g/cm^3. The specific scattering coefficient of small particles to light, $\beta^0{}_{\mathrm{snap}}(\lambda)$, can be derived from the spectral correlation of suspended small particles of matter as [10]

$$\beta^0_{snap}(\lambda) = 1.151302\left(\frac{400}{\lambda}\right)^{1.7}, \tag{7.19}$$

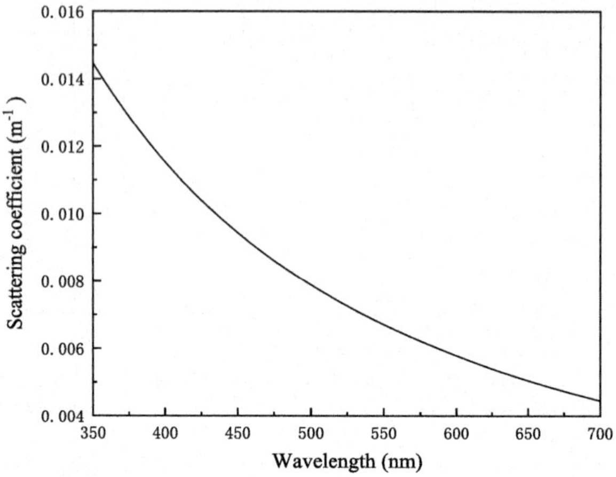

Fig. 7.7 Plot of the scattering coefficient of small suspended particles versus different wavelengths [10]

where the concentration of small suspended particles C_s is mainly influenced by the concentration of chlorophyll C_c in seawater. The expression for this correlation is [10]

$$C_S = 0.01739C\exp\left(0.11631\left(\frac{C}{C_c^0}\right)\right), \tag{7.20}$$

where C is the chlorophyll concentration affecting the suspended particles, and $C_c{}^0$ is the chlorophyll concentration in seawater. Because Eq. (7.26) can be used to calculate the concentration C_s of small particles when the chlorophyll concentration $C_c{}^0 = 1\ \text{mg/m}^3$, the scattering coefficient of small suspended particles to light is obtained by substituting the visible wavelength band λ (350–700 nm) into Eq. (7.20), as shown in Fig. 7.7. From Fig. 7.7, it can be seen that the scattering characteristics of suspended particles decrease with increasing wavelength; however, from the entire visible-light band, the scattering coefficient of small suspended particles changes very little, decreasing only by 0.009 m^{-1}.

Scattering Characteristics of Large Light Particles

Particles suspended in seawater with a diameter greater than 5 μm are called large particles. The density of large suspended particles is 1 g/cm^3 and the refractive index is 1.03. The specific scattering coefficient of large particles to light, $\beta^0{}_{\text{lnap}}(\lambda)$, can be derived from the spectral correlation of suspended particles of large particulate matter as [10]

$$\beta_{lnap}^0(\lambda) = 0.341074\left(\frac{400}{\lambda}\right)^{0.3}, \tag{7.21}$$

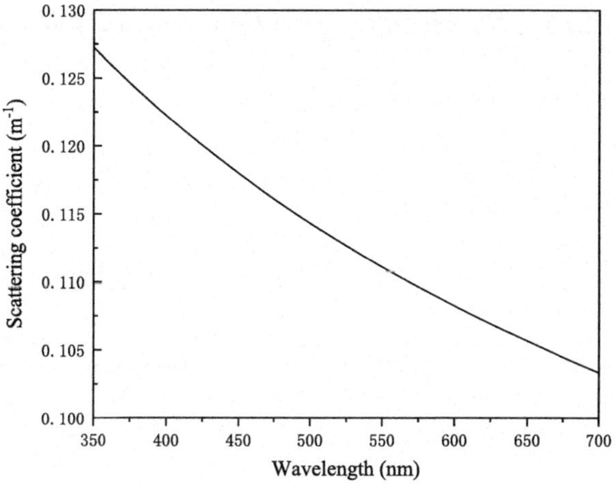

Fig. 7.8 Plot of the scattering coefficient of large suspended particles versus different wavelengths

where the concentration of large suspended particles C_l is mainly influenced by the concentration of chlorophyll in seawater C_c. The expression for this correlation is [10]

$$C_l = 0.76284C \exp\left(0.03092\left(\frac{C}{C_c^0}\right)\right), \tag{7.22}$$

where C is the chlorophyll concentration affecting the suspended particles, and C_c^0 is the chlorophyll concentration in seawater. Equation (7.22) can be calculated from the chlorophyll concentration of $C_c^0 = 1\,\text{mg/m}^3$, when the concentration of large particles is C_l. The visible-light band λ (350–700 nm) is substituted in Eq. (7.22) to obtain the relationship between the scattering coefficient of large suspended particles and the light, as shown in Fig. 7.8.

The scattering characteristics of large suspended particles in seawater decrease with the increasing wavelength of the light wave (Fig. 7.8). Compared with the scattering coefficients of small particles in Fig. 7.8, when the wavelength of the light wave is fixed, the large suspended particles in seawater scatter light significantly more than the small suspended particles.

7.2 Underwater Optical Direct-Vision Transmission

The different scattering coefficients of seawater at different depths change the characteristics of blue-green light transmissions in seawater at different depths. In this section, the Monte Carlo method is used to simulate the scattering transmission of photons in seawater and investigate the normalized received optical power and bit-error rate (BER) of photodetectors for the horizontal, vertical, and slant-range line-of-sight (LOS) transmission of blue-green light in seawater at different depths.

7.2.1 Received Optical Power and BER

7.2.1.1 Received Optical Power

If a photon reaches the photodetector after subscattering, the received optical power of the photodetector is

$$P_s = \frac{Mh\nu P_r}{t}, \tag{7.23}$$

where M is the number of photons emitted by the laser, h is Planck's constant, ν is the light frequency, and t is the unit time. The normalized received optical power of the photodetector is

$$P_N = 10\log_{10}\frac{P_s}{P_T}, \tag{7.24}$$

where P_T is the emitted optical power of the laser. The expression for P_T is

$$P_T = \frac{Mh\nu}{t}. \tag{7.25}$$

Substituting Eqs. (7.23) and (7.25) into Eq. (7.24) yields

$$P_N = 10\log_{10} P_r. \tag{7.26}$$

7.2.1.2 Bit-Error Rate

The output signal-to-noise ratio of the photodetector at the receiving end is

$$SNR = \frac{P_s}{P_n}, \tag{7.27}$$

where P_n denotes the noise power. Assuming that a PIN photodetector is used at the receiving end, the noise consists mainly of thermal, scattering-grain, and dark-current noises [15]. The thermal noise of the PIN photodetector is [15]

$$i_t^2 = \frac{4k_B T_{\mathrm{PIN}} B}{R}, \tag{7.28}$$

where B is the bandwidth of the photodetector, assuming that the bandwidth of the photodetector is twice the information transmission rate R_b, i.e., $B = 2R_b$. R is the load resistance; based on engineering experience, $R = 50\,\Omega$. T_{PIN} is the absolute temperature at which the PIN photodetector operates. The scattering noise of the PIN photodetector is [15]

$$i_{sh}^2 = 2eI_sB. \tag{7.29}$$

The scattering noise generated by background light is neglected in Eq. (7.29), where I_s represents the photocurrent generated by the signal light arriving at the photodetector, with the expression

$$I_s = \frac{e\eta_0}{hv}P_s, \tag{7.30}$$

where e is the electron charge, η_0 is the quantum efficiency and $\eta_0 = 0.6$. The dark-current noise of the PIN photodetector is [14]

$$i_{dark}^2 = 2eI_DB, \tag{7.31}$$

where I_D denotes the dark current. The typical value of the dark current of a Si-based PIN photodetector is 1–10 nA [15]; the typical value of the dark current of a Ge-based PIN photodetector is 50–500 nA; and the typical value of the dark current of an InGaAs-based PIN photodetector is 1–10 nA [15]. $I_D = 10\text{nA}$. The total noise of the PIN photodetector is

$$i_n^2 = i_t^2 + i_{sh}^2 + i_{dark}^2. \tag{7.32}$$

The output signal-to-noise ratio of the PIN photodetector is

$$SNR = \frac{e^2\eta_0^2P_s^2}{h^2v^2i_n^2}. \tag{7.33}$$

Let the optical signal from the laser be modulated by an external modulator to be an on–off keying (OOK) signal, and the bit error rate P_e of the photodetector be

$$P_\mathrm{e} = \frac{1}{2}\mathrm{erfc}\left(\sqrt{\frac{e^2\eta_0^2P_s^2}{4h^2v^2i_n^2}}\right), \tag{7.34}$$

where erfc(·) is the complementary error function.

7.2.2 Simulation Analysis

In this section, the Monte Carlo method is used to simulate the scattering transmission of photons in seawater. The normalized received optical power and BER of the photodetector are simulated to study the transmission of 450-nm blue light and 520-nm green light in pure seawater at depths of 600–700 m in a sea area at 24.5° N, 23.8° W. The seawater has a chlorophyll concentration of 0.25 mg/m^3 for horizontal, vertical, and diagonal-range LOS transmissions.

7.2.2.1 Horizontal LOS Transmissions

The laser and photodetector were placed at a depth of 650 m. The transmitting-elevation angle of the laser and receiving-elevation angle of the photodetector were set to study the normalized received optical power of 450-nm blue light and 520 nm green light transmitted by the photodetector in pure seawater. The normalized-received optical power of the photodetector and simulation results are shown in Fig. 7.9.

When the transmission distance is less than approximately 10 m, the transmission attenuation of the 450-nm blue light in pure seawater is less than that of the 520-nm green light, and in seawater, it is greater than that of the 520-nm green light (Fig. 7.9). When the transmission distance is greater than approximately 10 m, the transmission attenuation of the 450-nm blue light in both pure seawater and seawater is greater than that of the 520-nm green light.

The emitted optical power of the laser is $P_T = 0.1$ W. Using a 20-m horizontal-LOS transmission as an example, the BER at the receiving end is analyzed under different information transmission rates R_b. The results are shown in Fig. 7.10.

When the blue and green lights are transmitted 20 m horizontally in seawater, the BER at the receiving end increases with an increase in the information transmission rate. Moreover, the BER of the 450-nm blue light is larger than that of the 520-nm green light under the same information transmission rate. According to the simulation conditions, under the premise that the BER is less than 10^{-6}, both the 450-nm blue light and 520-nm green light can realize horizontal LOS communication in seawater with a 20-m transmission distance and an information transmission rate of 100 Mbps.

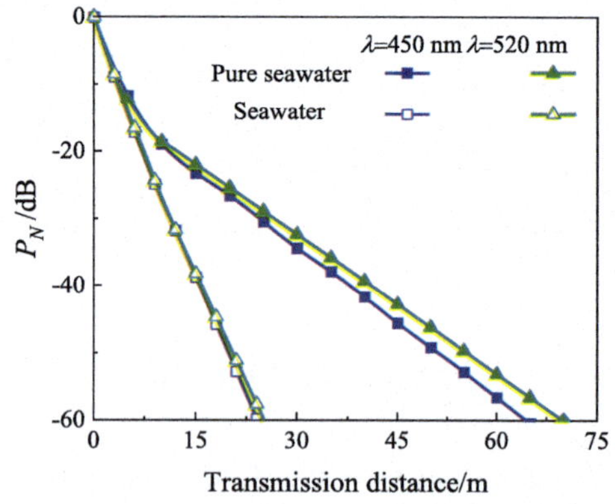

Fig. 7.9 Normalized received optical power of the photodetector for blue-green optical-level LOS transmissions [16]

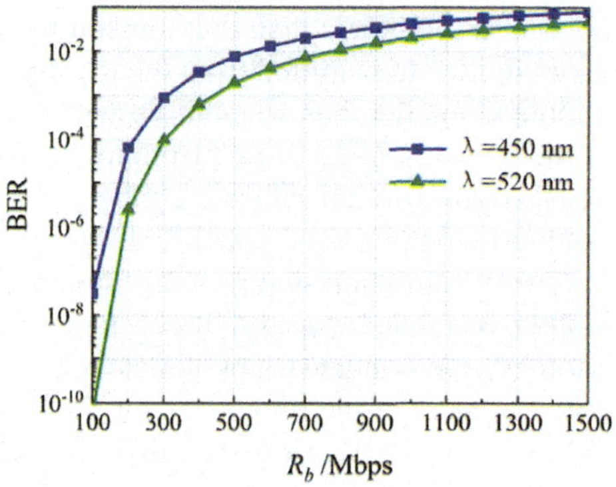

Fig. 7.10 BER of blue-green optical level LOS transmissions at different information transmission rates [16]

7.2.2.2 Vertical LOS Transmissions

According to the simulation flow given in Fig. 7.9, the normalized received optical power of the photodetector for blue-green light was studied in pure seawater and seawater. The chlorophyll concentration is 0.25 mg/m^3 at a depth of 620 m transmitted vertically from the top to the bottom LOS and at a depth of 680 m transmitted vertically from the bottom to the top LOS in a sea area at 24.5° N, 23.8° W. The simulation results are shown in Fig. 7.11.

The attenuation of the 450-nm blue light is less than that of the 520-nm green light in pure seawater for a vertical LOS transmission, when the distance is less than approximately 10 m. When the transmission distance is greater than 10 m, the attenuation of blue light is greater than that of green light. In a seawater vertical LOS transmission, the attenuation of blue light is greater than the attenuation of green light.

When the transmission distance is greater than approximately 10 m, the attenuation of blue-green light in pure seawater and seawater transmitted vertically from

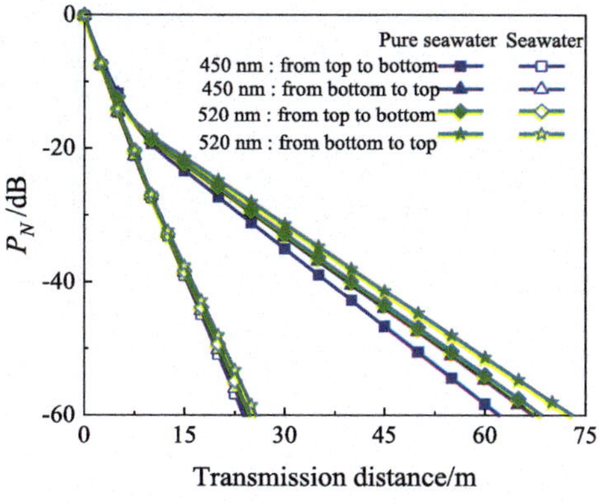

Fig. 7.11 Normalized received optical power of blue-green optical vertical LOS transmissions [16]

bottom to top is less than that transmitted from top to bottom. Because when blue-green light is transmitted from the top down, the initial transmission step size of the photon is smaller than when it is transmitted from the bottom up rendering the value of the transmission $|r_n' - r_r|$ from top to bottom larger than that from bottom to top, i.e., the transmission values p_{1n} and p_{2n} from top to bottom are smaller than that from bottom to top.

The transmission step r_l of the photon given in Eq. (7.34) is considerably smaller than $|r_n' - r_r|$, compared to the distance from the scattering point of the n scattered photons of seawater to the photodetector $|r_n' - r_r|$. Therefore, the normalized received optical power of the photodetector for blue-green light transmitted vertically from the top-down LOS is less than that vertically transmitted from the bottom-up LOS, after single scattering and multiple scattering.

We set the transmit optical power of the laser to $P_T = 0.1$ W. P_T and the normalized received optical power P_N of a vertical LOS transmission photodetector are substituted into Eq. (7.24) to determine the photodetector received optical power P_S. P_S can be introduced into Eq. (7.34) to obtain the BER at the receiving end. Using a vertical LOS transmission of 20 m as an example, we analyzed the BER at the receiving end under different information transmission rates, R_b, and the results are shown in Fig. 7.12.

From Fig. 7.12, it can be observed that the BER at the receiving end of the vertical LOS transmission of blue-green light in seawater increases with an increase in the information transmission rate, and the BER of the vertical bottom-up LOS transmission of blue-green light at the same information transmission rate is smaller than that of the top-down transmission.

Fig. 7.12 BER of blue-green optical vertical LOS transmission at different information transmission rates [16]

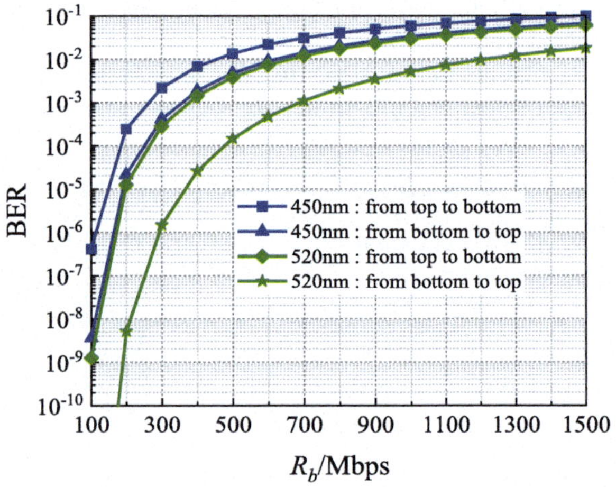

7.2.2.3 Inclined LOS Transmissions

The simulation results for the normalized received optical power of the photodetector for a slanted-range top-down LOS transmission at a depth of 620 m and a slanted-range bottom-up LOS transmission at a 680-m depth are shown in Fig. 7.13, where the angle between the line connecting the laser and photodetector and the horizontal line is 30°. From Fig. 7.13, it can be observed that the attenuation of the 450-nm blue light is less than that of the 520-nm green light when the pure-seawater oblique-range LOS transmission distance is less than approximately 10 m. The attenuation of the blue light is greater than that of the green light when the transmission distance is greater than approximately 10 m.

The attenuation of blue light transmitted in the oblique LOS range in seawater is greater than that of green light. When the transmission distance is greater than approximately 10 m, the attenuation of blue-green light in pure seawater and seawater in an oblique transmission from bottom to top is smaller than that from top to bottom, for the same reason as that of the blue-green light with a vertical LOS transmission.

The transmission optical power of the laser is $P_T = 0.1$ W. P_T and the normalized receiving optical power P_N of the photodetector of the oblique-range LOS transmission are introduced into Eq. (7.50) to obtain the receiving optical power of the photodetector. P_S is introduced into Eq. (7.60) to obtain the BER of the receiving end. Using a 20-m slant-range LOS transmission as an example, we analyzed the BER at the receiving end under different information transmission rates R_b; the results are shown in Fig. 7.14.

From Fig. 7.14, it can be observed that the BER of the blue-green light in the seawater diagonal-range LOS transmission at the receiving end increases with an increase in the information-transmission rate, and the BER of the blue-green light diagonal-range bottom-up LOS transmission at the same information-transmission rate is smaller than that of a top-down transmission.

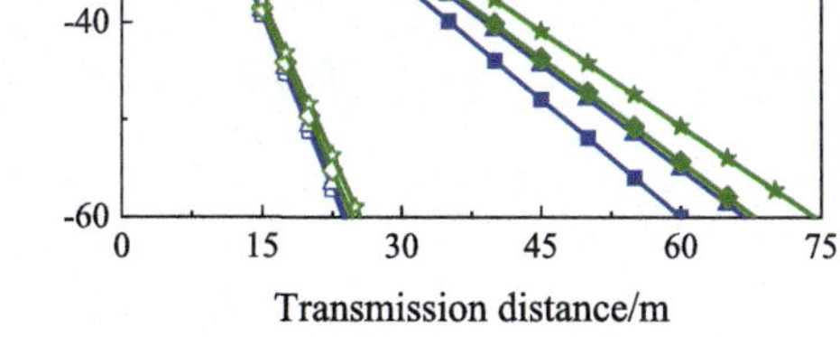

Fig. 7.13 Normalized received optical power of a blue-green oblique-range LOS transmission photodetector [16]

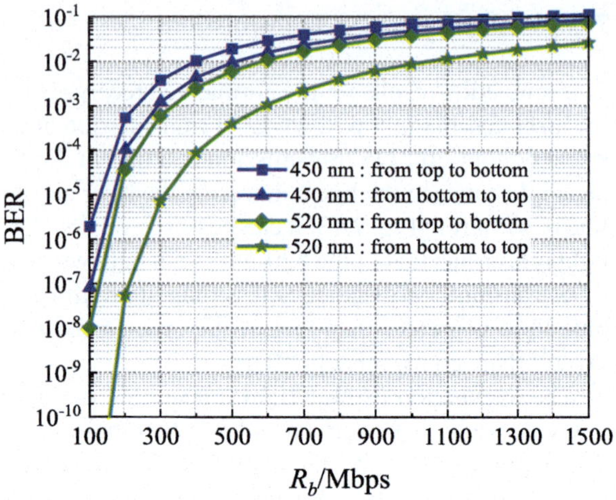

Fig. 7.14 BER of a blue-green oblique-range LOS transmission at different information transmission rates [16]

7.3 Light Transmission in Ocean Turbulence

Light transmission in seawater, in addition to absorption and scattering, is also affected by ocean turbulence, which results in light-intensity scintillation and beam drift. In this section, the scintillation index, beam drift, and BER of Gaussian beams were investigated for transmission in ocean turbulence.

7.3.1 Theoretical Calculation of the Scintillation Index

The scintillation index of a Gaussian beam in weak turbulence is [17]

$$\sigma_I^2(\vec{r}, L) = \sigma_{I,r}^2(\vec{r}, L) + \sigma_{I,l}^2(\vec{r}, L), \tag{7.35}$$

where $\sigma_{I,r}^2(\vec{r}, L)$ is the scintillation index of the radial component of the Gaussian beam and $\sigma_{I,l}^2(\vec{r}, L)$ is the scintillation index of the axial component of the Gaussian beam. The expressions for $\sigma_{I,r}^2(\vec{r}, L)$ and $\sigma_{I,l}^2(\vec{r}, L)$ are, respectively [17],

$$\sigma_{I,r}^2(\vec{r}, L) = 8\pi^2 k^2 L \int_0^1 \int_0^\infty \kappa \Phi_n(\kappa) \exp\left(\frac{-\Lambda L \kappa^2 \xi^2}{k}\right) [I_0(2\Lambda r \xi \kappa) - 1] d\kappa d\xi \tag{7.36}$$

$$\sigma_{I,l}^2(\vec{r}, L) = 8\pi^2k^2L\int_0^1\int_0^\infty \kappa\Phi_n(\kappa)\exp\left(\frac{-\Lambda L\kappa^2\xi^2}{k}\right)\left\{1-\cos\left[\frac{L\kappa^2\xi(1-\bar{\Theta}\xi)}{k}\right]\right\}d\kappa d\xi, \tag{7.37}$$

where $\Lambda = \frac{2L}{kW^2}$ is the receiving-plane Fresnel ratio, W is the spot radius of the receiving plane, L is the transmission distance of the beam, r is the off-axis distance, k is the wavenumber, κ is the spatial wavenumber, $\Phi_n(\kappa)$ is the spatial power-spectral function of the ocean turbulence refractive-index fluctuation, $\overline{\Theta}$ is the normalization parameter of the Gaussian beam, $\xi = 1 - \frac{z}{L}$ is the normalized distance variable, and $I_0(\cdot)$ is the first type of the corrected zero-order Bessel function. For the scintillation index of the radial component of the Gaussian beam $\sigma_{I,r}^2(\vec{r}, L)$, because [18]

$$I_0(2\Lambda r\xi\kappa) - 1 = \sum_{n=1}^{\infty}\frac{(\Lambda r)^{2n}\kappa^{2n}\xi^{2n}}{(n!)^2}, \tag{7.38}$$

$$\sigma_{I,r}^2(\vec{r}, L) = 8\pi^2k^2L\int_0^1\int_0^\infty \kappa\Phi_n(\kappa)\exp\left(\frac{-\Lambda L\kappa^2\xi^2}{k}\right)\sum_{n=1}^{\infty}\frac{(\Lambda r)^{2n}\kappa^{2n}\xi^{2n}}{(n!)^2}d\kappa d\xi. \tag{7.39}$$

Substituting Eq. (7.35) into Eqs. (7.39) and (7.37), respectively, yields [18]

$$\sigma_{I,r}(\vec{r}, L) \approx 1.949\Lambda^2r^2W^2L\varepsilon^{-\frac{1}{3}}\frac{\chi_T}{\omega^2}\big[(3.641\times10^{-6}\omega^2 - 2.323\times10^{-6}\omega + 6.728\times10^{-6})k^2W^2 - (\omega^2 - 5.421\omega + 770.381)L^2\big] \tag{7.40}$$

$$\begin{aligned}\sigma_{I,l}^2(\vec{r}, L) \approx{}& 0.388\times10^{-8}\pi^2\varepsilon^{-\frac{1}{3}}\frac{\chi_T}{\omega^2}\left[4L^3\left(\frac{1}{3}-\frac{\overline{\Theta}}{2}+\frac{\overline{\Theta}^2}{5}\right)(7.245\times10^7\omega^2\right.\\&- 4.184\times10^8\omega + 8.136\times10^{10}) - 8k^{-2}W^{-2}L^5\left(\frac{1}{5}-\frac{\overline{\Theta}}{3}+\frac{\overline{\Theta}^2}{7}\right)\\&\times\left(3.457\times10^{14}\omega^2 - 4.16\times10^{15}\omega + 4.663\times10^{19}\right) + 8k^{-4}W^{-4}L^7\\&\left.\times\left(\frac{1}{7}-\frac{\overline{\Theta}}{4}+\frac{\overline{\Theta}^2}{9}\right)(2.905\times10^{21}\omega^2 - 7.211\times10^{22}\omega + 4.513\times10^{28})\right].\end{aligned} \tag{7.41}$$

Introducing Eqs. (7.40) and (7.41) into Eq. (7.35) yields the scintillation index of a Gaussian beam transmitted through ocean turbulence.

7.3.2 *Simulation Analysis of Scintillation Indices*

In this section, we discuss the simulation of the transmission of a Gaussian beam in ocean turbulence under weak undulations, and the effects of three characteristic parameters of ocean turbulence on the scintillation index at different transmission distances. In the simulation, $r = 0.1$ m, $W = 0.1$ m, and $\overline{\Theta} = 1$.

7.3.2.1 Effect of Temperature-Induced to Salinity-Induced Ratio on Scintillation Indices

When simulating the effect of the temperature-induced to salinity-induced ratio on the scintillation index of a Gaussian beam transmitted through ocean turbulence at different transmission distances, we set $\varepsilon = 10^{-3}\,\mathrm{m}^2/\mathrm{s}^3$ and $\chi_T = 10^{-8}\,\mathrm{K}^2/\mathrm{s}$. The simulation results are shown in Fig. 7.15.

Figure 7.15 shows that the Gaussian beam transmitted through ocean turbulence is influenced by the ratio of temperature-induced to salinity-induced light. The scintillation indices of blue and green light are equal at a transmission distance of 45.5 m. The transmission distance is constant at the same scintillation indices of blue and green light at different ratios of temperature-induced to salinity-induced light.

When the transmission distance is less than 45.5 m, the scintillation index of blue light is greater than that of green light at the same transmission distance; when the transmission distance is greater than 45.5 m, the scintillation index of blue light is less than that of green light at the same transmission distance. It can be concluded that green light can be used to minimize the effect of ocean turbulence on the performance of the communication system when communicating within a close range (transmission distance less than 45.5 m) and blue light can be used to minimize the effect of ocean turbulence on the performance of the communication system when communicating over a long range (transmission distance greater than 45.5 m).

The scintillation index of a Gaussian beam transmitted in ocean turbulence, influenced by the temperature-induced to salinity-induced ratio at the same transmission

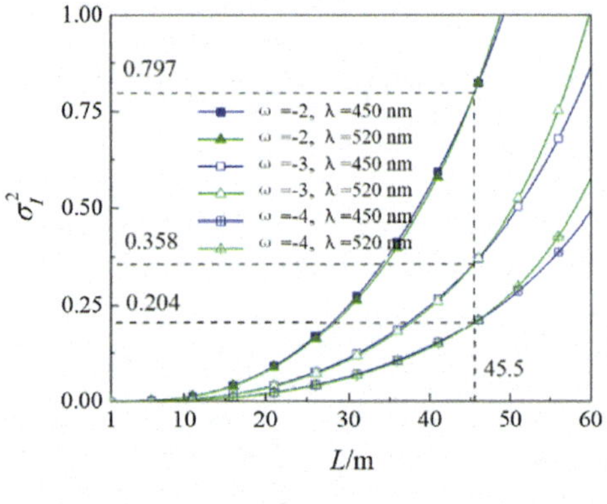

Fig. 7.15 Effect of the temperature-induced to salinity-induced ratio on the scintillation index [16]

distance, increases with an increase in the temperature-induced to salinity-induced ratio, indicating that salinity changes have a larger effect on the transmission of Gaussian beams in ocean turbulence than temperature changes.

7.3.2.2 Influence of the Kinetic-Energy Dissipation Rate on the Scintillation Index

When simulating the effect of the kinetic-energy dissipation rate on the scintillation index of a Gaussian beam transmitted through ocean turbulence, $\omega = -3$ and $\chi_T = 10^{-8}\ \text{K}^2/\text{s}$. The simulation results are shown in Fig. 7.16.

As can be seen in Fig. 7.16, the Gaussian beam transmitted in ocean turbulence is affected by different kinetic-energy dissipation rates, and the scintillation indices of blue and green light are also equal at a transmission distance of 45.5 m, i.e., changes in the kinetic-energy dissipation rate do not affect the transmission distance when the scintillation indices of blue and green light are equal.

Similarly, green light can be used to minimize the effect of ocean turbulence on the performance of a communication system for near-distance (transmission distance less than 45.5 m) communication, whereas blue light can be used to minimize the effect of ocean turbulence on the performance of a communication system for long-distance (transmission distance greater than 45.5 m) communication. When a Gaussian beam transmitted through ocean turbulence is affected by the kinetic-energy dissipation rate at the same transmission distance, the scintillation index decreases with an increase in the kinetic-energy dissipation rate, indicating that the smaller the kinetic energy of seawater, the smaller is the effect on the transmission of the Gaussian beam through ocean turbulence.

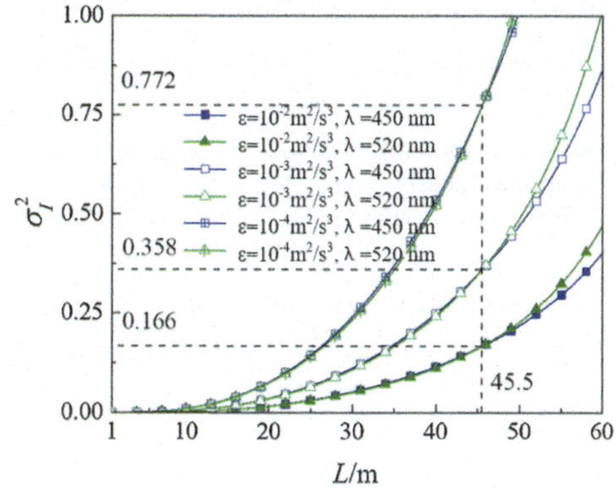

Fig. 7.16 Effect of the kinetic-energy dissipation rate on the scintillation index [16]

7.3.2.3 Influence of the Temperature Dissipation Rate on the Scintillation Index

When simulating the effect of the temperature dissipation rate on the scintillation index of a Gaussian beam transmitted through ocean turbulence, we consider $\omega = -3$ and $\varepsilon = 10^{-3}\ \mathrm{m}^2/\mathrm{s}^3$. The simulation results are shown in Fig. 7.17.

As can be seen in Fig. 7.17, a Gaussian beam transmitted through ocean turbulence is affected by different temperature dissipation rates. The scintillation indices of the blue and green lights are equal at a transmission distance of 45.5 m, i.e., the change in the temperature dissipation rate does not affect the transmission distance when the scintillation indices of the blue and green lights are equal. Similarly, green light can be used to minimize the effect of ocean turbulence on the performance of a communication system within a close range (transmission distance less than 45.5 m) and blue light can be used to minimize the effect of ocean turbulence on the performance of a communication system over a long range (transmission distance greater than 45.5 m).

When a Gaussian beam is transmitted through ocean turbulence, the scintillation index increases with an increase in the temperature difference dissipation rate at the same transmission distance. This indicates that the larger the temperature difference dissipation of seawater, the greater is the effect on the transmission of the Gaussian beam through ocean turbulence.

7.3.3 Beam Drift and BER

Because of ocean turbulence, laser transmissions in seawater are subject to beam drift. Based on the general model proposed by Andrews, the beam drift is expressed as [17]

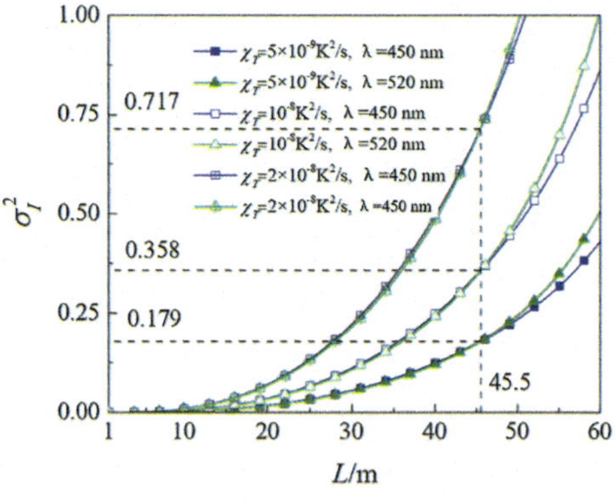

Fig. 7.17 Effect of the temperature dissipation rate on the scintillation index [16]

$$\langle r_c^2\rangle = W^2T_{LS} = 4\pi^2k^2W^2\int_0^L\int_0^\infty \kappa\phi_n(\kappa)H_{LS}(\kappa,z)\left[1-\exp\left(\frac{-\Lambda Lk^2\xi^2}{k}\right)\right]d\kappa dz, \tag{7.42}$$

where $\langle\cdot\rangle$ denotes statistical averaging. $H_{LS}(\kappa, z)$ is a large-scale filter function with the expression [17]

$$H_{LS}(\kappa,z) = \exp\left[-\kappa^2W^2(z)\right] = \exp\left\{-\kappa^2W_0^2\left[\left(\Theta_0+\overline{\Theta}_0\xi\right)^2+\Lambda_0^2(1-\xi)^2\right]\right\}, \tag{7.43}$$

where W_0 is the beam-waist radius of the Gaussian beam, $W(z)$ is the spot radius of a Gaussian beam transmitted to the z-point, Θ_0 is the curvature parameter of the beam at the emitting end, $\Theta_0 = 1-\overline{\Theta}_0$, and Λ_0 is the Fresnel ratio at the emitting end. Using the geometrical-optics approximation yields [17]

$$1-\exp\left(\frac{-\Lambda L\kappa^2\xi^2}{k}\right)\approx\frac{\Lambda L\kappa^2\xi^2}{k},\ \frac{L\kappa^2}{k}\ll 1. \tag{7.44}$$

where $\ll$ symbol means much less than.

For a collimated Gaussian beam $\Theta_0 = 1$, the beam drift of the collimated Gaussian beam can be obtained by substituting Eqs. (7.36), (7.43), and (7.44) into Eq. (7.42):

$$\begin{aligned}\langle r_c^2\rangle = {} & 1.552\times10^{-8}\frac{\pi^2kW^2\Lambda L\chi_T\varepsilon^{-\frac{1}{3}}}{\omega^2}\int_0^L\int_0^\infty \kappa^{-\frac{2}{3}}\xi^2\exp\left(-\kappa^2W_0^2\right)\\ & \times\left[1+2.35(\kappa\eta)^{\frac{2}{3}}\right]\left(\omega^2e^{-A_T\delta}+e^{-A_S\delta}-2\omega e^{-A_{TS}\delta}\right)d\kappa dz.\end{aligned} \tag{7.45}$$

Integrating and organizing Eq. (7.45) yields [16]

$$\begin{aligned}\langle r_c^2\rangle = {} & 0.517\times10^{-8}\pi^2kW^2L\Lambda\varepsilon^{-\frac{1}{3}}\frac{\chi_T}{\omega^2}\times\left\{\omega^2\left[37.9244\eta^4A_T^2\left(W_0^2+12.978A_T\eta^2\right)^{-\frac{11}{6}}\right.\right.\\ & +15.2043\eta^{\frac{8}{3}}A_T^2\left(W_0^2+12.978A_T\eta^2\right)^{-\frac{3}{2}}-9.03014\eta^{\frac{8}{3}}A_T\left(W_0^2+12.978A_T\eta^2\right)^{-\frac{7}{6}}\\ & -4.67544\eta^{\frac{4}{3}}A_T\left(W_0^2+12.978A_T\eta^2\right)^{-\frac{5}{6}}+2.08263\eta^{\frac{4}{3}}\left(W_0^2+12.978A_T\eta^2\right)^{-\frac{1}{2}}\\ & \left.+2.78316\left(W_0^2+12.978A_T\eta^2\right)^{-\frac{1}{6}}\right]-2\omega\left[37.9244\eta^4A_{TS}^2\left(W_0^2+12.978A_{TS}\eta^2\right)^{-\frac{11}{6}}\right.\\ & +15.2043\eta^{\frac{8}{3}}A_{TS}^2\left(W_0^2+12.978A_{TS}\eta^2\right)^{-\frac{3}{2}}-9.03014\eta^{\frac{8}{3}}A_{TS}\left(W_0^2+12.978A_{TS}\eta^2\right)^{-\frac{7}{6}}\\ & -4.67544\eta^{\frac{4}{3}}A_{TS}\left(W_0^2+12.978A_{TS}\eta^2\right)^{-\frac{5}{6}}+2.08263\eta^{\frac{4}{3}}\left(W_0^2+12.978A_{TS}\eta^2\right)^{-\frac{1}{2}}\\ & \left.+2.78316\left(W_0^2+12.978A_{TS}\eta^2\right)^{-\frac{1}{6}}\right]+\left[37.9244\eta^4A_S\left(W_0^2+12.978A_S\eta^2\right)^{-\frac{11}{6}}\right.\end{aligned}$$

$$+ 15.2043\eta^{\frac{8}{3}}A_S^2\left(W_0^2 + 12.978A_S\eta^2\right)^{-\frac{3}{2}} - 9.03014\eta^{\frac{8}{3}}A_S\left(W_0^2 + 12.978A_S\eta^2\right)^{-\frac{7}{6}}$$
$$- 4.67544\eta^{\frac{4}{3}}A_s\left(W_0^2 + 12.978A_s\eta^2\right)^{-\frac{5}{6}} + 2.08263\eta^{\frac{4}{3}}\left(W_0^2 + 12.978A_s\eta^2\right)^{-\frac{1}{2}}$$
$$+ 2.78316\left(W_0^2 + 12.978A_s\eta^2\right)^{-\frac{1}{6}}\Big]\Big\} \tag{7.46}$$

The collimated Gaussian beam is transmitted to point z, and the light-intensity distribution on the beam cross section is

$$I(z, x) = I_0\frac{W_0^2}{W^2(z)}\exp\left[\frac{-2x^2}{W^2(z)}\right], \tag{7.47}$$

where x denotes the beam drift $\langle r_c^2\rangle$.

Assuming that the Gaussian beam emitted by the laser is modulated by an external modulator to be an OOK signal, and assuming that the optical power when transmitting the "0" code is 0, the optical power when transmitting the "1" code is P_1, and the efficiency of the transmitting antenna is η_1, the optical power of the Gaussian beam emitted by antenna P_2 is [19]

$$P_2 = P_1\eta_1 = \eta_1\int_0^{W_0}\int_0^{2\pi} I(0, x)xdxd\varphi. \tag{7.48}$$

Introducing Eq. (7.47) into Eq. (7.48) yields [19]

$$I_0 = \frac{2\eta_1 P_1}{0.865\pi W_0^2}. \tag{7.49}$$

Assuming that the aperture radius of the receiving antenna is $R_2 \ll W(z)$ and the efficiency of the receiving antenna is η_2, it can be assumed that the intensity of the light entering the receiving antenna is $I(z, x)$. Then, the optical power entering the photodetector through the receiving antenna $P_s(z)$ is

$$P_s(z) = \pi\eta_2R_2^2I(z, x) = \pi\eta_2R_2^2I_0\frac{W_0^2}{W^2(z)}\exp\left[\frac{-2x^2}{W^2(z)}\right]. \tag{7.50}$$

The output optical power of the photodetector $P_{ss}(z)$ is

$$P_{ss}(z) = \left(\frac{e\eta_0}{hv}\right)^2P_s^2(z)R. \tag{7.51}$$

Assuming that the photodetector remains a PIN photodetector, the BER at the receiving end of the system is

$$P_{e2} = \frac{1}{2}\text{erfc}\left(\sqrt{\frac{r_{SNR}}{4}}\right) = \frac{1}{2}\text{erfc}\left[\sqrt{\frac{e^2\eta_0^2 P_s^2(z)}{4h^2v^2 i_n^2}}\right]. \tag{7.52}$$

7.3.4 *Beam Drift and BER Analysis*

In this section, we discuss the simulation of a collimated Gaussian-beam transmission through ocean turbulence and the effects of three characteristic parameters of ocean turbulence (temperature-induced to salinity-induced ratio ω, kinetic-energy dissipation rate ε, and temperature-differential dissipation rate χ_T) on the beam drift and BER. For the simulation, $W_0 = 0.05$m, $W = 0.5$m, $\eta = 10^{-3}$m, $\Lambda = 0.5$m, $\eta_0 = 0.6$, $\eta_1 = 0.6$, $\eta_2 = 0.6$, $R_2 = 0.1$ m, $T = 288.15$K, $B = 2$GHz, $P_1 = 0.1$ mW, $R = 50\,\Omega$, and $L = 60$m.

7.3.4.1 Effect of the Temperature-Induced to Salinity-Induced Ratio on the Beam Drift and Bit-Error Rate

When simulating the effect of the temperature-induced to salinity-induced ratio on the beam drift and BER of a collimated Gaussian beam transmitted through ocean turbulence, $\varepsilon = 10^{-6}\,\text{m}^2/\text{s}^3$ and $\chi_T = 10^{-6}\,\text{K}^2/\text{s}$. The simulation results are shown in Fig. 7.18.

As can be seen in Fig. 7.18, the beam drift and BER of collimated Gaussian beams transmitted through ocean turbulence, influenced by the temperature-induced to salinity-induced ratio, increases with the ratio. For the same transmission distance,

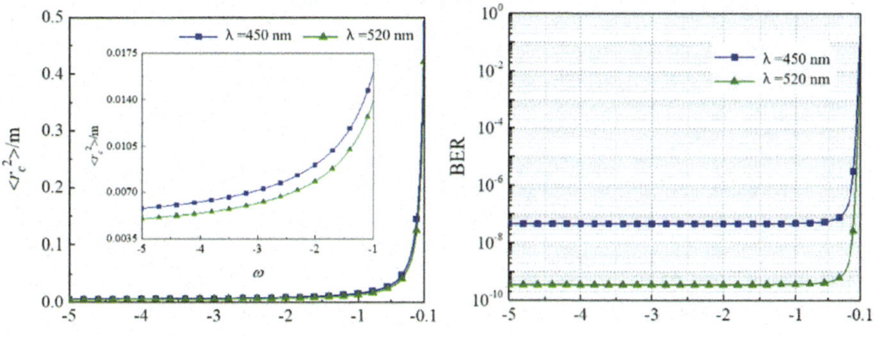

Fig. 7.18 Effect of the temperature-induced to salinity-induced ratio on the beam drift and BER [16]

kinetic-energy dissipation rate, and temperature-dissipation rate, the beam drift and BER of a 450-nm blue light are larger than those of a 520-nm green light, and the beam drift and BER of collimated Gaussian beams are larger for transmissions in ocean turbulence, where salinity induction dominates. It can be concluded that salinity changes have a greater effect than temperature changes on the transmission of collimated Gaussian beams in ocean turbulence.

7.3.4.2 Effect of the Kinetic-Energy Dissipation Rate on the Beam Drift and BER

When simulating the effect of the kinetic-energy dissipation rate on the beam drift and BER of a collimated Gaussian beam transmitted through ocean turbulence, we consider $\omega = -3$ and $\chi_T = 10^{-6}\mathrm{K}^2/\mathrm{s}$. The simulation results are shown in Fig. 7.19.

As shown in Fig. 7.19, the beam drift and BER decrease with increasing kinetic-energy dissipation rate when a collimated Gaussian beam is transmitted through ocean turbulence, influenced by the kinetic-energy dissipation rate. For the same transmission distance, temperature-induced to salinity-induced ratio, and temperature-dissipation rate, the beam drift and BER of a 450-nm blue light are larger than those of a 520-nm green light. Moreover, the beam drift and BER of collimated Gaussian beams transmitted through ocean turbulence with large kinetic-energy dissipation are small, indicating that the smaller kinetic energy of seawater has less effect on the transmission of collimated Gaussian beams through ocean turbulence.

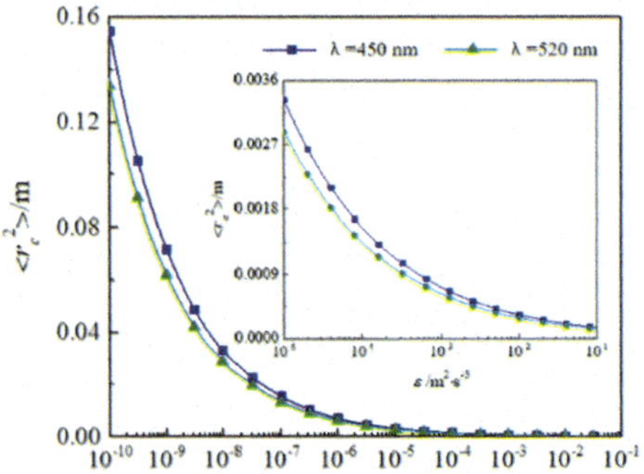

(a) Beam drift

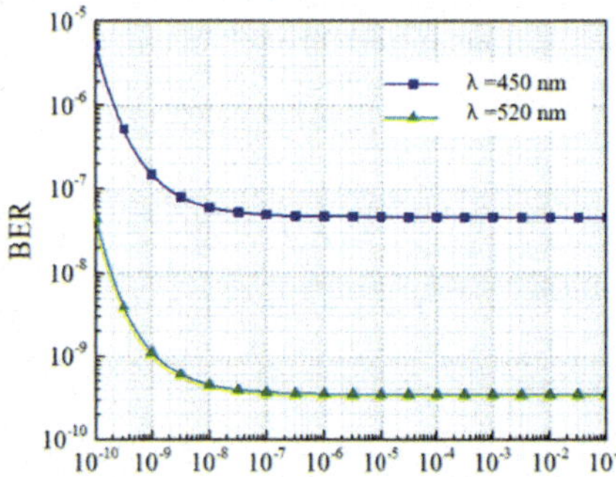

(b) BER

Fig. 7.19 Effect of the kinetic-energy dissipation rate on the beam drift and BER [16]

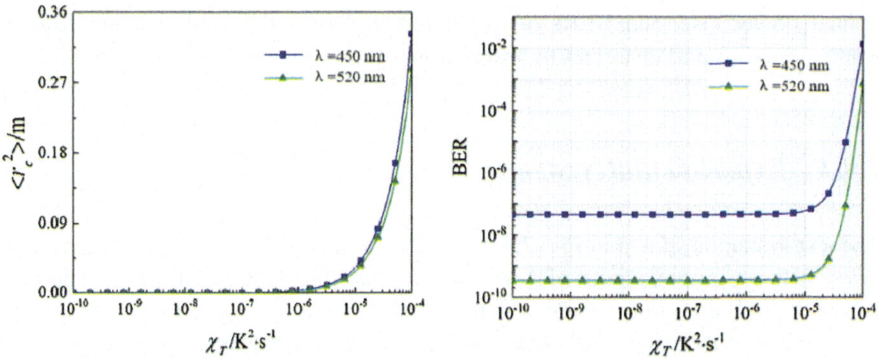

(a) Beam drift (b) BER

Fig. 7.20 Effect of the temperature-dissipation rate on the beam drift and BER [16]

7.3.4.3 Effect of the Temperature-Dissipation Rate on the Beam Drift and BER

When simulating the effect of the temperature-dissipation rate on the beam drift and BER of a collimated Gaussian beam transmitted through ocean turbulence, we consider $\omega = -3\ \varepsilon = 10^{-5}\ \mathrm{m}^2/\mathrm{s}^3$. The simulation results are shown in Fig. 7.20.

As shown in Fig. 7.20, the beam drift and BER increase with the temperature-dissipation rate when the collimated Gaussian beam transmitted through ocean turbulence is influenced by the temperature-dissipation rate. For the same transmission distance, temperature-induced to salinity-induced ratio, and kinetic-energy dissipation rate, the beam drift and BER of the 450-nm blue light were larger than those of the 520-nm green light. Moreover, the beam drift and BER of collimated Gaussian beams transmitted through ocean turbulence with a large temperature-differential dissipation were large, indicating that the larger the temperature-differential dissipation, the greater is the effect on the transmission of collimated Gaussian beams through ocean turbulence.

7.4 Underwater Optical-Wireless Communication Experiment

The effect of underwater turbulence on a laser transmission can significantly decrease the accuracy and transmission distance of laser communications and can even interrupt communication in severe cases. In an experiment, measurements were made in the Chan River and Ba River in Xi'an, Shaanxi Province, and the First Sea Bathing Beach in Yantai, Shaanxi Province, to analyze the skewness and steepness, underwater structural constants, probability-density distributions, scintillation indices, optical power loss, and BERs of the underwater optical samples under different water velocities and distances in different waters.

7.4.1 Experimental Systems

The test system is shown in Fig. 7.21. The transmitting end is a 532-nm-wavelength laser, with an output power of 1 W and an operating voltage of 4.2 V. The laser is transmitted through the underwater channel and then passes through the double-layer compound-eye lens [20]. The curved lens in the second layer of the double-layer compound-eye lens converges and superimposes the converged beams from the lens arrays, to combine the light passing through the various small lenses on the surface of the detector. The detector is then combined with the filter to measure the instantaneous optical signal, which is converted into electrical signals and sent to the computer for data acquisition and signal processing. The detector model was a PD300-UV with a spectral response range of 200–1100 nm, a 10 mm $\times$ 10 mm aperture size, and an accuracy of $\pm$ 1.1%. The filter band used was 513–545 nm, and the aperture 20 mm.

During the experiment, a probe-type electronic thermometer was used to measure the underwater temperature, and the current water-flow rate was averaged using the float method for several measurements.

The experimental system is shown in Fig. 7.21. During the experiment, the transmitter and receiver were aligned, and the laser passed through the double-layer compound-eye lens. After the laser was operating stably, the receiver was used to collect the optical signal using the PD300-UV detector. In addition, the temperature values during the experiment were recorded, and finally, the optical-signal optical power values were analyzed and processed to obtain the undulation characteristics of the optical-signal samples and the optical-power loss under different turbulence intensities.

The research group conducted underwater laser-transmission experiments at multiple locations in the Chan River and Ba River in Xi'an City, Shaanxi Province, and at the First Sea Bathing Beach in Yantai, Shaanxi Province, from November 2022 to June 2023. The link consisted of a transmitter, channel, and receiver, and the parameter measurements of some of the data during the experimental period are listed in Tables 7.1, 7.2, and 7.3. Moreover, to avoid the error caused by wind speed in the measured results, the experimental link was kept 10 cm from the water surface [21].

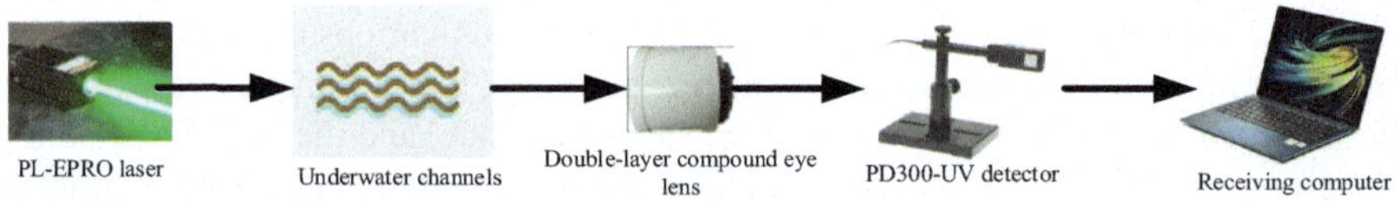

Fig. 7.21 Experimental-system diagram

Table 7.1 Basic experimental conditions at different water velocities

Location	Measurement samples	Date	Link distance (m)	Wind speed (m/s)	Water temperature (°C)	Water velocity (m/s)
Chan River	T1	2023.2.17	0.5	2	5–5.2	0.34
	T2	2023.2.17	0.5	2	5.2–5.4	0.47
	T3	2023.2.17	0.5	2	5.4–5.6	0.53
	T4	2023.2.17	0.5	2	5.6–5.8	0.64
Ba River	T5	2023.3.5	1.5	2	7.9–8.1	0.43
	T6	2023.3.5	1.5	2	8.1–8.3	0.52
	T7	2023.3.5	1.5	2	8.3–8.5	0.61
	T8	2023.3.5	1.5	2	8.5–8.7	0.70

Table 7.2 Basic conditions for different water-depth experiments

Location	Measurement samples	Date	Link distance (m)	Wind speed (m/s)	Water temperature (°C)	Water direction
Chan River	T9	2023.2.19	0.5	2	5–5.2	Smooth
	T10	2023.2.19	0.5	2	5.2–5.4	Adverse
	T11	2023.2.19	0.5	2	5.4–5.6	Horizontal
Ba River	T12	2023.3.6	1.5	1	8–8.2	Smooth
	T13	2023.3.6	1.5	1	8.2–8.4	Adverse
	T14	2023.3.6	1.5	1	8.4–8.6	Horizontal

Table 7.3 Basic conditions for experiments at different distances

Location	Measurement samples	Date	Wind speed (m/s)	Water temperature (°C)	Link distance (m)
Yantai First Sea Bathing Area	T15	2023.4.2	3.3	8.5–8.8	1
	T16	2023.4.2	3.3	8.8–9.1	3
	T17	2023.4.2	3.3	9.1–9.4	5
	T18	2023.4.2	3.3	9.4-9.7	6.5

For multiple sets of measurements of the same water velocity over the range of temperature variations measured in the experiment, the sample variance of the optical signal remained essentially constant; therefore, the effect of temperature on the results can be ignored for all subsequent experiments.

7.4.2 Statistical Analysis of Data

Moment estimation involves estimating the overall moments using sample moments, where the first-order origin moments of the measured signal sequence X denote the expectation of the measured signal sequence and the second-order center moments denote the variance of the measured signal sequence. The expression for the skewness-statistics data-distribution skewed direction and degree of skewness by the third-order centroid distance to the third power of the sample variance is [22]

$$S_X = \frac{\langle (X - \langle X \rangle)^3 \rangle}{\sigma_X^3}. \tag{7.53}$$

The steepness measures the degree of data clustering at the center and is represented by the fourth-order central moment with variance, [22]

$$K_X = \frac{\langle (X - \langle X \rangle)^4 \rangle}{\sigma_X^4}. \tag{7.54}$$

Figure 7.22 shows the relationship between the skewness and steepness of the measured samples and the variance in different waters, where the horizontal coordinate is the variance of the measured light-signal samples, and the vertical coordinate is their skewness and steepness. Figure 7.22a, b show the statistical distribution of the measured light-signal samples under different water velocities in the Chan and Ba Rivers, respectively, and Fig. 7.22c shows the statistical distribution of the measured light-signal samples at different distances in the Yantai First Sea Bathing Beach.

Figure 7.22a–d show that the variance of the measured optical-signal samples increases with an increase in water velocity, and Fig. 7.22e, f show that the variance of the measured optical-signal samples increases with an increase in distance.

The distribution range of the variance of the light-signal samples was not the same for different waters, resulting in different ranges of increased skewness and steepness. In Fig. 7.22, the variances of the measured optical-signal samples of Chan River, Ba River, and Yantai First Sea Bathing Beach are within the range of 10^{-3}, 10^{-2}, and 10^{-1} orders of magnitude, respectively.

Among the corresponding measured optical-signal samples, the maximum values of skewness and kurtosis of the Chan River optical-signal samples are 1.362 and 3.5, respectively; those of the Ba River optical-signal samples are 1.43 and 3.8, respectively; and those of the First Sea Bathing Beach optical-signal samples are 1.49 and 4.3, respectively.

The maximum values of the skewness and kurtosis of the Chan River light-signal samples are 1.49 and 4.3, respectively, while those of the Ba River light-signal samples are 1.43 and 3.8, respectively, and those of the First Beach light-signal samples are 1.49 and 4.3, respectively. Thus, the Chan River light-signal samples have the smallest skewness and kurtosis, the Ba River light-signal samples have the second highest, and those of the First Beach light-signal samples have the largest.

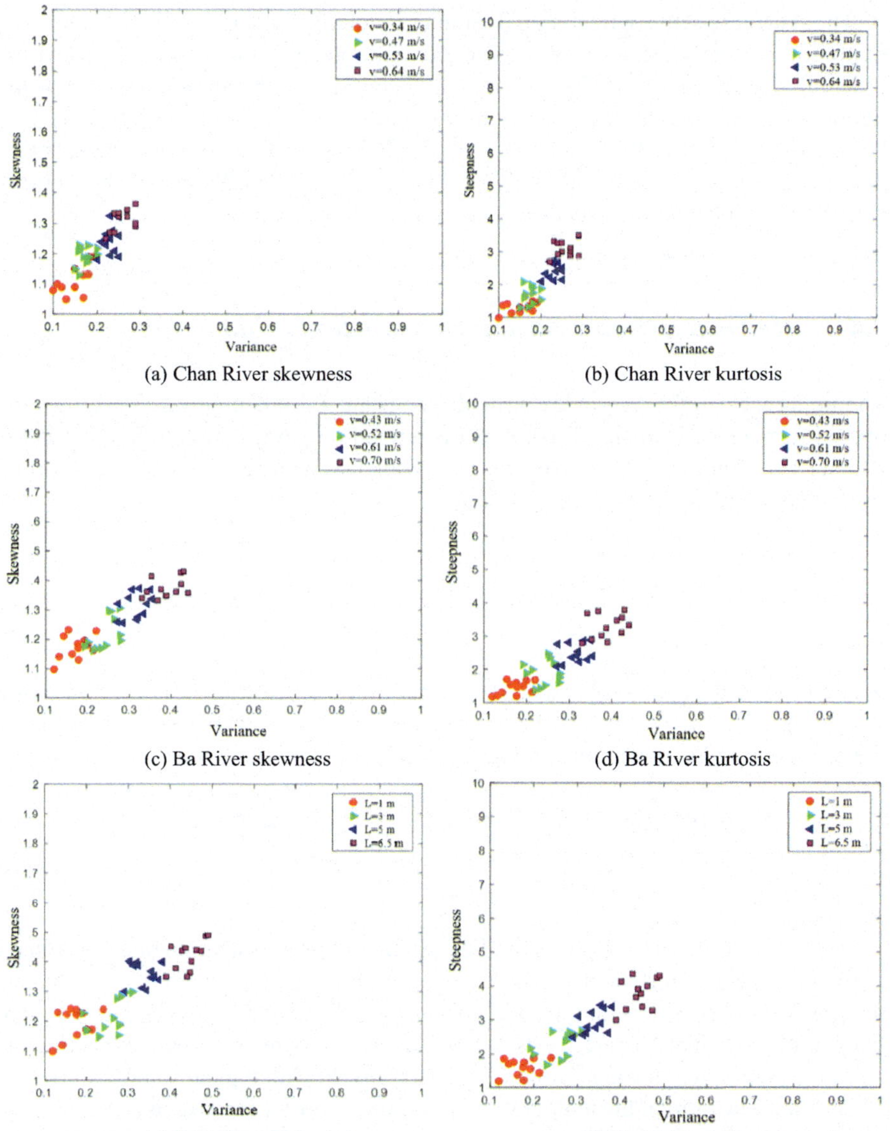

(a) Chan River skewness (b) Chan River kurtosis

(c) Ba River skewness (d) Ba River kurtosis

(e) Yantai Beach No. 1 skewness (f) Yantai Beach No. 1 kurtosis

Fig. 7.22 Degree of skewness and steepness versus variance for samples measured in different watersheds

The skewness of the optical-signal samples gradually increases with their variance, implying that the greater the water velocity and distance, the less symmetrical the probability distribution of the optical-signal sample power values compared to the normal distribution. The steepness of the optical-signal samples gradually increases, but the distribution of the optical signal sample power values became increasingly dispersed, compared to the normal distribution, as the distribution of the optical-signal sample power values became increasingly larger.

7.4.3 Refractive-Index Structural Constants Under Water

The underwater refractive-index structure constant C_{n}^2 quantitatively describes the optical turbulence intensity. Baykal characterized the atmospheric refractive-index structure constant C_{n}^2 in terms of oceanic turbulence parameters using the modified Rytov theory, which is expressed as [23]

$$C_n^2 = 16\pi^2 k^{-7/6} L^{-11/6} \int_0^L \int_0^\infty \kappa \phi_n(\kappa) \cdot \left\{1 - \cos[\kappa^2 \zeta (L - \zeta)/(Lk)]\right\} d\kappa d\zeta, \quad (7.55)$$

where ζ is the distance of the Gaussian beam from the lens mirror at the end of the laser transmitter; $k = 2\pi/\lambda$ is the wavenumber; and λ is the wavelength. $\phi_n(\kappa)$ is the refractive-index power-spectral density function with the following expression:

$$\begin{aligned}\Phi_n(\kappa) &= 0.388 \times 10^{-8} \varepsilon^{-1/3} \kappa^{-11/3} \times \left[1 + 2.35(\kappa\eta)^{2/3}\right] \\ &\times \frac{\chi_T}{\omega^2}\left(\omega^2 e^{-A_T\delta} + e^{-A_S\delta} - 2\omega e^{-A_{TS}\delta}\right),\end{aligned} \quad (7.56)$$

where $\delta = 1.5C_1^2(\kappa\eta)^{\frac{4}{3}} + C_1^3(\kappa\eta)^2$ and the kinetic-energy dissipation rate ε is defined as the rate at which turbulent kinetic energy per unit mass is converted to molecular-thermal kinetic energy, with typical values of $(10^{-10}$–$10^{-1})$ $\mathrm{m^2/s^3}$. The mean-square temperature-dissipation rate χ_T is defined as the amount of turbulence acting on the temperature field through molecular-heat transfer, with typical values of $(10^{-10}$–$10^{-4})$ $\mathrm{K^2/s}$. The ratio of the temperature to salinity contribution is defined as the ratio of ocean turbulence due to temperature and ocean turbulence due to salinity, with typical values of [– 5, 0]. κ indicates the magnitude of the spatial frequency, $\mathrm{A_T} = 1.863 \times 10^{-2}$, $\mathrm{A_S} = 1.9 \times 10^{-4}$, and $\mathrm{A_{TS}} = 9.41 \times 10^{-3}$.

The value change with different water bodies, water velocities, etc., which is measured in this section using the scintillation method [24], has the following expression:

$$C_{\mathrm{n}}^2 = 4.48 C D_t^{7/3} L^{-3} \sigma_x^2, \quad (7.57)$$

where L is the link length, $C = D_r/D_t$, where D_r and D_t are the receiving and transmitting aperture sizes, respectively. σ_x^2 is the logarithmic amplitude variance, which can be expressed as [25]

$$\sigma_x^2 = \frac{1}{4}\ln\left[1 + \left(\frac{s}{\langle I\rangle}\right)^2\right], \tag{7.58}$$

where $\langle I\rangle$ is the mean value of the light intensity and s is the standard deviation.

To obtain the C_n^2 change rule under different water speeds and distances in different waters, the received light signals were normalized and divided into 200 groups. We find the mean $\langle I\rangle$ and standard deviation s of 500 light-intensity values for each group; then, we use Eq. (7.58) to obtain σ_x^2 and Eq. (7.57) to obtain C_n^2. The change curve is shown in Fig. 7.23.

To quantitatively describe the pattern of values of C_n^2 taken at different water velocities and distances in different waters, the maximum, minimum, and mean values of C_n^2 were also calculated, as shown in Table 7.4.

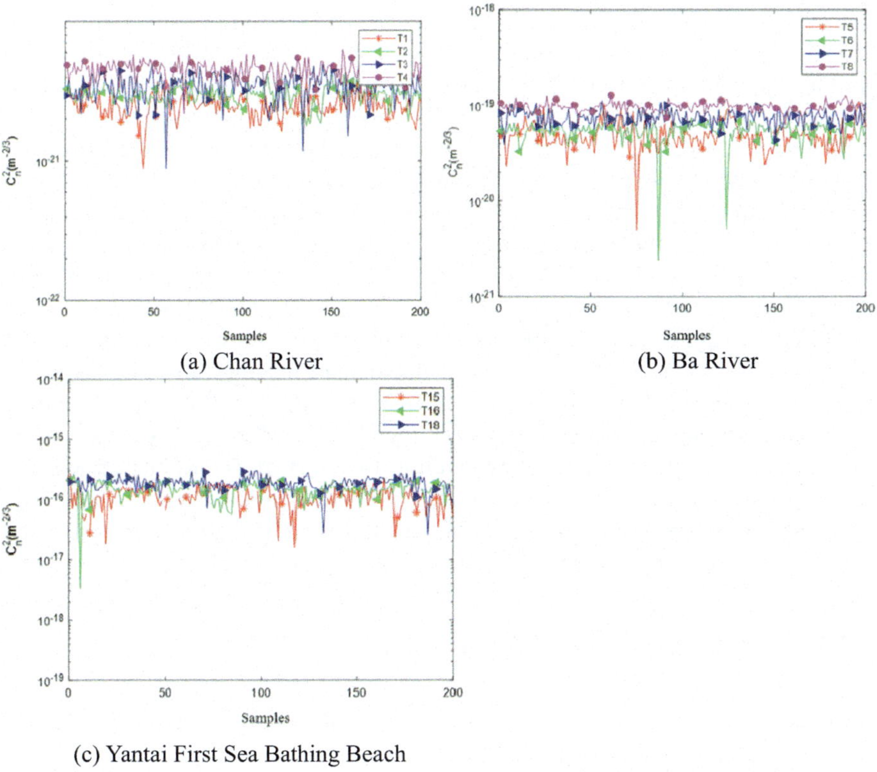

Fig. 7.23 Structural constants of measured samples of optical signals in different waters

Table 7.4 Structural constants of optical-signal samples

Measurement sample	Maximum value ($m^{-2/3}$)	Minimum value ($m^{-2/3}$)	Average value ($m^{-2/3}$)
T1	3.9475×10^{-21}	7.5496×10^{-22}	2.5071×10^{-21}
T2	4.6414×10^{-21}	1.1592×10^{-21}	3.0325×10^{-21}
T3	5.1554×10^{-21}	8.8539×10^{-22}	3.51×10^{-21}
T4	6.7502×10^{-21}	2.1776×10^{-21}	4.5620×10^{-21}
T5	1.0438×10^{-19}	4.9211×10^{-21}	4.9773×10^{-20}
T6	7.3957×10^{-20}	2.3715×10^{-21}	5.5249×10^{-20}
T7	1.2376×10^{-19}	3.9042×10^{-20}	7.5235×10^{-20}
T8	1.5142×10^{-19}	4.3730×10^{-20}	9.5812×10^{-20}
T15	2.3255×10^{-16}	5.5792×10^{-19}	1.0531×10^{-16}
T16	3.1901×10^{-16}	3.4070×10^{-18}	1.4859×10^{-16}
T18	4.7438×10^{-16}	2.6121×10^{-17}	1.8970×10^{-16}

In experiments T1–T4, the Chan River measurement samples C_n^2 increased with increasing water velocity; in experiments T5–T8, the Ba River measurement samples C_n^2 increased with increasing water velocity; and in experiments T15, T16, and T18, the First Sea Bathing Beach measurement samples C_n^2 increased with increasing distance.

Considering Fig. 7.23 and Table 7.4, it can be seen that the C_n^2 of the Chan River spans two orders of magnitude and fluctuates randomly for different water velocity variations. Most of the values were distributed between 1.16×10^{-21} $m^{-2/3}$ and 5.75×10^{-21} $m^{-2/3}$. The Ba River C_n^2 was divided into three orders of magnitude, that is, between 2.3715×10^{-21} $m^{-2/3}$ and 1.5142×10^{-19} $m^{-2/3}$, and fluctuated randomly; most of them were between 5.65×10^{-20} $m^{-2/3}$ and 1.21×10^{-19} $m^{-2/3}$.

The first bathing area had four orders of magnitude, between 5.5792×10^{-19} $m^{-2/3}$ and 4.7438×10^{-16} $m^{-2/3}$, which fluctuated randomly; most of them were distributed between 3.41×10^{-18} $m^{-2/3}$ and 2.93×10^{-16} $m^{-2/3}$. According to the Davis standard, the turbulence intensity of the Chan and Ba rivers is weak, and the turbulence intensity of the First Sea Bathing Beach is moderate; the stronger the turbulence, the more unstable is the signal. The fluctuation of the light-signal sample in the First Sea Bathing Beach was the largest, followed by the Ba and Chan Rivers.

7.4.4 Probability Density Function

The light intensity in underwater turbulence is characterized by random fluctuations. The probability density distribution is the most basic method used to describe the statistical characteristics of real underwater turbulence. Currently, many studies have been conducted on the probability density distribution of light-intensity fluctuations.

It is generally believed that, in a weak-undulation region, the probability density function of the light-intensity undulation follows a lognormal distribution, and its expression is [26]

$$f_I(I) = \frac{1}{I\sigma_{\ln I}\sqrt{2\pi}} \exp\left\{ -\frac{\left[\ln(I) + 0.5\sigma_{\ln I}^2\right]^2}{2\sigma_{\ln I}^2} \right\}, \tag{7.59}$$

where $\sigma_{\ln I}^2$ is the logarithmic light intensity variance of the optical signal. σ_I^2 can be expressed as the scintillation index under weak-turbulence conditions, I represents the light intensity, and $\langle\cdot\rangle$ represents the systematic mean.

The gamma–gamma light-intensity fluctuation probability-distribution model is a two-parameter model that is often used to describe light-intensity fluctuation statistics in the medium and strong turbulence regions [27]. The light intensity at the receiving end can be regarded as a multiplicative modulation process from the large- to the small-scale light-intensity fluctuation. When both the large-scale and small-scale light intensity fluctuations follow the gamma distribution and are statistically independent, the light-intensity fluctuation follows the gamma distribution; in this case, its probability density function is [28]

$$f_I(I) = \frac{2(\alpha\beta)^{(\alpha+\beta)/2}}{\Gamma(\alpha)\Gamma(\beta)} I^{(\alpha+\beta)/2-1} K_{\alpha-\beta}(2\sqrt{\alpha\beta I}), \tag{7.60}$$

where I is the normalized irradiance, α and β are used to describe the outer-scale and inner-scale parameters of the light-intensity scintillation, respectively, $\Gamma(\cdot)$ is the gamma function, and $K_\nu(\cdot)$ is the second type of modified ν-order Bessel function. α and β are obtained from the following equation, when a Gaussian beam is used for transmission and an aperture is used for reception at the receiving end:

$$\alpha = \left[\exp\left(\frac{0.196\sigma_l^2}{(1 + 0.18d^2 + 0.186\sigma_l^{12/5})^{7/6}}\right) - 1\right]^{-1} \tag{7.61}$$

$$\beta = \left[\exp\left(\frac{0.204\sigma_l^2(1 + 0.23\sigma_l^{12/5})^{-5/6}}{1 + 0.9d^2 + 0.207d^2\sigma_l^{12/5}}\right) - 1\right]^{-1}, \tag{7.62}$$

where $d = \sqrt{kD^2/4L}$, $\sigma_l^2 = 1.23C_n^2k^{7/6}L^{11/6}$ is the Rytov variance under strong turbulence. Where $k = 2\pi/\lambda$, λ is the wavelength of light, D is the size of the receiving aperture, and C_{n}^2 is the refractive-index structure constant. Figure 7.24 shows the probability density function (PDF) of the measured samples from different waters.

From Fig. 7.24a, b, it can be seen that the light intensity I on the receiving surface is normalized, and the interval of the value of I increases with the increase of the water velocity. In addition, it deviates to the left, has weak turbulence intensity, and the distribution of P(I) is in line with the characteristics of the lognormal distribution.

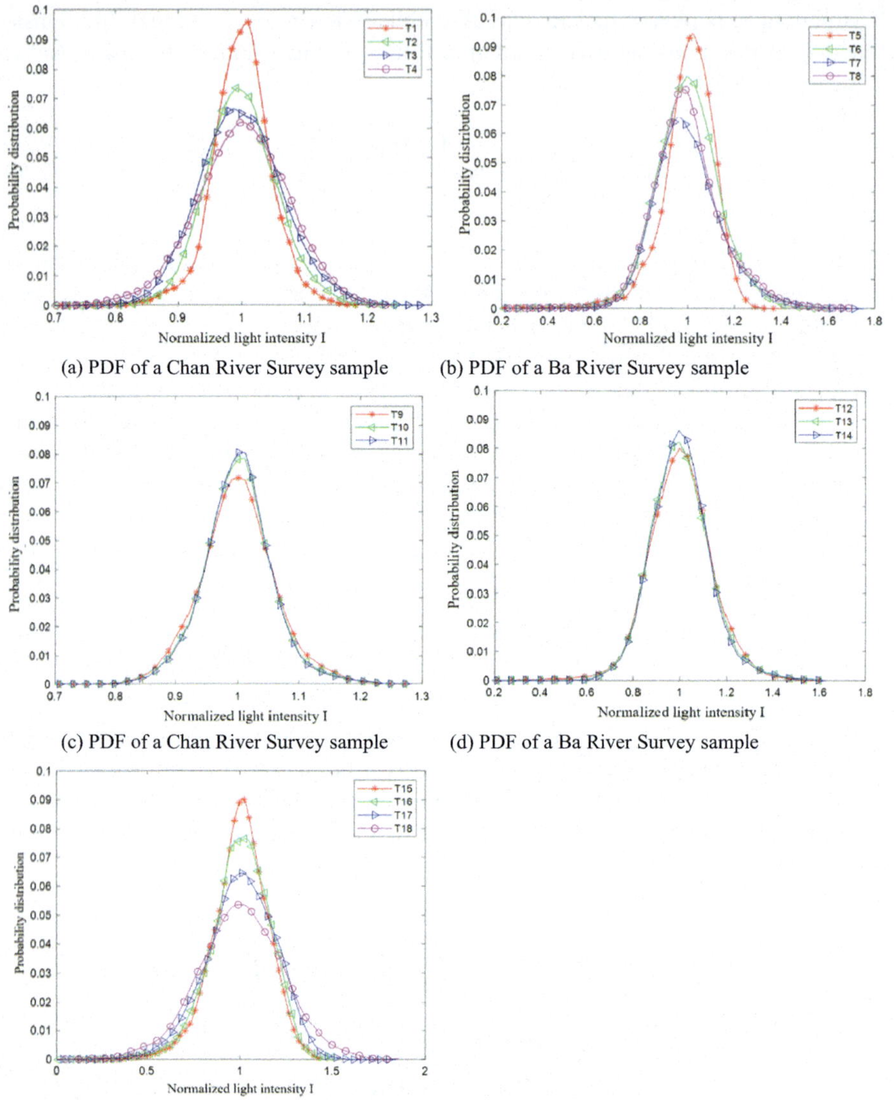

(a) PDF of a Chan River Survey sample

(b) PDF of a Ba River Survey sample

(c) PDF of a Chan River Survey sample

(d) PDF of a Ba River Survey sample

(e) PDF of samples taken at different distances from Yantai No. 1 beach

Fig. 7.24 PDF of measurement samples from different watersheds

It can be seen that the increase in water velocity increases the turbulence intensity, resulting in a greater degree of random fluctuation in the atmospheric refractive index, reducing the stability of the received optical signal.

From Fig. 7.24c, d, it can be seen that the range of values of I under different water directions remains basically the same, the probability density curves overlap, and the water direction has no effect on the change in the probability density function. It can be seen that the size of the turbulence intensity is irrelevant to the water direction when the laser is transmitted horizontally underwater.

From the measurement results at different distances in the Yantai First Sea Bathing Beach, i.e., Fig. 7.24e, it is found that the range of values of I increases with an increase in the communication distance, while deviating to the left. Moreover, the turbulence intensity is moderate, which is in line with the characteristics of the gamma–gamma distribution. Therefore, an increase in distance increases the turbulence intensity, which leads to greater instability in the optical signal received by the entire link.

7.5 Analysis of Experimental Results

Underwater turbulence affects the performance of communication systems and can even cause system interruptions in severe cases. This section further evaluates the performance of an underwater communication system by investigating the scintillation index, optical power loss, and BER of optical signals at different signal-to-noise ratios in different waters at different water speeds and distances.

7.5.1 Scintillation Index

Light-intensity scintillation is usually characterized statistically in time by the scintillation index, expressed as [29]

$$\sigma_I^2 = \frac{\left(\langle I^2 \rangle - \langle I \rangle^2\right)}{\langle I \rangle^2} = \frac{\langle I \rangle^2}{\langle I \rangle^2} - 1, \tag{7.63}$$

where $\langle \cdot \rangle$ denotes the systematic average and I is the beam intensity. The normalized light intensity obtained by processing multiple sets of data measured in the underwater laser-transmission experiment is shown in Fig. 7.25.

Figure 7.25 shows that the normalized light-intensity values in the Chan River fluctuate randomly in the range 0.9–1.1, and those in the Ba River fluctuate randomly in the range 0.8–1.2. The range of light-intensity fluctuations increases with the increase in water velocity. The increase in water velocity leads to a greater intensity of underwater turbulence and also causes a greater degree of flickering of the received optical signal, reducing the stability of the received optical signal.

The normalized light-intensity values of the Yantai First Sea Bathing Beach fluctuate randomly within the range 0.5–1.5, and the fluctuation range increases with the increase in communication distance. The intensity of underwater turbulence increases

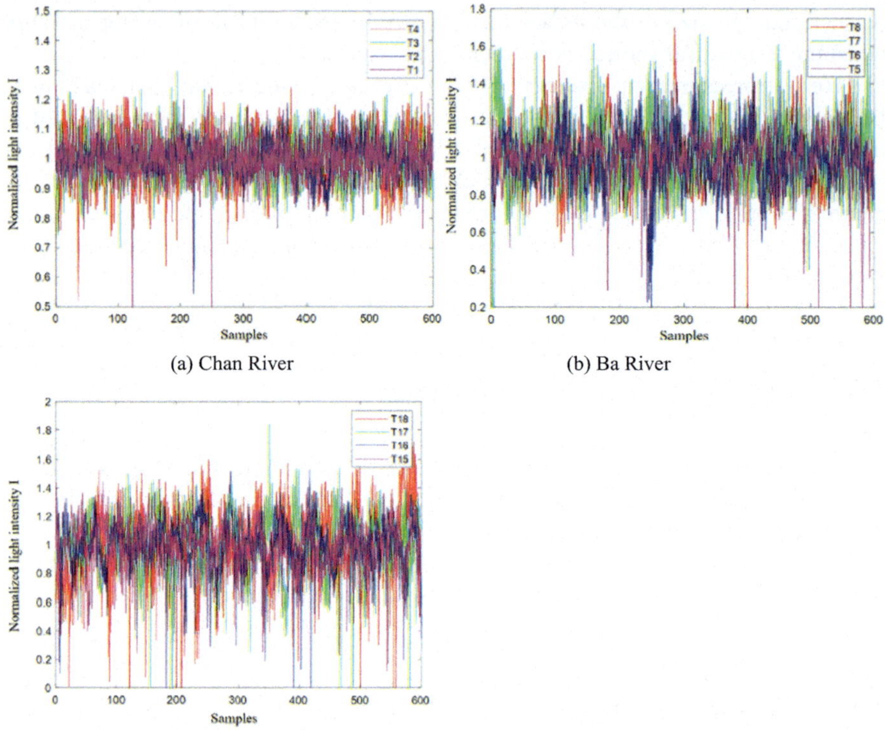

(a) Chan River

(b) Ba River

(c) Yantai Beach No. 1

Fig. 7.25 Light-intensity fluctuation of light-signal samples in different waters

with an increase in the communication distance and causes an increase in the flickering degree of the received optical signals, which makes the received optical signals more unstable.

To further illustrate from a quantitative point of view that the degree of light-intensity fluctuation at high water velocities is greater than that at low water velocities, and that the degree of light-intensity fluctuation at long distances is greater than that at short distances, the scintillation index was calculated from the standard deviation of each sample set.

The scintillation indices of the Chan River under different water velocities from low to high are 0.0032, 0.0041, 0.0050, and 0.0065, which are all in the 10^{-3} order of magnitude. The scintillation indices of the Ba River under different water velocities from low to high are 0.045, 0.052, 0.060, and 0.068, which are all in the 10^{-2} order of magnitude. The quantification of the light-intensity samples from the Chan and Ba Rivers can confirm that the fluctuation of light intensity under high water velocities is greater than that under low water velocities.

The scintillation indices of the light-intensity samples in the Chan and Ba Rivers confirm that the fluctuation of light intensity under a high water velocity is larger than that under a low water velocity, and the light-intensity fluctuation under a high water velocity is larger than that under a low water velocity. The scintillation indices at different distances of the First Sea Bathing Beach in Yantai are 0.27, 0.38, 0.52, and 0.61, which are all in the order of 10^{-1}. The quantized scintillation indices of the optical signal samples from the First Sea Bathing Beach show that the intensity of the underwater turbulence increases with an increase in the communication distance.

7.5.2 Optical Power Loss

Based on Beer's theorem, the change in light power in seawater can be simplified as an exponential attenuation [30]:

$$I(L) = I_0 \exp(-c(\lambda)L), \tag{7.64}$$

where L is the distance that light travels in water, I_0 is the initial light intensity, $I(L)$ is the received light intensity, and $c(\lambda)$ is the attenuation coefficient.

Figure 7.26 shows the power loss due to turbulence under different water conditions. Figure 7.26a, b show the attenuation curves of the received optical power with the scintillation index at different water velocities for the Chan and Ba Rivers, respectively, and Fig. 7.26c shows the attenuation curves of the received optical power with the scintillation index at different distances for the First Sea Bathing Beach. The loss of optical power increases with an increase in the scintillation index.

Comparing the measurements of the Chan River, Ba River, and First Sea Bathing Beach, it can be seen that for different degrees of scintillation index, the degree of optical-signal loss is different, and the larger the scintillation index, the greater is the degree of optical-signal loss. The scintillation indices at different water speeds in the Chan River are all of the order of 10^{-3}, and the optical-signal loss is 1.9 dB. The scintillation indices at different water speeds in the Ba River are all of the order of 10^{-2}, and the optical-signal loss is 5.9 dB. The scintillation indices at different communication distances in the First Sea Bathing Beach are all of the order of 10^{-1}, and the maximum loss of optical signals is 50 dB. A farther distance will increase the loss of the received optical signal.

The maximum allowable transmission distances of the Chan River, Ba River, and First Sea Bathing Beach were 1 m, 1.5 m, and 6.5 m, respectively. The attenuation coefficient of the First Sea Bathing Beach was 2.13, whereas the attenuation coefficients of the Chan and Ba Rivers were considerably greater than 2.13. By comparing the maximum allowable transmission distances of the lasers for different water qualities, we found that the maximum distance of the laser in water was determined by the attenuation coefficient, i.e., turbidity, of the water. A laser can be transmitted across a long distance in clear seawater, whereas the transmission distance is shorter in turbid water. The water quality of the Chan River was the worst, followed by that of the Ba River. The water quality of the Yantai First Sea Bathing Beach was the best, which is consistent with the actual situation.

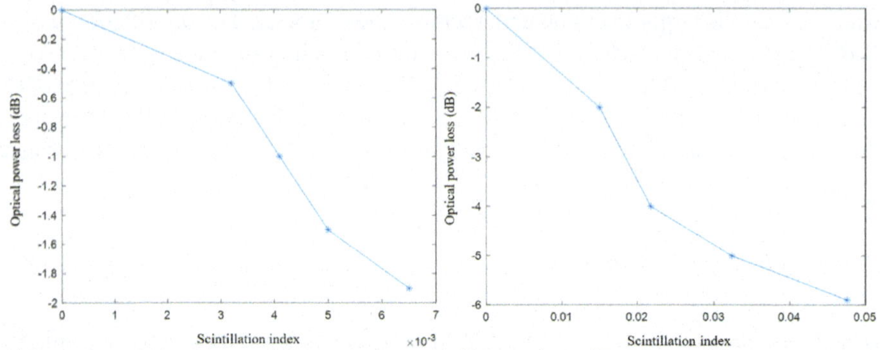

(a) Power loss of Chan River at different water speeds; (b) Power loss of Ba River at different water speeds

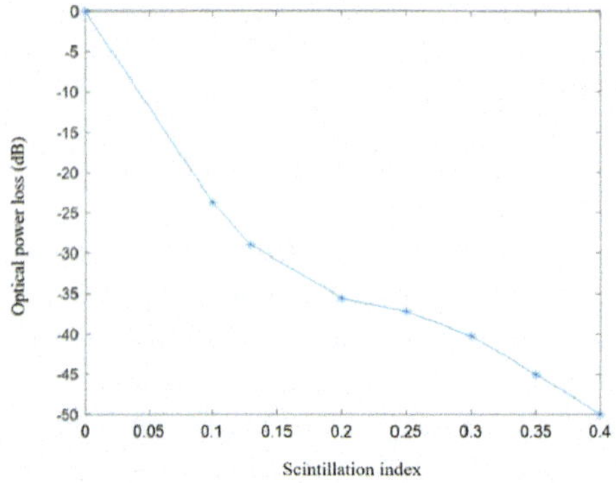

(c) Power loss of the Yantai No. 1 Sea Bathing Beach at different distances

Fig. 7.26 Power loss due to turbulence in different water areas

7.5.3 BER with Different SNR

The probability-density function [31] $f_\mu(\mu)$ of the instantaneous signal-to-noise ratio of the received signal can be expressed as

$$f_\mu(\mu) = \frac{1}{2\mu\sigma_I\sqrt{2\pi}}\exp\left\{-\frac{[\ln(\mu/\overline{\mu}) + \sigma_I^2]^2}{8\sigma_I^2}\right\}. \tag{7.65}$$

Typically, the average BER under fading channels can be calculated by multiplying and summing the instantaneous BER with the probability-density function $f_\mu(\mu)$ of the instantaneous SNR of the received signal [32]:

$$\overline{P_\mathrm{e}} = \int_0^\infty f_\mu(\mu)P_e d\mu, \tag{7.66}$$

where the instantaneous bit-error rate $P_e = \frac{1}{2}erfc\left(\frac{\sqrt{\mu}}{2}\right)$ is expressed by the average bit-error rate formula [33]:

$$\overline{P_e} = \int_0^{\infty} \frac{1}{2\mu\sigma_I\sqrt{2\pi}} \exp\left[-\frac{(\ln(\mu/\overline{\mu}) + \sigma_I^2)^2}{8\sigma_I^2}\right] \times \frac{1}{2}erfc\left(\frac{\sqrt{\mu}}{2}\right)d\mu, \tag{7.67}$$

where μ is the instantaneous signal-to-noise ratio, $\overline{\mu}$ is the average signal-to-noise ratio, and μ_{th} is the critical threshold of the SNR.

By introducing the probability-density function of the gamma–gamma distribution into the above equation, we obtain [34]

$$\overline{P_e} = \frac{2^{\alpha+\beta-3}}{\pi^{3/2}\Gamma(\alpha)\Gamma(\beta)} \times G_{5,2}^{2,4}\left[\frac{4\overline{\mu}}{\alpha^2\beta^2}\Big|_{0,\frac{1}{2}}^{\frac{1-\alpha}{2},\frac{2-\alpha}{2},\frac{1-\beta}{2},\frac{2-\beta}{2},1}\right]. \tag{7.68}$$

Figure 7.27 shows the BER under different S/N ratios for different waters. The BER decreases with increase of the S/N ratio. Meanwhile, from Fig. 7.27a, b, when the signal-to-noise ratio is 10 dB, the BERs corresponding to the four water velocities in the Chan River are 0.0133, 0.0134, 0.0135, and 0.0139, and the BERs corresponding to the four water velocities in the Ba River are 0.0154, 0.0166, 0.0184, and 0.0210, from small to large, respectively. When the signal-to-noise ratio is fixed, the BER of the measurement samples with a high water velocity was larger than that of the measurement samples with a low water velocity.

In Fig. 7.27c, when the signal-to-noise ratio is 10 dB, the BERs corresponding to communication distances of 1, 3, 5, and 6.5 m in the First Sea Bathing Beach are 0.0297, 0.0344, 0.0390, and 0.0449, respectively. This shows that the larger the communication distance, the higher the BER, when the signal-to-noise ratio is constant.

When the BER is 10^{-2}, the corresponding signal-to-noise ratios for different water speeds in the Chan River were 11, 11.05, 11.1, and 11.15, and the corresponding signal-to-noise ratios for different water speeds in the Ba River were 13.5, 14, 14.5, and 15, indicating that, for a certain BER, the larger the water speed, the larger the required signal-to-noise ratio. When the BER is 0.15, the corresponding signal-to-noise ratios for different distances from the First Sea Bathing Beach are 5, 6, 7, and 8, respectively; with a certain BER, the farther the distance, the larger is the required signal-to-noise ratio.

Figure 7.27 shows that the difference in the BER between different water speeds in the Ba River is larger than that in the Chan River. The BER curves corresponding to different water speeds in the Chan River have smaller variations, whereas those in the Ba River have larger variations.

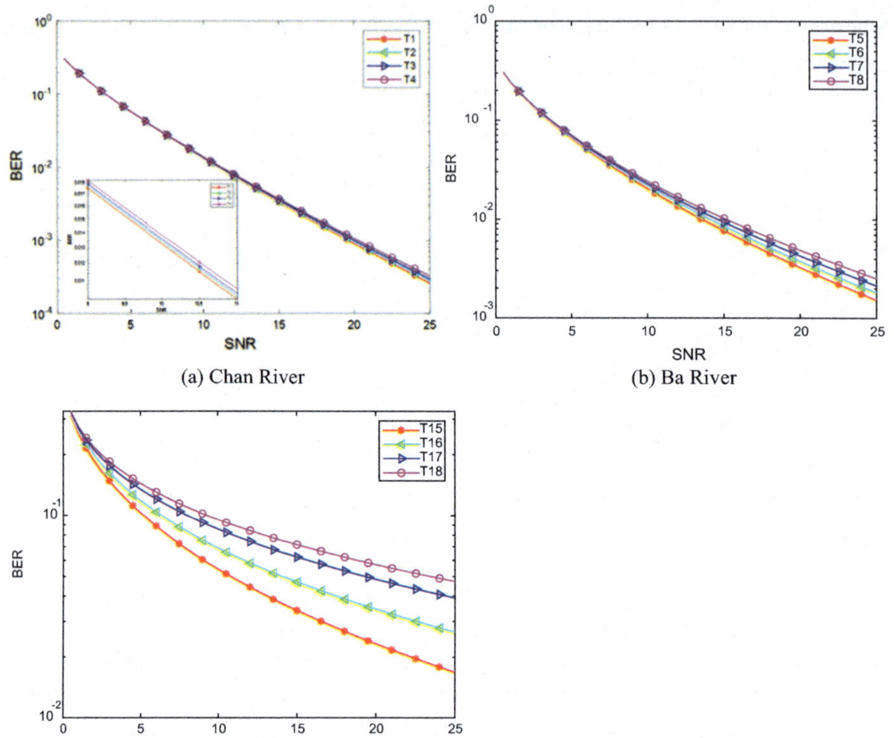

(a) Chan River

(b) Ba River

(c) First Bathing Beach

Fig. 7.27 Bit-error rates at different signal-to-noise ratios for different water areas

7.6 Transmedia Optical Transmission Characterization

At the sea–air interface, laser light is transmitted across the medium from a light-dense to a light-sparse medium, and the random fluctuation of the sea surface leads to the expansion, scintillation, and drift of the spot at the receiving surface, resulting in changes in the communication performance. Underwater transmedia optical communication is not only affected by the absorption and scattering of water in the center of mass of the spot but also by the wind speed at the sea surface. Spot images from different waters were collected, and the spatial intensity distribution of the spot, as well as the offset of the spot center of mass for transmedia transmission, were investigated and analyzed.

7.6.1 Optical Transmedia Transport Modeling

Owing to the influence of wind on the sea surface, the sea surface randomly undulates with time. When the laser is irradiated on the sea surface, the different refractive indices of the atmosphere and seawater medium cause the transmitted laser light to refract. Figure 7.28 shows the refractive reflection of light from the air–sea interface, assuming that the refractive index of the medium in which the incident light is located is n_1, and the refractive index of the medium in which the refracted light is located is n_2.

The relationship between the incidence angle θ_i, reflection angle θ_r, and refraction angle θ_t in the local sea-surface coordinate system satisfies the reflection law and Snell's law

$$\theta_i = \theta_r \tag{7.69}$$

$$\frac{\sin\theta_i}{\sin\theta_t} = \frac{n_1}{n_2}. \tag{7.70}$$

The proportion R of the power of the incident light reflected by the air–sea interface to the power of the incident light is called reflectivity. The reflectivity of vertically polarized light is R_s and the reflectivity of parallel-polarized light is R_p:

$$R_s = \frac{\sin^2(\theta_i - \theta_t)}{\sin^2(\theta_i + \theta_t)} \tag{7.71}$$

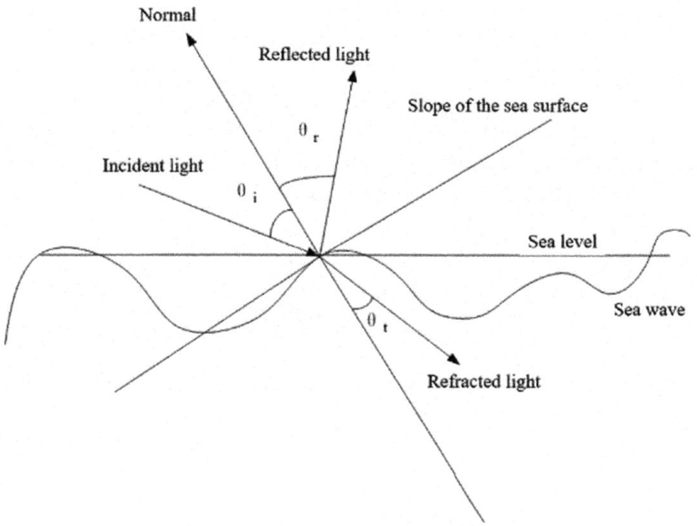

Fig. 7.28 Simplified model of the air–sea interface for light transmission

$$R_p = \frac{\tan^2(\theta_i - \theta_t)}{\tan^2(\theta_i + \theta_t)}. \tag{7.72}$$

The ratio of the power of the refracted light to that of the incident light is called the transmittance T:

$$R = \frac{1}{2}(R_s + R_p) \tag{7.73}$$

$$T = 1 - R. \tag{7.74}$$

Considering the cross-medium light-transmission background, a model of the center shift of the light spot caused by the underwater cross-medium turbulence using a Gaussian function was proposed. The probability-density function is defined as follows:

$$f(I) = A \times \frac{1}{\sqrt{2\pi} \times \sigma} \times e^{\left|-\frac{(x-\mu)^2}{2\times\sigma^2}\right|}, \tag{7.75}$$

where A is the normalized correction coefficient and the values are in the range 0–1. $\int_0^\infty f(I) = 1$, so that μ is the mean value of the spot-center offset in the current turbulent state, and σ is the dithering variance of the spot-center offset in the current turbulent state.

The Gaussian function was used to fit the cumulative-distribution function of the offset of the spot center caused by cross-medium turbulence:

$$F(I) = \int_0^I f(x)dx = \int_0^I A \times \frac{1}{\sqrt{2\pi} \times \sigma} \times e^{\left|-\frac{(x-\mu)^2}{2\times\sigma^2}\right|} dx. \tag{7.76}$$

When the signal-to-noise ratio of the cross-medium optical communication system is less than γ_{th} the communication system is interrupted, so the interrupt probability of the system can be expressed as:

$$P_{out} = F(\gamma_{th}) = \int_0^{\gamma_{th}} A \times \frac{1}{\sqrt{2\pi} \times \sigma} \times e^{\left|-\frac{(x-\mu)^2}{2\times\sigma^2}\right|} dx. \tag{7.77}$$

Therefore, the general expression of the bit-error rate for the different binary modulation modes is

$$P_{BER} = \frac{q^p}{2\Gamma(p)} \int_0^\infty \exp(-qx)x^{p-1}F(x)dx, \tag{7.78}$$

where the different values of parameters p and q represent different modulation methods, and $p = 1$ and $q = 1$ represent differential phase-shift keying.

7.6.2 *Experiments*

7.6.2.1 Spatial Distribution of the Spot Light Intensity

To obtain the law of the light-intensity spatial distribution under different water conditions, several transmedia laser transmissions were carried out in the Chan and Ba Rivers in Xi'an City, Shaanxi Province, and the upper reaches of the Qianyou River in Shangluo City, Zhashui County, Shaanxi Province. The temperature was measured by an electronic thermometer, and the water velocity was measured several times by the floating method to find the average value. Some of the experimental results are selected, and the experimental conditions are listed in Table 7.5.

To obtain the law of the light-intensity spatial distribution in different waters, spot images captured by a CCD camera were normalized by gray values. Typical spot images measured in the experiments are shown in Fig. 7.29.

As shown in Fig. 7.29, the light-spot intensity distribution of the underwater optical-wireless transmedia transmission in different waters follows a Gaussian distribution. The spatial distribution of the light intensity changes with different water qualities. It can be seen from the figure that the light-spot image of the Qianyou River is more aggregated, whereas that of the Chan River is more diffuse.

The distortion and dispersion of light spots ranged from large to small for the Chan, Ba, and Qianyou Rivers, respectively. This corresponds with the water quality of the three rivers in reality, which ranges from poor to good for the Chan, Ba, and Qianyou Rivers, respectively. The worse the water quality, the greater is the degree of absorption and scattering, and consequently, the greater the degree of spot distortion and dispersion.

To further analyze the specific differences in the light-intensity distribution of light spots in different waters, the two-dimensional distribution of light intensity was obtained by projecting a grayscale image along the x/y axis. The results are shown in Fig. 7.30.

As shown in Fig. 7.30, the spatial distribution of the spot intensity undulate randomly and the morphology and degree of energy attenuation of the received spots differ from one water body to another. The spot-intensity distribution in the

Table 7.5 Environmental conditions of the experiments

Water	Date	Water temperature (°C)	Wind speed (m/s)	Water speed (m/s)
Chan River	2023.5.9	20–21.5	1	0.45
Ba River	2023.5.13	21.3–22.9	1	0.47
Qianyou River	203.6.10	19.1–20.3	2	0.5

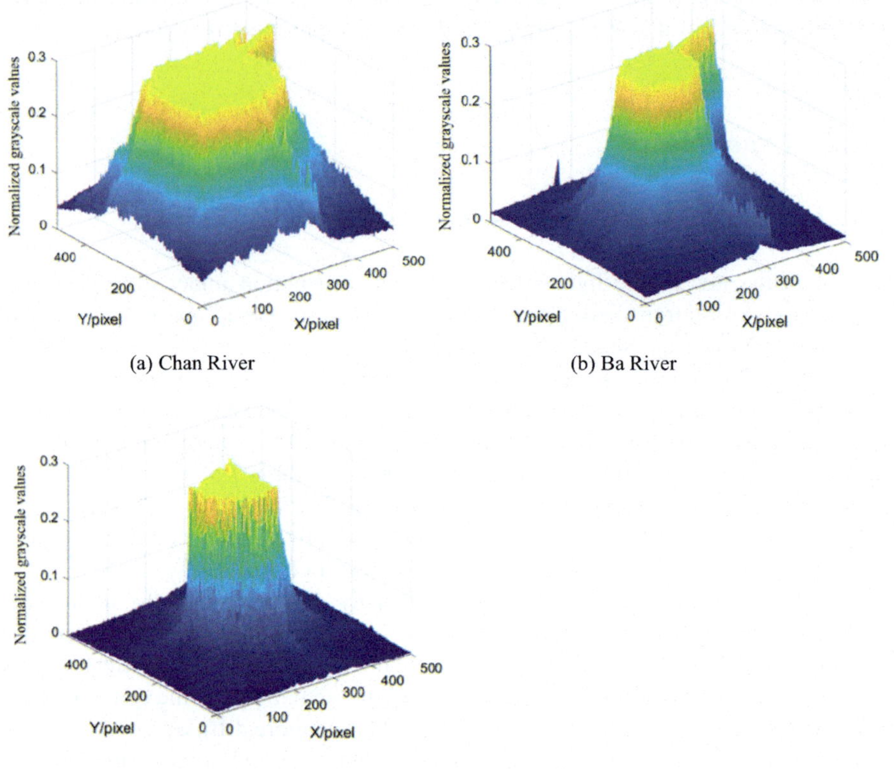

(a) Chan River

(b) Ba River

(c) Qianyou River

Fig. 7.29 Three-dimensional images of light intensities in different waters

Qianyou River was the steepest, followed by the Ba River, and the Chan River is the smoothest. As the water quality deteriorates, the attenuation coefficient of the water quality increases. The photons are absorbed and scattered by particles in seawater more often, the spot area that finally reaches the receiving surface increases, and the spot spread becomes serious. Owing to the increase in water turbidity, the absorption and scattering of photons by water increase, and the spotlight intensity decrease.

The maximum normalized gray values of the received spots in the Qianyou, Ba, and Chan Rivers were 0.2550, 0.2318, and 0.2125, respectively, indicating that the attenuation of the light intensity was the smallest in the Qianyou River, followed by the Ba River, and the largest in the Chan River. On the light-intensity distribution curve, it can be seen that the light-intensity distribution alternates between weak and strong, which is caused by the laser-diffraction phenomenon in the transmission process, resulting in the appearance of a bright and dark halo on the receiving surface.

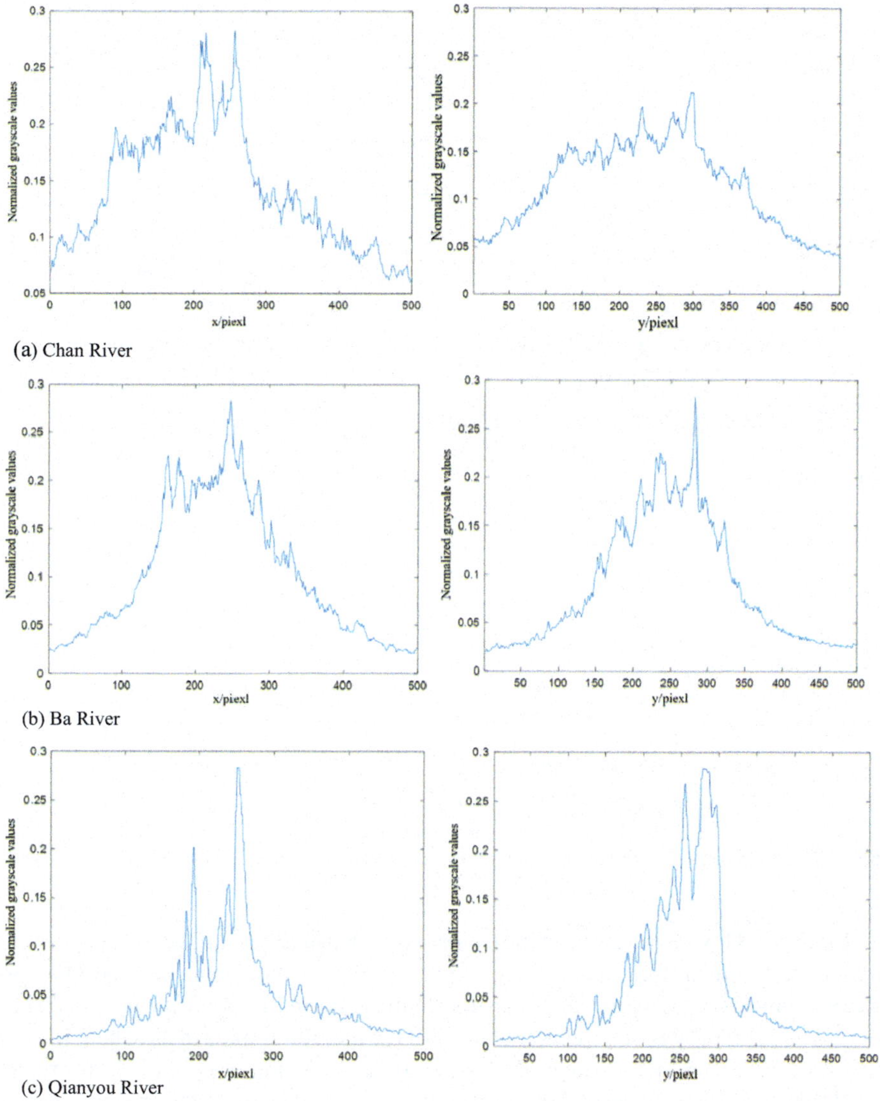

Fig. 7.30 Two-dimensional projection of light intensity for different waters

7.6.2.2 Spot-Center Offset

By taking the center of mass of multiple spot images measured over a long period of time under different water conditions and graphing them, the probability density function of the spot center-of-mass offset under different water conditions can be obtained (Fig. 7.31).

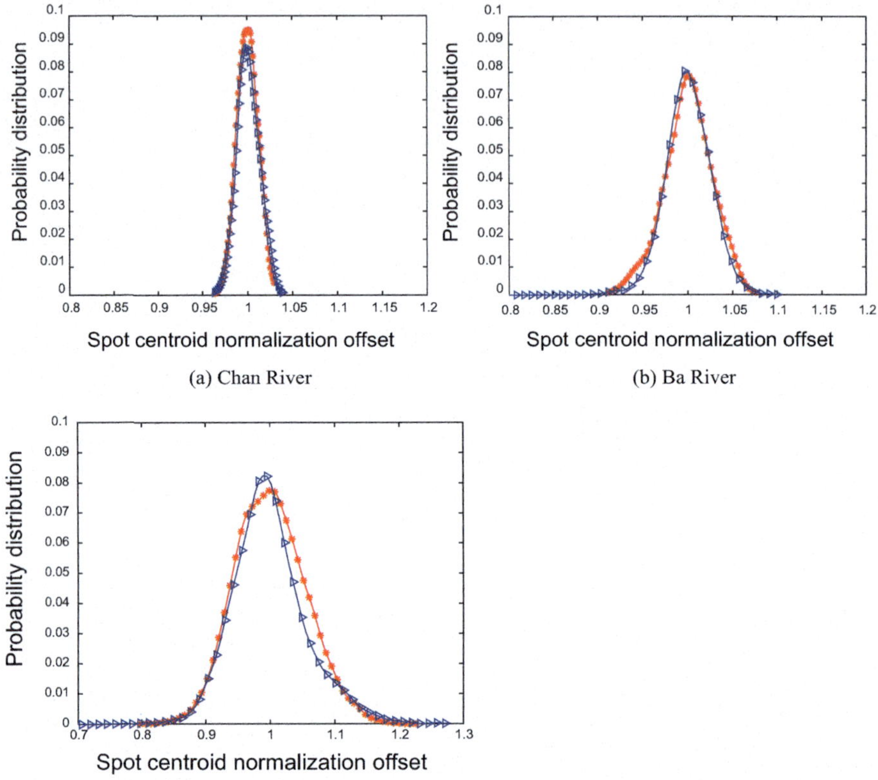

(a) Chan River (b) Ba River

(c) Qianyou River

Fig. 7.31 Probability-density distribution of the centroid offset of the light spot in different waters

Figure 7.31 shows the probability density distributions of the center-of-mass offsets of the measured spots in the upper reaches of the Chan, Ba, and Qianyou Rivers, which all satisfy the Gaussian distribution. From Fig. 7.31, it can be seen that the normalized offsets of the center of mass of the measured spots in the Chan, Ba, and Qianyou Rivers are in the ranges 0.96–1.04, 0.92–1.08, and 0.85–1.15, respectively.

With the intensification of the water-surface wind speed, the water-surface random fluctuation becomes more intense, resulting in the emergence of a larger sea-surface tilt angle. Photons are projected out of the sea surface; the direction is farther away from the center of the receiving surface position, and the light-field distribution is increasingly dispersed. Coupled with the reception of the field-of-view angle limitations, or even falling outside the reception of the field-of-view angle, the number of photons on the surface are reduced. Therefore, they are not included in the statistics of the results. The position of the weight in the center of the receiving surface results in a gradual decline in the weight at the center of the surface.

7.7 Summary

According to the above measurement model, the following experimental results can be obtained:

1. The spatial distribution of the spot light intensity follows the Gaussian distribution in different waters, and the spot image received in low-quality water has a higher degree of aberration and dispersion;
2. The center-of-mass (COM) offsets of transmedia optical communication spots are consistent with the Gaussian function model;
3. The probability-density distributions of the COM offsets are different in different waters.

Under the experimental conditions, the higher the wind speed, the faster is the water speed, and the more severe the water-surface undulation; consequently, the more significant the offsets in the center-of-mass of the spots received in the waters.

References

1. Nikishov VV, Nikichov VI (2000) Spectrim of turbulent fluctuations of the seawater refraction index. Int J Fluid Mech Res 27(1):82–98
2. Smith RC, Baker KS (1981) Optical properties of the clearest natural waters. Appl Opt 20(2):177–184
3. Morel A, Maritorena S (2001) Bio-optical properties of oceanic waters: a reappraisal. J Geophys Res: Oceans 106(4):7163–7180
4. Prieur L, Sathyendranath S (1981) An optical classification of coastal and oceanic waters based on the specific spectral absorption curves of phytoplankton pigments, dissolved organic matter, and other particulate materials 1. Limnol Oceanogr 26(4):671–689
5. Morel A, Prieur L (1977) Analysis of variations in ocean color 1. Limnol Oceanogr 22(4):709–722
6. Wu YS, Zhang SK, Zhang SQ et al (2002) Experimental study on the light absorption properties of sea yellow matter. Oceans Lakes 33(04):402–406
7. Bricaud A, Morel A, Prieur L (1981) Absorption by dissolved organic matter of the sea (yellow substance) in the UV and visible domains. Limnol Oceanogr 26(1):43–53
8. Pegau WS, Cleveland JS, Doss W et al (1995) A comparison of methods for the measurement of the absorption coefficient in natural waters. J Geophys Res: Oceans 100(7):13201–13220
9. Gilerson A, Zhou J, Hlaing S et al (2007) Fluorescence component in the reflectance spectra from coastal waters. Dependence on water composition. Opt Express 15(24):15702–15721
10. Haltrin VI (1999) Chlorophyll-based model of seawater optical properties. Appl Opt 38(33):6826–6832
11. Roesler CS, Boss E (2003) Spectral beam attenuation coefficient retrieved from ocean color inversion. Geophys Res Lett 30(9):1468–1469
12. Poulin C, Zhang X, Yang P et al (2018) Diel variations of the attenuation, backscattering and absorption coefficients of four phytoplankton species and comparison with spherical, coated spherical and hexahedral particle optical models. J Quant Spectrosc Radiat Transfer 217(01):288–304
13. Sun Y (2022) Experimental research on underwater blue-green optical communication system. Xi'an University of Technology, Xi'an

14. Stramski D, Bricaud A, Morel A (2001) Modeling the inherent optical properties of the ocean based on the detailed composition of the planktonic community. Appl Opt 40(18):2929–2945
15. Xu F, Khalighi MA, Bourennane S (2011) Impact of different noise sources on the performance of PIN- and APD-based FSO receivers. In: 11th International conference on telecommunications, Graz, Austria, 2011, pp 211–218
16. Li G (2023) Characterization of blue-green light transmission in the sea. Xi'an University of Technology, Xi'an
17. Andrews LC, Phillips RL (2005) Laser beam propagation through random media. SPIE Press, Bellingham, pp 202–203, 263
18. Lu L (2016) Effect of ocean turbulence on laser beam transmission. University of Science and Technology of China, Anhui Hefei, pp 13, 16–17, 115
19. Kun Yu (2008) BER characterization of wireless laser communication system. Henan Normal University, Zhengzhou, Henan
20. Ke XZ, Yang SJ, Sun Y et al (2022) Underwater blue-green LED communication using a double-layered, curved compound-eye optical system. Opt Express 30(11):18599–18616
21. Li J, Luo J, Yuan X (2021) Impact of water surface wave perturbation on optical-wireless communication. J Opt 41(07):47–54
22. Jiang Y, Chen Z, Tao J et al (2006) Numerical simulation of non-Gaussian stochastic processes with specified power spectral density, skewness and skewness values. J Syst Simul 18(5):1127–1130
23. Baykal Y (2016) Expressing oceanic turbulence parameters by atmospheric turbulence structure constant. Appl Opt 55(6):1228–1231
24. Arnold T (2008) Optical turbulence parameters characterized via optical measurements over a 2.33 km free-space laser path. Opt Express 16(19):14645–14654
25. Wu X, Wang H, Song B et al (2015) Measurement and transmission characterization of light intensity undulation in different environments. Optoelectron-Laser 26(6):1138–1145
26. Pang WN, Wang P, Guo LX et al (2019) Performance investigation of UWOC system with multiuser diversity scheduling schemes in oceanic turbulence channels. Opt Commun 441(1):138–148
27. Ke XZ, Deng LJ (2016) Optical-wireless communications. Science Press, Beijing, pp 107–109
28. He F, Du Y, Zhang J et al (2019) BER study of pulse position modulated optical-wireless communication under anisotropic turbulence in Gamma-Gamma ocean. Phys Lett 68(16):236–244
29. Ji X, Yin H, Jing L et al (2022) Performance analysis of underwater optical-wireless communication system for strong turbulent channel based on strong fluctuation theory. J Opt 42(18):9–16
30. Huang A-P, Zhang Y-L, Tao L-W (2017) Channel characterization of underwater laser communication by Monte Carlo simulation. Infrared Laser Eng 46(04):226–231
31. Zhu XM, Kahn JM (2002) Free-space optical communications through atmospheric turbulence channels. IEEE Trans Commun 50(8):1293–1300
32. Chen Q, Yu X, Cui W et al (2019) Visible light communication based on short- and medium-range interplanetary links and performance analysis. J Opt 39(10):102–112
33. Qin Y, Xie M, Xu Z et al (2022) Performance analysis of free-space optical communication system based on Airy light. Opt Commun Technol 46(04):46–50
34. Fu Y, Duan Q, Zhou L (2020) Performance study of underwater optical-wireless communication system under Gamma-Gamma strong ocean turbulence and aiming error. Infrared Laser Eng 49(02):110–117

Chapter 8
Noise Model of Ultraviolet Optical Communication

In recent years, with the rapid development of ultraviolet (UV) detection technology, ultraviolet optical communication has gradually become a research hotspot because of its inherent advantages, such as low solar-background noise, non-direct vision, strong confidentiality, no acquisition, pointing, and tracking (APT), and good concealment. Owing to its strong scattering properties in the atmosphere, the ultraviolet band makes it ideal for non-line-of-sight (NLOS) communication.

8.1 Theoretical Model of Ultraviolet Background Light

The atmosphere is a layer of gas on the surface of the earth. With changes of altitude and geographical location, the atmosphere is divided into the troposphere, stratosphere, mesosphere, thermosphere, dispersion layer, etc. Air contains aerosols and various molecules, the presence of which affects the transmission of light.

Owing to the absorption of the ozone layer, the intensity of ultraviolet radiation in the 200 nm–280-nm band near the earth's surface is very low; it is called the "solar blind zone" [1]. At present, most studies have focused on the central band of the solar blind zone, with low background noise. To reduce the influence of background noise on communication quality, and achieve more stable high-altitude ultraviolet communication and wider communication bandwidths, it is necessary to study the background noise of ultraviolet light in the solar blind zone.

8.1.1 Theoretical Model of Solar–Terrestrial Radiation

The total solar-radiation calculation models can be divided into three categories: physical, empirical, and machine learning [2]. The empirical model is one of the

X. Ke and C. Ke, *Noise Models in Optical-Wireless Communication Systems*, Optical Wireless Communication Theory and Technology,
https://doi.org/10.1007/978-981-97-7550-7_8

most widely used because of its high precision and low complexity. Angstrom [3] first established the total solar-radiation calculation model, based on the correlation between the sunshine percentage and total solar radiation/sunny radiation. However, because of the complexity of calculating the radiation in a clear sky, Prescott proposed to use astronomical radiation to replace the clear-sky radiation. He established a model based on the Angstrom model, which is also called the A-P model [4].

8.1.1.1 Photon Energy

The sun transmits energy in the form of light throughout the solar system and beyond. The sun's photons carry the energy of the sun, known as quantum light. Its rest mass is zero, it has no electric charge, and its energy is the product of Planck's constant and the electromagnetic radiation frequency, which can be expressed by the following formula [5]:

$$E = \frac{hc}{\lambda} = hf, \tag{8.1}$$

where h is Planck's constant, $h = 6.6260693 \times 10^{-34}$ J s, c is the speed of light, $c = 2.998 \times 10^{8}$ m/s, and f is the frequency of light at different wavelengths. Light composed of high-energy photons has a shorter wavelength and light composed of low-energy photons has a longer wavelength.

Photon energy can be described using joules for the unit, but also can be expressed as electron-volts. Electron-volts represent the energy required to raise an electron by one volt, so a photon has the energy of 1 eV $= 1.602 \times 10^{-19}$ J. The above constant hc can therefore be recharacterized in eV [6]:

$$hc = (1.99 \times 10^{-25}\,\mathrm{J\,m}) \times (1\,\mathrm{eV}/1.602 \times 10^{-19}\,\mathrm{J}) = 1.24 \times 10^{-6}\,\mathrm{eV\,m}. \tag{8.2}$$

The photon-energy equation is transformed into eV and μm, and a common expression is obtained, which represents the relation between photon energy and the wavelength, as shown below:

$$E(\mathrm{eV}) = \frac{1.24}{\lambda(\mu m)}. \tag{8.3}$$

The photon energies vary from band to band.

Table 8.1 gives the frequency, wavelength, and photon energies of the photons that do not pass through the band. As can be seen from the table, as the wavelength decreases, the energy carried by the photon gradually increases; that is, the shorter the wavelength, the greater the energy of the photon. The minimum energy of a microwave is 1.2×10^{-6} eV–1.2×10^{-2} eV, and the maximum energy of a gamma ray is more than 1.2×10^{5} eV.

Table. 8.1 Photon energies for different wavebands [7]

Band	Frequency f/Hz	Wavelength λ/μm	Photon energy hf/eV
Microwave	3×10^8–3×10^{12}	10^6–10^2	1.2×10^{-6}–1.2×10^{-2}
Infrared light	8.8×10^{11}–4.3×10^{14}	3.4×10^2–7×10^{-1}	3.6×10^{-3}–1.7
Visible light	4×10^{14}–7.5×10^{14}	7.6×10^{-1}–4×10^{-1}	1.6–3
Ultraviolet light	7.5×10^{14}–3×10^{16}	4×10^{-1}–10^{-2}	3–120
X-rays	3×10^{16}–3×10^{20}	10^{-2}–10^{-6}	1.2×10^2–1.2×10^6
Gamma rays	$> 3 \times 10^{19}$	$> 10^{-5}$	$> 1.2 \times 10^5$

8.1.1.2 Blackbody Radiation Model

Blackbody radiation is a type of thermal radiation; that is, the radiation spectrum of an object at different temperatures changes in intensity and frequency. Its theoretical origin can be traced back to the early nineteenth century. Planck and Wien proposed the blackbody radiation law, which provided the basis for subsequent studies [8].

It is assumed that the black body is composed of many simple harmonic oscillators with different vibrational frequencies and energy levels, and that each harmonic oscillator can emit or absorb radiation of a certain frequency (energy). If the energy of a harmonic oscillator in an energy state is E, then the number of harmonic oscillators in that state is $n(E)$ [9], according to the Boltzmann distribution function:

$$n(E) = \frac{g(E)}{\exp(\frac{E}{\kappa_B T}) - 1}, \tag{8.4}$$

where $g(E)$ is the number of harmonic oscillators in the black body at each energy level, κ_B is the Boltzmann constant, $\kappa_s = 1.380 \times 10^{-23}$ J/K, and T is the temperature of the black body. At very high temperatures, the $-\ 1$ term in the denominator can be ignored, yielding

$$n(E) \approx \frac{g(E)}{\exp(\frac{E}{\kappa_B T})}. \tag{8.5}$$

Within a certain frequency range, the number of all harmonic oscillators located between energy E and $E + \Delta E$ can emit all the electromagnetic waves with energy ΔE; therefore, all electromagnetic waves at that frequency should have the energy ΔE. Let this frequency range be v to $v + \Delta v$; then, the energy flow through this band in unit time (i.e., the radiation-flux density) can be expressed as

$$I_v(v, T) = \frac{g(E)}{\exp(\frac{E}{\kappa_B T}) - 1} \cdot \frac{hv^2}{\lambda^3}. \tag{8.6}$$

According to electromagnetic-wave theory, converting the frequency v to wavelength λ yields

$$I_{\lambda}(\lambda, T) = \frac{hc^2}{\lambda^5.\exp(\frac{hc}{\lambda \kappa_B T}) - 1}, \tag{8.7}$$

where λ is the wavelength of light, T is the blackbody temperature in K, and h is Planck's constant, $h = 6.626 \times 10^{-34}$ J s. c is the vacuum speed of light; the constant is 3×10^8 m/s.

This formula shows the relationship between the radiant-flux density at each frequency of a black body at different temperatures. It is one of the basic formulas that must be used in the fields of cosmology, astrophysics, and infrared technology.

8.1.1.3 Solar-Radiation Model

Figure 8.1 shows a schematic representation of solar radiation from the sun to the Earth's surface, where β is the ecliptic angle. We set the direction of the vernal equinox to 0° longitude. α is called the yellow meridian of the sun, and is the angle between the current line of the Earth and the Sun to the line of the earth and the vernal equinox. ν_0 is the meridian at perihelion. (X, Y, Z) represents the coordinate system of the equator. It can be transformed to the zodiacal coordinate system (X′, Y′, Z′) according to Eq. (8.8) [10]:

$$\begin{bmatrix} X \\ Y \\ Z \end{bmatrix} = \begin{bmatrix} 1 & 0 & 0 \\ 0 & \cos(\beta) & -\sin(\beta) \\ 0 & \sin(\beta) & \cos(\beta) \end{bmatrix} \begin{bmatrix} X' \\ Y' \\ Z' \end{bmatrix}. \tag{8.8}$$

The position of the sun can be expressed as P_s^y. It can be determined by the celestial longitude α of the day; namely, $P_s^y = \left[r\cos(\alpha)\ r\sin(\alpha)\ 0 \right]^T$. r is the distance between the sun and the earth, which can be expressed as

$$r = A(1 - \frac{e(\cos(\alpha + \nu_0))}{1 + e\cos(\alpha + \nu_0)}), \tag{8.9}$$

where A is the semi-major axis of the Earth's orbit, $A = 1.496 \times 10^{11}$ m and e is the eccentricity of the solar orbit, $e = 0.016722$. The ecliptic α for a given date can be determined by the following formula:

$$\frac{\int_0^{\alpha} \frac{1}{2} r^2 d\alpha}{\pi A \sqrt{1 - e^2}} = \frac{n}{365}, \tag{8.10}$$

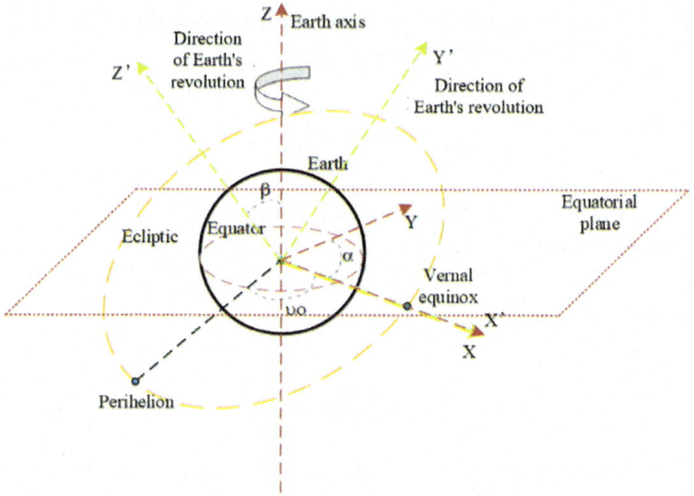

Fig. 8.1 Sun–earth radiation schematic

where n is the cumulative day, that is, the number of days from the vernal equinox. $\int_0^{\alpha} \frac{1}{2} r^2 d\alpha$ is the area swept by the earth–sun line from the vernal equinox to the current date. $\pi A \sqrt{1 - e^2}$ is the area of the ecliptic ellipse.

When there is an angle γ between the solar radiation and the surface horizontal normal vector, as shown in Fig. 8.2, the total surface solar-radiation intensity can be expressed as [11]

$$E_0 = \frac{M_0}{4\pi r^2} \cos \gamma, \tag{8.11}$$

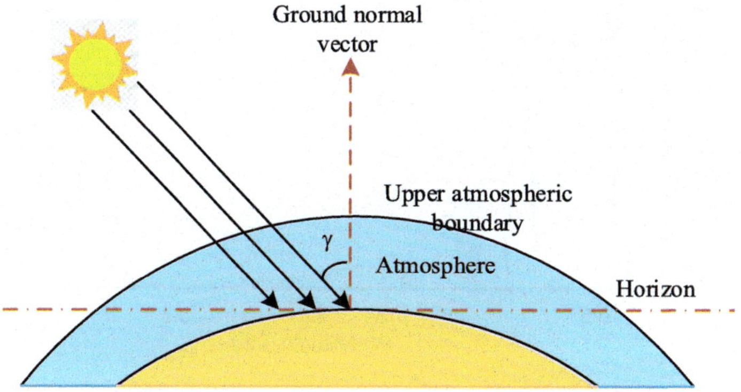

Fig. 8.2 Solar radiation on the ground

where M_0 is the solar external radiant power with a constant of 3.9×10^{26} W. $\cos\gamma$ can be determined by the following equation:

$$\cos\gamma = \frac{P_s^r n_g^r}{\left\| P_s^r \right\|}, \tag{8.12}$$

where P_s^r is the position of the Sun in the equatorial system, and can be derived from the transformation matrix in Eq. (8.8). n_g^r is the surface normal vector, expressed as $n_g^r = [\cos\theta\cos\phi\ \sin\theta\sin\phi\ \sin\phi]^T$, ϕ is the local latitude, and θ is the right ascension. Then, the solar-radiation intensity can be expressed as [12, 13]

$$E_0 = \frac{M_0}{4\pi r^2 \left\| P_s^r \right\|} P_s^r n_g^r. \tag{8.13}$$

Figure 8.3 shows the solar spectrum of the Earth's surface at a zenith angle of 60° in clear weather conditions [14].

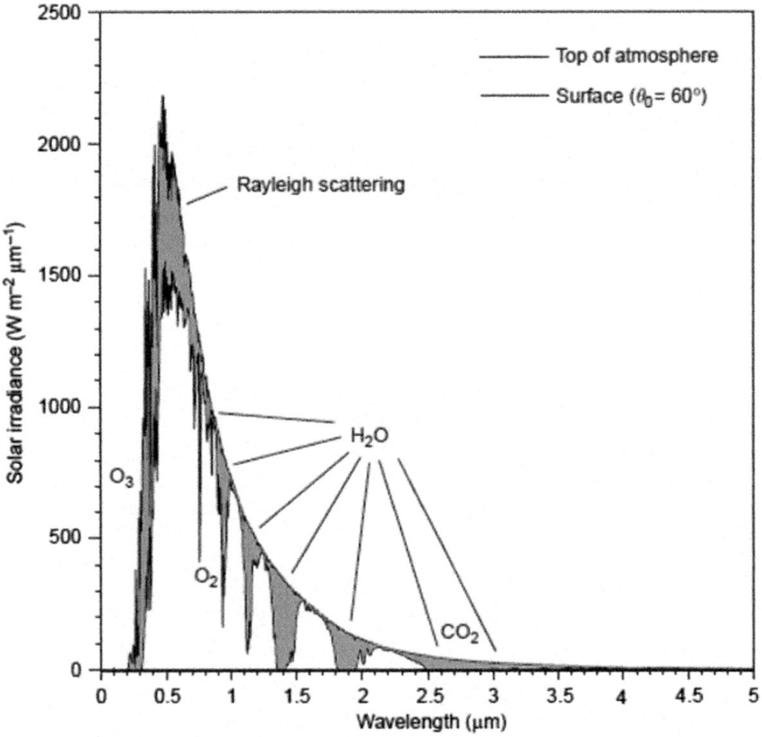

Fig. 8.3 Sun spectrum on a clear day [14]

8.1.2 Atmospheric Parameters

Atmospheric transparency is an important feature of a radiative climate and a decisive factor in the Earth's surface receiving direct ultraviolet radiation. It objectively reflects the state of atmospheric transparency and the physical state of atmospheric light radiation, and is an important parameter to characterize the degree of atmospheric attenuation of UV radiation [15].

At a certain solar altitude, atmospheric transparency and cloud changes are the main factors determining the intensity of UV radiation [16]. Direct radiation is mainly weakened by the reflection and absorption of clouds, which are related to the cloud shape. However, the effect of clouds on effective long-wave radiation is mainly through two aspects. One is that the cloud's shielding effect on direct short-wave radiation decreases the surface heating, which weakens long-wave radiation on the ground. The second is that the presence of clouds increases inverse atmospheric radiation, reducing the effective radiation on the ground to a greater extent [17].

According to the Bouguer–Beer–Lambert theorem, when solar radiation of wavelength A passes through the dm-thick atmosphere, its radiation decay decreases by the following amount [18]:

$$dI_\lambda = -\alpha_\lambda I_0 \lambda dm, \tag{8.14}$$

where α_λ is the atmospheric-extinction coefficient. Owing to the absorption of the ozone layer, the intensity of ultraviolet radiation in the 200–280 nm band is very low; this band is called the solar blind zone [1]. For the solar-blind-zone band, the entire band is integrated from 200 to 280 nm; then,

$$S_m = \int_{0.2}^{0.28} S_0 \lambda P_\lambda^m d\lambda. \tag{8.15}$$

It can be simplified to

$$S_m = S_0 P_\lambda^m, \tag{8.16}$$

or expressed as

$$P_\lambda = \sqrt[m]{S/S_0}, \tag{8.17}$$

where P_λ is the compound transparency coefficient, m is the atmospheric mass, and S_0 is the value of the solar constant corrected by the distance between the sun and the earth; $S_0 = 1381.6\,\mathrm{W\,m^{-2}}$. S_m indicates the solar-radiation flux on the ground perpendicular to the surface of the sun's rays when the atmospheric mass is m.

Equation (8.17) is an exact theoretical equation, not an empirical relationship. P_λ^m increases with m, a phenomenon known as the Forbes effect. The application of this effect greatly limits P_λ. To overcome the Forbes effect, atmospheric transparency needs to be adjusted to a given atmospheric quality. P can be divided into six grades: Very turbid ($P < 0.625$), turbid ($P = 0.625$–0.675), low ($P = 0.676$–0.725), normal ($P = 0.726$–0.775), high ($P = 0.776$–0.825), and very transparent ($P \geq 0.826$) [19].

According to the measurement data of meteorological stations in various places, China has established an empirical look-up table for correcting P_λ to P. The corrected formula deduced from the look-up table is [19]

$$P = (1.91/S_0)^{1/2}(S_h/1.91)^{(\sinh+0.15)/1.3}, \tag{8.18}$$

where h is the solar altitude angle and S_h is the direct solar radiation perpendicular to the solar-ray plane.

8.1.3 Solar-Blind-Area Ultraviolet Background-Radiation Model

The energy of solar radiation is mainly concentrated in the wavelength range of 0.15–4 μm. In this wavelength range, it can be divided into three main regions, namely, the shorter ultraviolet region, the longer infrared region, and the visible region between them. The energy of solar radiation is mainly distributed in the visible and infrared regions, the former accounting for 50% of the total solar radiation and the latter accounting for 43%. The ultraviolet region accounts for only seven percent of the Sun's energy. Solar radiation reaches its highest level at a wavelength of 0.475 μm.

The radiation capacity of the sun at various wavelengths can be described by the blackbody radiation model, and the radiation capacity can be expressed as $w(\lambda)$[20]:

$$w(\lambda) = \frac{c^2 h}{\lambda^5}\left\{\frac{1}{\exp[hc/(k\lambda T)] - 1}\right\}, \tag{8.19}$$

where c is the vacuum speed of light, 3×10^8 m/s; h is Planck's constant, $h = 6.626 \times 10^{-34}$ J s; T is the temperature in Kelvin, and k is Boltzmann constant, $k = 1.380 \times 10^{-23}$ J/K. Then, the percentage of the overall radiation capacity of a particular band of ultraviolet $F_{(\lambda_1-\lambda_2)}$ can be expressed as [21]

$$F_{(\lambda_1-\lambda_2)} = \frac{\int_{\lambda_2}^{\lambda_1} w(\lambda)d\lambda}{\int_0^{\infty} w(\lambda)d\lambda} = \frac{1}{\sigma T^4}\int_{\lambda_2}^{\lambda_1} w(\lambda)d\lambda = \frac{1}{\sigma T^4}\left[\int_0^{\lambda_1} w(\lambda)d\lambda - \int_0^{\lambda_2} w(\lambda)d\lambda\right]. \tag{8.20}$$

For a solar-blind ultraviolet communication system, the sensitive wavelength of background noise is in the range of 200–300 nm, and the main source is sunlight [22].

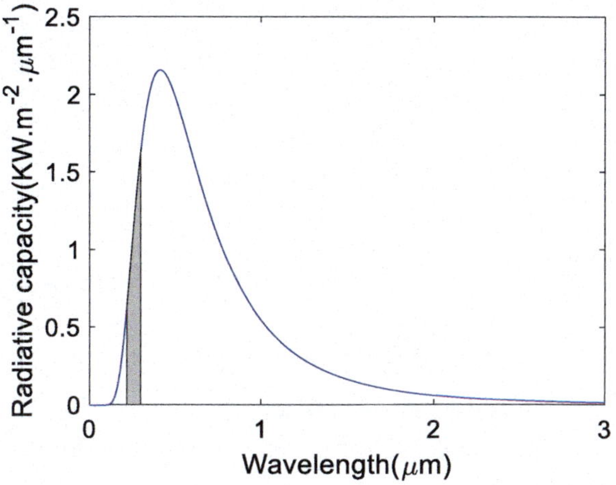

Fig. 8.4 Solar-radiation capacity at each wavelength

Figure 8.4 shows the blackbody radiation map of the Sun when its surface temperature is 5800 K. The gray area is the ultraviolet background-radiation spectrum, and its integral sum is the sum of the solar background noise in this band.

At latitude ϕ and longitude θ, for a receiver radius of $D_r/2$, the filter transmittance is η_B, receiving field angle is $\theta_{r,signal}$, and the ultraviolet background-light power at the receiving end is [23]

$$P_B = F_{(\lambda_1-\lambda_2)}.\eta_B\pi\left(\frac{D_r}{2}\right)^2 E_0.[P^m + \frac{1-P^m}{2(1-1.4\ln P)}].\frac{\pi}{4}\theta_{r,signal}^2, \tag{8.21}$$

where $F_{(\lambda_1-\lambda_2)}$ represents the percentage of a certain band to the overall radiation capacity, η_B is the filter transmittance, D_r is the receiver diameter, E_0 is the surface solar radiation intensity; P is the atmospheric transparency, and m is the air quality.

Suppose the measurement time is from 6 a.m. to 6 p.m. on a sunny day, and the measurement location is Xi'an, China. The measurement wavelength is 300 nm, the receiver diameter is 10 mm, and the receiving field of view is 20 mrad. The filter transmittance η_B is 0.24 and the bandwidth is $\pm$ 10 nm. Xi'an was simulated on spring equinox March 21, summer solstice June 21, and autumn equinox September 23, 2023, and winter solstice December 22, 2022 [24].

The variation of the UV background light with the season under sunny conditions is shown in Fig. 8.5. It can be seen that the background-light radiation value of the summer solstice is the largest, those of the autumn equinox and spring equinox are similar, and the winter solstice is the smallest. The daily-radiation peak also changes with the seasons. Figure 8.6 shows the comparison of the measured and theoretical values of the winter solstice, spring equinox, and summer solstice. It can be seen that the winter solstice, spring equinox, and summer solstice conform to the theoretical changes from day to day.

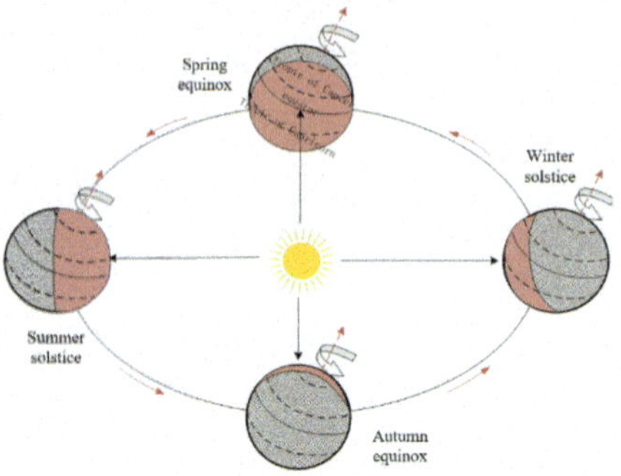

(a) Diagram of different solar terms

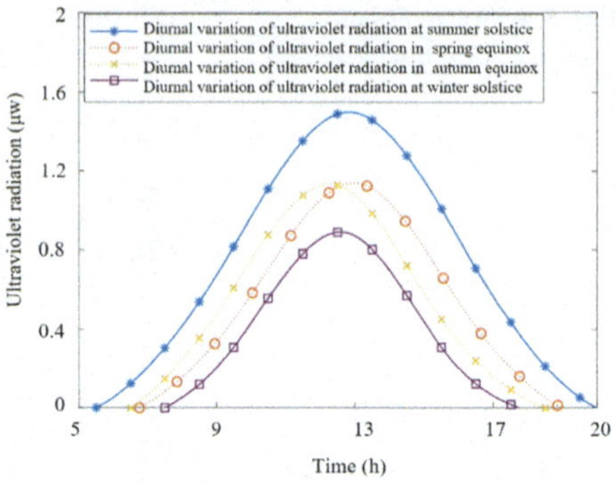

(b) Theoretical value of diurnal variation of ultraviolet radiation intensity in different seasons

Fig. 8.5 Theoretical values of the diurnal variation of UV radiation intensity in different seasons

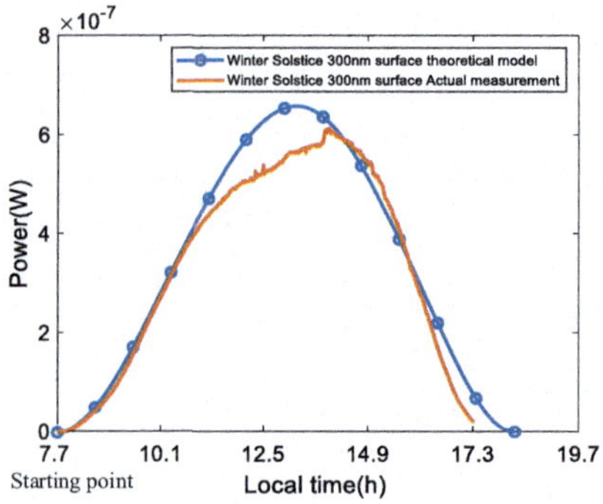

(a) Theoretical and measured winter-solstice values

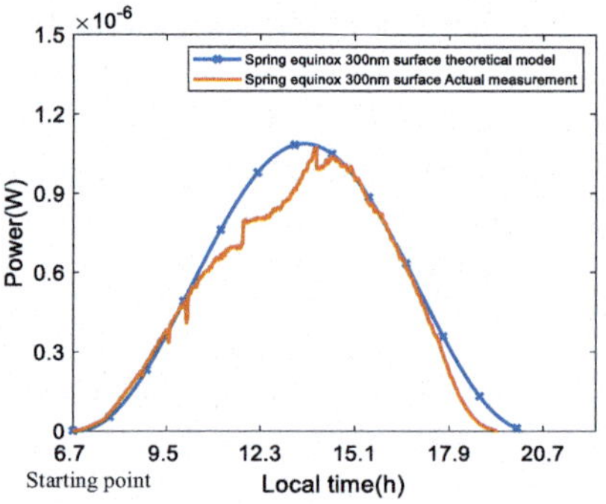

(b) Theoretical and measured values of the spring equinox

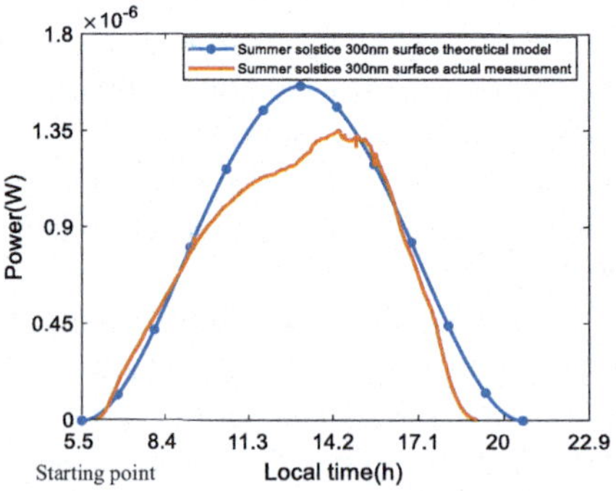

(c) Theoretical and measured values of the summer solstice

Fig. 8.6 Theoretical and measured values of the winter solstice, spring equinox, and summer solstice

8.2 UV Communication Path-Loss Models

8.2.1 Single-Scattering Model

The single-scattering model was proposed by Luegtten et al. [25] in 1991, which is an extension of the single-scattering transceiver-axis collinear model proposed by Reilly and Warde [26] in 1979. The essence is a single-scattering coplanar direct-integral model; the transceiver axis is coplanar. For ease of calculation, ellipsoidal coordinates are established, as shown in Fig. 8.7, where each point can be uniquely determined by ξ, the diameter corresponding angle η, and the azimuth angle ϕ.

As shown in Fig. 8.7, F_1 and F_2 are the two focal points of the ellipsoid, r is the focal length of the ellipsoid, r_1 and r_2 are the two focal radii of the ellipsoid, and ψ_1 and ψ_2 are the two focal angles of the ellipsoid.

The transformation relationship between any point in the Cartesian coordinate system (x, y, z) to the ellipsoid coordinate system (ξ, η, φ) is

$$\xi = (r_1 + r_2)/r(1 \leq \xi \leq \infty) \tag{8.22}$$

$$\eta = (r_1 - r_2)/r\,(- 1 \leq \eta \leq 1) \tag{8.23}$$

$$\phi = \arctan(y, x)\,(- \pi \leq \xi \leq \pi) \tag{8.24}$$

$$\theta_s = \psi_1 + \psi_2. \tag{8.25}$$

From Eqs. 8.22 and 8.23, we can solve r_1 and r_2:

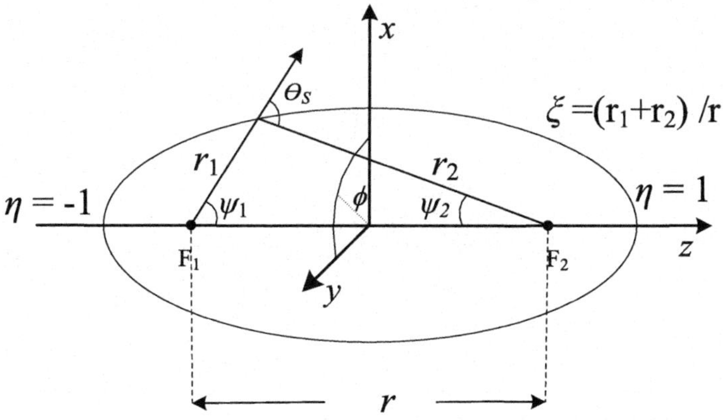

Fig. 8.7 Single-scattering ellipsoidal coordinates

$$r_1 = \frac{r(\xi + \eta)}{2} \tag{8.26}$$

$$r_2 = \frac{r(\xi - \eta)}{2}. \tag{8.27}$$

In ellipsoidal coordinates, the focal angle and the scattering angle can be expressed as

$$\cos \psi_1 = (1 + \xi\eta)/(\xi + \eta) \tag{8.28}$$

$$\sin \psi_1 = \sqrt{(\xi^2 - 1)(1 - \eta^2)}/(\xi + \eta) \tag{8.29}$$

$$\cos \theta_s = (2 - \xi^2 - \eta^2)/(\xi^2 - \eta^2). \tag{8.30}$$

The single-scattering model simplifies the model by assuming that photons are emitted from the transmitting end and scattered only once during their arrival at the receiving end. The single-scatter coplanar model is shown in Fig. 8.8. The receiver and transmitter are located on the two focal points of the ellipsoid, F_1 and F_2, respectively, θ_R is the receiving field-of-view half-angle, θ_T is the transmit field-of-view half-angle, β_R is the receiving-elevation angle, β_T is the transmission elevation angle, θ_s is the scattering angle, and $\theta_s = \beta_T + \beta_R$.

r is the distance between the transmitter and receiver, r_1 is the distance from the transceiver to the effective scattering point, and r_2 is the distance from the effective scattering point to the receiver. The shadow part is the effective scattering volume V; that is, after one scattering, only the atmospheric light signal located in the volume V can be received by the receiver.

It is assumed that the emitting and receiving optical axes are in the same plane, and the effect of atmospheric turbulence is not considered. At time $t = 0$, the emitter emits the ultraviolet light pulse with response energy E_T, and leaves the emitter at the

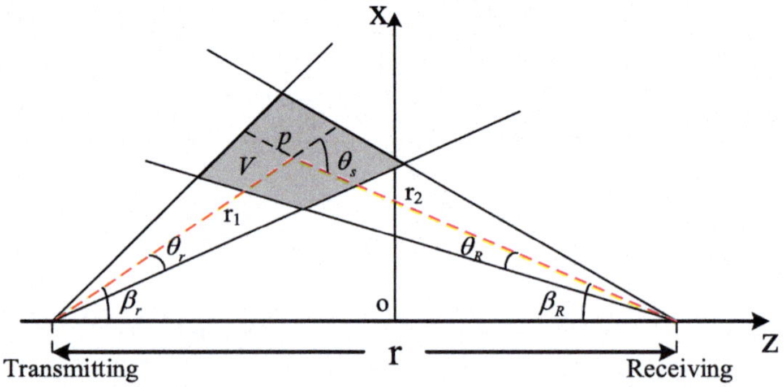

Fig. 8.8 Single-scattering NLOS model

emission elevation angle β_T. After being scattered and absorbed by the atmosphere, it reaches the point $P = (\xi, \eta, \phi)$ at a distance r_1 from the emitter. The energy at point P is

$$H_p = \frac{E_T \exp(-k_e r_1)}{\Omega_r (r_1^2)}, \tag{8.31}$$

where $\Omega_r = 2\pi(1 - \cos\theta_T)$ is the emission of the inherent cone angle and k_e is the atmospheric extinction coefficient.

The differential volume element dV can be regarded as a secondary light source and a spherical-point light source, which emits energy of

$$dR_p = k_s \frac{E_T \exp(-k_e r_1)}{\Omega_r (r_1^2)} . \frac{P(\cos\theta_s)}{4\pi} dV. \tag{8.32}$$

Energy dR_p reaches the receiver after the transmission distance r_2; then, the energy received per unit area of the receiver is

$$\begin{aligned} dH_R &= dR_p . \frac{\cos(\xi)\exp(-k_e r_2)}{r_2^2} \\ &= \frac{E_T k_s \cos(\xi)\exp(-k_e(r_1 + r_2))}{4\pi\,\Omega_r (r_1 r_2)} P(\cos\theta_s) dV, \end{aligned} \tag{8.33}$$

where k_s is the atmospheric scattering coefficient, ξ is the angle between the line between the scattering point S, the center of the receiving surface at the receiving end, and the central axis of the receiving field-of-view cone. $P(\cos\theta_s)$ is a scattering-phase function. $\cos(\xi)$ is the effective area of the detection and can be expressed as

$$\cos(\xi) = \sin\beta_R \sin\psi_1 \cos\phi + \cos\beta_R \cos\psi_1. \tag{8.34}$$

In ellipsoidal coordinates, dV can be expressed as

$$dV = \frac{r^3}{8}(\xi^2 - \eta^2) d\xi d\eta d\phi. \tag{8.35}$$

Equations (8.26), (8.27), and (8.35) are added to Eq. (8.32) and produce the following:

$$dH_R = \frac{E_T k_s \cos(\xi)\exp(-k_e r \xi)}{2\pi\,\Omega_r r(\xi^2 - \eta^2)} P(\cos\theta_s) d\xi d\eta d\phi. \tag{8.36}$$

The time for the UV pulse to reach the receiver is $t = (r_1 + r_2)/c$, and $\xi = (r_1 + r_2)/r$.

$$dE = \frac{E_T c k_s \cos(\xi) \exp(-k_e r \xi)}{2\pi \Omega_r r^2 (\xi^2 - \eta^2)} P(\cos\theta_s) d\eta d\phi \tag{8.37}$$

At time $t = \xi r/c$, Eq. (8.36) is integrated to obtain the acceptable irradiance in units, which can be expressed as

$$E = \begin{cases} \frac{E_T c k_s \exp(-k_e r \xi)}{2\pi \Omega_r r^2} \int\limits_{\eta_1(\xi)}^{\eta_2(\xi)} \int\limits_{\phi_1(\xi,\eta)}^{\phi_2(\xi,\eta)} \frac{\cos(\xi) P(\cos\theta_s)}{\xi^2 - \eta^2} d\eta d\phi \ (\xi_{\min} < \xi < \xi_{\max}) \\ 0 \qquad (otherwise). \end{cases} \tag{8.38}$$

From the symmetry of the ellipsoid coordinates $\phi_2(\xi, \eta) = -\phi_1(\xi, \eta)$, for the ϕ integral, Eq. (8.37) can be simplified to

$$E = \begin{cases} \frac{E_T c k_s \exp(-k_e r \xi)}{2\pi \Omega_r r^2} \int\limits_{\eta_1(\xi)}^{\eta_2(\xi)} \frac{2g[\phi_2(\xi,\eta)] P(\cos\theta_s)}{\xi^2 - \eta^2} d\eta \ (\xi_{\min} < \xi < \xi_{\max}) \\ 0 \qquad (otherwise), \end{cases} \tag{8.39}$$

where $g[\phi_2(\xi, \eta)] = \phi_2(\xi, \eta) \cos\beta_R \cos\psi_1 + \sin \sin\psi_1 \sin[\phi_2(\xi, \eta)]$. Integrating the above equation gives the energy density J/m^2 at the receiver as

$$H_p = \int\limits_{t_{\min}}^{t_{\max}} E(ct/r) dt. \tag{8.40}$$

The energy received by the receiver is

$$H = A_r . H_R, \tag{8.41}$$

where A_r is the area of the receiving aperture of the receiver. The path loss is

$$L = 10 \lg \frac{E_T}{H}, \tag{8.42}$$

where E_T is the pulse energy emitted by the emitter.

8.2.2 *Multiple-Scattering Model*

The multiple-scattering propagation model of non-line-of-sight non-coplanar ultraviolet light is shown in Fig. 8.9. We set the transmitting end at the origin of the coordinate system (x, y, z) and the receiving end on the positive semi-axis of the x-axis. The distance between Rx and Tx is *d*. C_t and C_r are cones representing the emitted beam and the receiving field of view, respectively. θ_t and θ_t represent

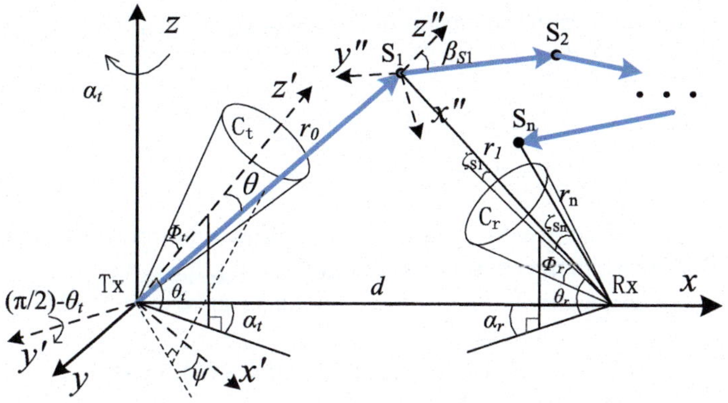

Fig. 8.9 Multiple-scattering propagation model of non-coplanar non-line-of-sight ultraviolet light

the elevation angle and divergence half-angle of the transmitting end, respectively. θ_r and ϕ_r represent the receiving elevation angle and the field-of-view half-angle, respectively.

α_t is the deflection angle of C_t and α_r is the deflection angle of C_r. S_n is the scattering point of the nth scatter. r_0 is the distance from T_X to S_1, r_1 is the distance from S_1 to R_X, and r_n is the distance to S_n. ζ_{sn} is the angle between the line of S_n and Rx the central axis. β_{sn} is the angle between the direction of photon incidence at the point and the direction of propagation after the photons scatter.

To conveniently describe the direction and distance of a photon transmission, the coordinate system (x, y, z) needs to be transformed. The point coordinates based on the coordinate system $\left(x',y',z'\right)$ and the point coordinates based on the coordinate system (x, y, z) are transformed according to Eq. (8.43).

$$\begin{bmatrix} x' \\ y' \\ z' \end{bmatrix} = R_{y'}(\frac{\pi}{2} - \theta_t)R_z(\alpha_t)\begin{bmatrix} x \\ y \\ z \end{bmatrix}, \tag{8.43}$$

where

$$R_{y'}\left(\frac{\pi}{2} - \theta_t\right)R_z(\alpha_t) = \begin{bmatrix} \cos\left(\frac{\pi}{2} - \theta_t\right)\cos\alpha_t & \cos\left(\frac{\pi}{2} - \theta_t\right)\sin\alpha_t & -\sin\left(\frac{\pi}{2} - \theta_t\right) \\ -\sin\alpha_t & \cos\alpha_t & 0 \\ \sin\left(\frac{\pi}{2} - \theta_t\right)\cos\alpha_t & \sin\left(\frac{\pi}{2} - \theta_t\right)\sin\alpha_t & \cos\left(\frac{\pi}{2} - \theta_t\right) \end{bmatrix}. \tag{8.44}$$

Suppose a photon is uniformly emitted in the emitting cone Tx, and any single photon is taken. In the coordinate system $\left(x', y',z'\right)$, the angle between the propagation direction of the photon and the forward direction of the coordinate axis z′ is set to θ. The angle between the projection of the photon and the coordinate axis in the plane $\left(x',yz'\right)$ is set to ϕ. (θ, ϕ) specifies the direction of the photon. θ is uniformly distributed among $(0, \phi_t)$:

$$\begin{aligned}\cos\theta &= 1-\xi^{(\theta)}(1-\cos\phi_t)\\ \phi &= 2\pi\xi^{(\phi)},\end{aligned} \tag{8.45}$$

where $\xi^{(\theta)}$ and $\xi^{(\phi)}$ are random numbers that follow a uniform distribution in the [0, 1] interval. Thus, in the coordinate system $\left(x',y',z'\right)$, the directional cosine of the direction of the propagation of the emitted photons can be expressed as

$$\left(u_{x'},\ u_{y'},\ u_{z'}\right) = \left(\sin\theta\cos\phi,\ \sin\theta\sin\phi,\ \cos\theta\right). \tag{8.46}$$

From Eq. (8.43), the directional cosine of the photon-propagation direction in the coordinate system (x, y, z) can be expressed as

$$\begin{bmatrix} u_x \\ u_y \\ u_z \end{bmatrix} = \left[R_y\left(\frac{\pi}{2}-\theta_t\right)R_z(\alpha_t)\right]^{-1}\begin{bmatrix} u_{x'} \\ u_{y'} \\ u_{z'} \end{bmatrix}. \tag{8.47}$$

In the photon-transmission process, both Rayleigh scattering and Mie scattering occur. Thus, the scattering-phase function can be expressed as the weighted sum of Rayleigh scattering and Mie scattering:

$$P(\cos\beta_s) = \frac{k_s^R}{k_s}P^R(\cos\beta_s) + \frac{k_s^M}{k_s}P^M(\cos\beta_s). \tag{8.48}$$

In the photon-propagation process, after scattering point S_1, after the scattering of the transmission direction and the incident direction of the angle, the scattering angle is recorded as β_{s1}. β_{s1} can be obtained by the following formula:

$$\xi^{(S)} = 2\pi\int_{-1}^{\mu_{S1}} P(\mu)d\mu, \tag{8.49}$$

where $\mu_{s1} = \cos\beta_{s1}, \xi^{(S)}$ are random variables that are uniformly distributed between 0 and 1, and $P(\mu)$ can be derived from Eq. (8.48).

The probability that a photon will reach the receiving surface is

$$P_{1n} = \frac{A\cos\zeta_{sn}}{4\pi r_n^2}P(\cos\beta_{sn}), \tag{8.50}$$

where A is the area of the receiving aperture and $P(\cos\beta_{sn})$ is the phase function of the nth scattering. The probability that a photon will travel a distance r_n after scattering by point S_n is

$$P_{2n} = e^{-k_e r_n}. \tag{8.51}$$

Therefore, the probability that a photon will reach the receiving surface after the nth scattering is

$$P_n = W_n P_{1n} P_{2n}, \tag{8.52}$$

where W_n is the probability of survival before the photon arrives. W_n is defined as

$$W_n = (1 - P_{n-1})\mathrm{e}^{-k_a |\mathrm{s}_n - \mathrm{s}_{n-1}|} W_{n-1} \tag{8.53}$$

where $|\mathrm{s}_n - \mathrm{s}_{n-1}|$ is the distance from scattering point S_{n-1} to scattering point S_n.

The probability P_0 of photon scattering directly reaching the receiving end zero times is zero. As the amount of photon scattering increases, the probability of photon survival is smaller, so the maximum number of photons scattering is set to $N \leq 5$. During the motion of the photon, each scattering that occurs is independent of the others; then, the probability that a photon reaches S_n and S_n is in the receiving plane is

$$P_N = \sum_{n=1}^{N} P_n. \tag{8.54}$$

Suppose that the emitter emits a total of M photons. In this study, $M = 10^6$. The total probability that photon m can travel to the receiving surface by a maximum of N scattering times is $(P_N)_m$. The average probability that a photon from the emitter will reach the receiving surface is

$$P = \frac{\sum_{m=1}^{M} (P_N)_m}{M}, \tag{8.55}$$

where $(P_N)_m$ can be found using Eq. (8.54). P reflects the average probability that a photon from the emitting end will reach the receiving end. The emitting end emits m photons, which are independent of each other. Then, the number of photons that can reach the receiving end is $M \times P$. the path loss of a non-line-of-sight ultraviolet communication system can be calculated as

$$PL = 10 \log_{10}\left(\frac{M}{M \times P}\right) = 10 \log_{10}\left(\frac{1}{P}\right). \tag{8.56}$$

8.2.3 Path-Loss Analysis of Ultraviolet Optical Communications

As shown in Fig. 8.10, there are three modes of non-line-of-sight communication [27]:

Mode 1: Vertical transceiver mode ($\beta_T = \beta_R = 90°$);

Mode 2: Vertical-receive, oblique-transmit mode or vertical-transmit, oblique-receive mode ($\beta_R = 90°$, $\beta_T < 90°$ OR $\beta_T = 90°$, $\beta_R < 90°$); and.

Mode 3: Oblique transmit-receive mode ($\beta_R < 90°$, $\beta_T < 90°$).

Assuming that the emitting and receiving optical axes are in the same plane, and UV light is scattered once on the way to the receiving plane, the three modes are simulated. The parameters are set as follows: the half-angle of the emitting side is $\theta_T = 45°$, the half-angle of the receiving side is $\theta_R = 30°$, wavelength $\lambda = 260$ nm, Rayleigh scattering coefficient $\kappa_s^{Rsy} = 0.24km^{-1}$, Mie scattering coefficient $\kappa_s^{Mie} = 0.25\,\mathrm{km}^{-1}$, atmospheric absorption coefficient $\kappa_a = 0.9\,\mathrm{km}^{-1}$, asymmetry factor $g = 0.72$, and the model parameters in the scattering phase function are $r = 0.017$ and $f = 0.5$. The simulation results are shown in Fig. 8.11.

At the same transmission distance, the path loss of the vertical transceiver mode is the largest, followed by the vertical-receive and oblique-transmit mode, and the path loss of the oblique transmit and receive mode is the smallest. Therefore, when designing a transceiver system, we should choose the oblique-transceiver mode, which is more advantageous to ultraviolet communication under the same conditions.

In the third mode, the path loss will differ with different receiving and transmitting elevation angles, and the parameter setting will not change. We set the divergence half-angle of the transmitter to $\theta_T = 10°$ and the field-of-view half-angle of the receiver to $\theta_R = 30°$. When the angle of the receiver is fixed to $\beta_R = 45°$, we change the elevation angle of the transmitter to obtain the path loss. When the angle of the transmitter is fixed at $\beta_T = 45°$, the path loss is simulated by changing the elevation angle of the receiver. The simulation results are shown in Fig. 8.12.

The graph shows that the path loss increases with the increase of the transmit–receive angle, and the effect of changing the elevation angle of the transmitter is obviously greater than that of the receiver. The path loss increases with the increase of the elevation angle of the receiver; however, the increase is small, so the elevation

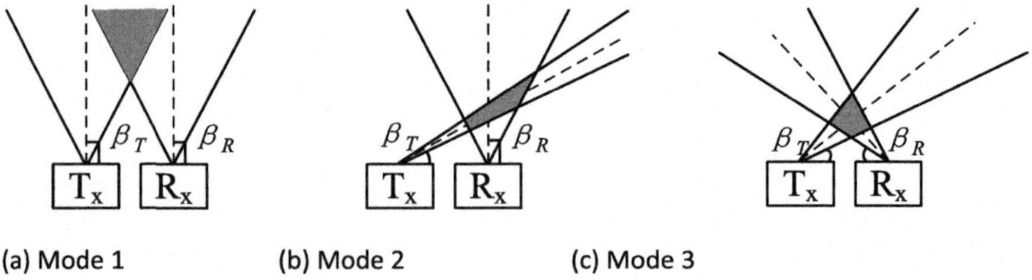

Fig. 8.10 Three modes of operation

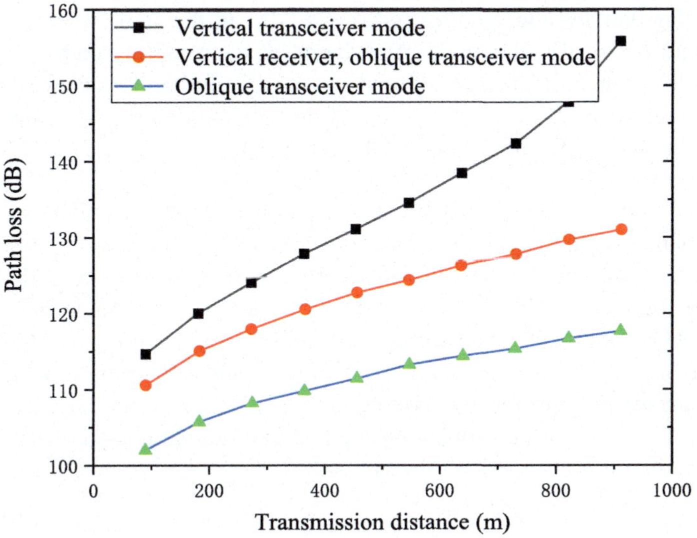

Fig. 8.11 Effect of the three working modes on the path loss

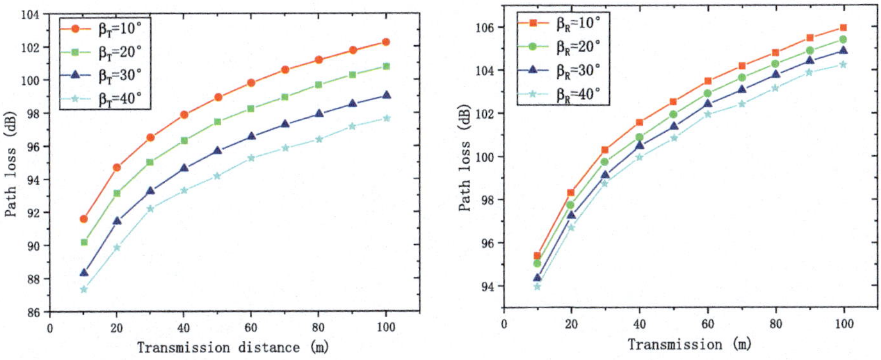

(a) Different transmit elevations (b) Different receive elevations

Fig. 8.12 Path loss at different receive and transmit elevation angles [27]

angle of the receiver should be fixed and the elevation angle of the transmitter be reduced to reduce the path loss.

In an ultraviolet communication system, the non-line-of-sight non-coplanar single-scattering model is more ideal than the non-line-of-sight non-coplanar multiple-scattering model. The path loss of the non-line-of-sight non-coplanar multiple-scattering model is calculated using the ergodic-element method and the pointing-probability method.

The simulation parameters are set as follows: UV wavelength is 266 nm; the receiving-surface radius of the receiving end is 1.5×10^{-2} m, the distance between the receiving end and the sending end is $d = 100$ m, the angle of elevation of the sending end $\beta_T = 30°$, the angle of elevation of the receiving end $\beta_R = 60°$, the angle of divergence of the sending end $\theta_T = 10°$, the field-of-view angle of the receiving end $\theta_R = 20°$, the offset angle of the sending end $\alpha_t = 0°$, and the offset angles of the receiving end are $\alpha_r = \{0°, 5°, 10°, 15°, 20°, 25°, 30°, 35°, 40°, 45°, 50°\}$.

In the pointing-probability method, the number of photons sent is set to 10^6, the parameters of the scattering-phase function (γ, g, f) are set to (0.017, 0.72, 0.5), the atmospheric absorption coefficient $k_a = 0.74 \times 10^{-3}\,\text{m}^{-1}$, the Rayleigh scattering coefficient $k_S^{Ray} = 0.24 \times 10^{-3}\,\text{m}^{-1}$, and the Mie scattering coefficient $k_S^{Mie} = 0.25 \times 10^{-3}\,\text{m}^{-1}$. The simulation results are shown in Fig. 8.13 [27].

Figure 8.13 shows that the path loss increases and the volume of the common scatterer decreases with the increase of the offset angle at the end. The path loss calculated by the pointing-probability method and the path-loss energy calculated by the ergodic micro-element method fit well.

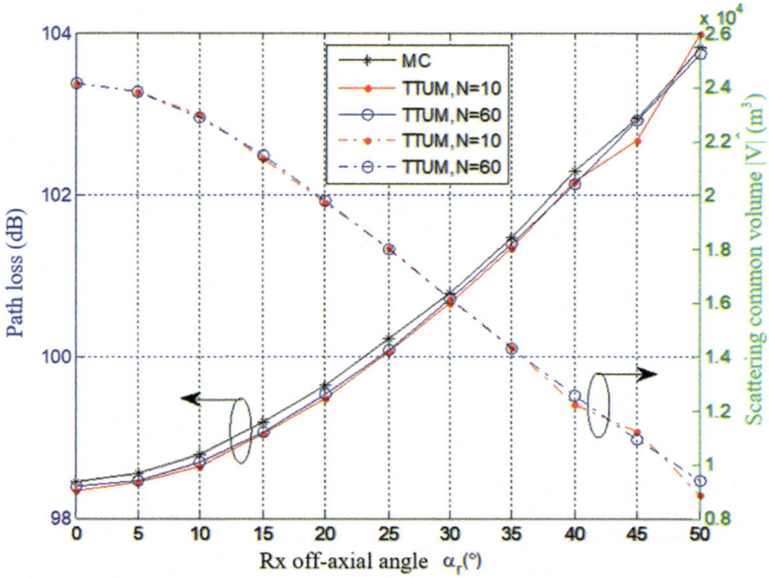

Fig. 8.13 Relationship between the path loss and volume of the common scatterer and offset angle of receiver [28]

8.3 UV-Light Scattering-Channel Characteristics Under Different Weather Conditions

8.3.1 UV-Light Scattering Channel Characteristics on Rainy Days

When the UV signal scatters with the raindrop particles in a rainy environment, the received UV signal will be attenuated to some extent, and the ultraviolet communication link will be interrupted. Therefore, to ensure the normal operation of the communication link, it is necessary to study the influence of rainfall particles with different sizes on the attenuation of ultraviolet light. At present, the main models used to describe the shape of raindrops are the approximate ellipsoid model, Beard–Chuang (BC) raindrop-shape model, and Pruppacher–Pitter shape model [29].

When the radius of raindrop particles is greater than 0.5 mm, the approximate model of the raindrop shape established by Pruppacher and Pitter is applicable. The model uses the ratio of the ellipsoid length to the short-axis radius a_r/b_r to approximate the deformation degree of the raindrop particles; the ellipsoid-shape diagram is shown in Fig. 8.14.

The ellipsoid is formed rotating on its long a_r and short axes b_r, and the corresponding ellipsoid shape is described as follows [29]:

$$r(\theta_p) = a_r(\sin^2\theta_p + \frac{a_r^2}{b_r^2}\cos^2\theta_p)^{-1/2}, \quad (8.57)$$

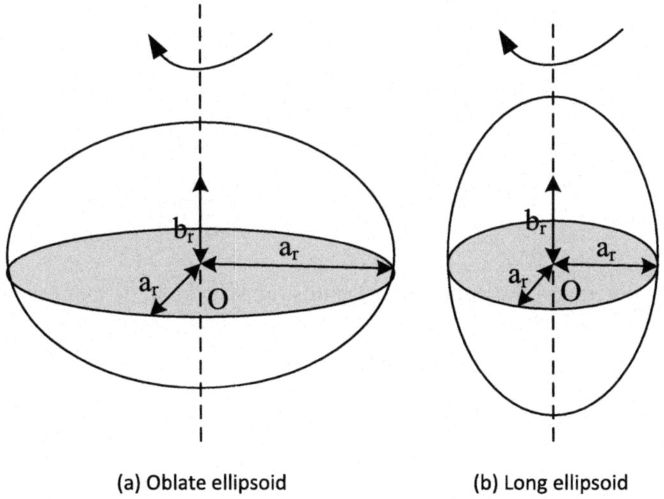

Fig. 8.14 Rotationally symmetric ellipsoids

where θ_p indicates the angle between the rotation angle and the longitudinal axis.

For deformed raindrop particles, the model proposed by Beard and Chuang, which applies the general Chebyshev particle to describe the shape of the particle, is suitable. The mathematical expression is [30]

$$r(\theta_p) = r_0[1 + \sum_{n_p=0}^{10} c_n \cos(\theta_p n_p)], \tag{8.58}$$

where $r(\theta_p)$ is the particle radius corresponding to different orientation angles θ_p and r_0 is the radius of an equal-area sphere. c_n is a constant. $\cos(\theta_p n_p) = T_n(\cos\theta_p)$, representing the Chebyshev polynomials of the n_p order.

The BC model for calculating the shape of raindrops can be approximated to an ellipsoid shape by the deformation of raindrop particles while falling, which is called the approximate-ellipsoid model. By considering the symmetry of the cross section of the raindrop model, the area of the cross section of the BC model is approximated by the volume equality. Because the ellipsoid cross section is symmetric along both the transverse and the longitudinal axes a_{pr}, the short axis c_{pr} of the approximate ellipsoid is solved by taking one half of the length of the corresponding longitudinal axis c_{pr} in the BC model. It can be expressed as [30]

$$c_{pr} = \frac{1}{2}[r(0^\circ) + r(180^\circ)] = r_v(1 + c_0 + c_2 + c_4 + c_6 + c_8 + c_{10}). \tag{8.59}$$

Because the cross-sectional area of the ellipsoid model is approximately equal to that of the BC model, the area of the vertical cross-sectional area of the BC model is expressed as follows:

$$\begin{aligned} A &= r_v^2 \int_0^\pi \left[1 + \sum_{n_p}^{10} c_n \cos(n_p\theta_p)\right]^2 \\ &= \pi r_v^2 \left[1 + 2c_0 + c_0^2 + \sum_{n_p=1}^{10} \frac{c_n^2}{2}\right]. \end{aligned} \tag{8.60}$$

The long axis a_{pr} of the approximate ellipsoid model can be expressed as

$$a_{pr} = (1.05 - 0.131 r_v)^{-\frac{1}{3}} r_v. \tag{8.61}$$

Because the raindrop particles in a rainy environment have different scales, it is necessary to consider raindrop particles with different sizes, in combination with their size distribution. A large number of measured data shows that the size distribution of raindrop particles can be modeled by the negative-exponential Marshall–Palmer (M-P) raindrop-spectrum distribution model. The expression is as follows:

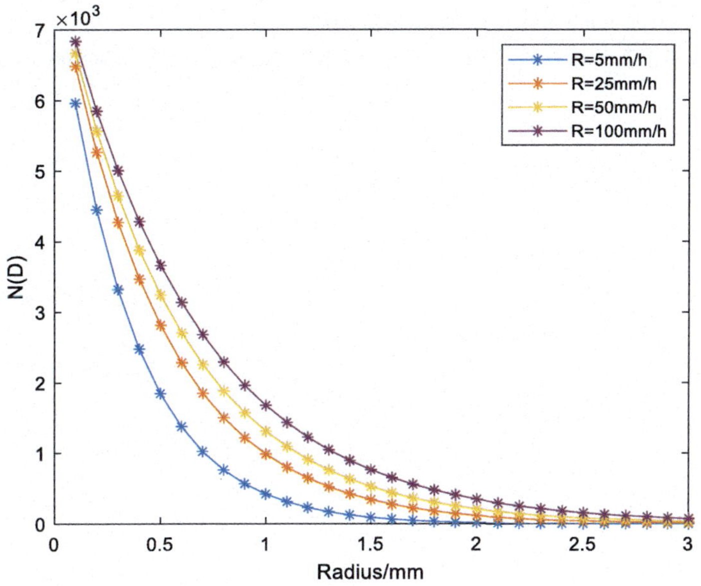

Fig. 8.15 Size distribution of raindrop particles [30]

$$N(D) = 8 \times 10^3 \exp(-4.1R^{-0.21}D), \tag{8.62}$$

where $N(D)$ is the unit volume, the number of raindrop particles per unit-radius interval; R is the rate of rainfall, with the unit mm/h; and D is the radius of the raindrop particles, with the unit mm. The simulation results are shown in Fig. 8.15. It can be seen from the graph that the larger the rainfall rate, the more the raindrop particles with the same raindrop-particle radius, and the number of raindrops increases exponentially with the size of the raindrops.

In practical applications, raindrop particles are randomly distributed in the atmosphere with different sizes and number densities, so the scattering effect of raindrop particles should be considered. The M-P spectrum distribution can be used to calculate the attenuation caused by a raindrop-particle swarm.

According to the relationship between the droplet-size distribution and particle-number density in Fig. 8.15, it can be concluded that the particle radius of a raindrop is mainly concentrated in the range of 0.1–3 mm. Depending on the intensity of the rainfall, the number of particles per unit volume can be determined. Combined with the above experimental conditions, a UV light path with 260-nm wavelength under different rainfall conditions was simulated and analyzed.

The experimental wavelength $\lambda = 260\,\text{nm}$, elevation angle $\beta_R = \beta_T = 10°$, divergence angle $\theta_T = 30°$, receiving field angle $\theta_R = 60°$, receiving aperture $A = 1.66\ \text{cm}^3$, and emitting light power $P_e = 1$ mW were assumed. The precipitation particle size, scattering coefficient, and absorption coefficient under different rainfall conditions are shown in Table 8.2.

Table. 8.2 Attenuation coefficients at different rainfall rates

Type of rainfall	Particle size (mm)	Scattering coefficient (m^{-1})	Absorption coefficient (m^{-1})
Drizzle	$0.1 < D < 1.5$	1.2×10^{-3}	0.906×10^{-3}
Widespread	$0.5 < D < 2$	3.4×10^{-3}	2.3×10^{-3}
Thunderstorm	$1 < D < 3$	9.3×10^{-3}	6.03×10^{-3}

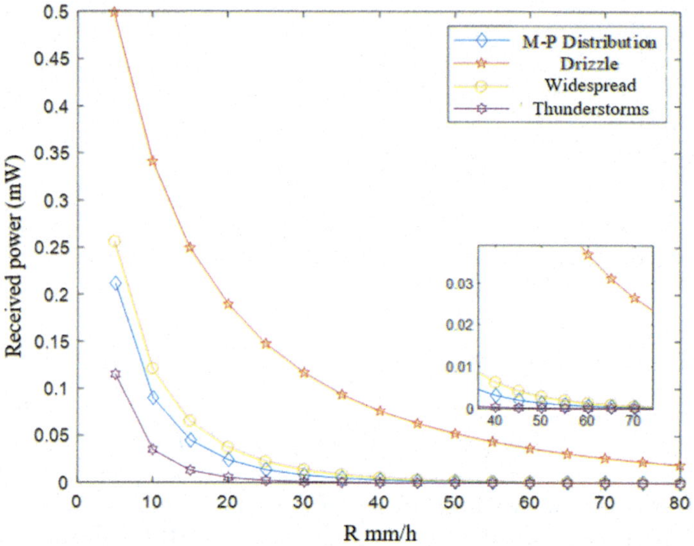

Fig. 8.16 Receiving power loss under different rainfall conditions [30]

Mie scattering is the main consideration for the scattering of UV light by raindrop particles. The results of the received-power simulation are shown in Fig. 8.16.

As shown in Fig. 8.16, the power of the receiver decreases with the increase of rainfall rate under different rainfall conditions, showing a negative-exponential trend. At the same rainfall rate, the transmission attenuation of ultraviolet light by thunderstorms is the largest, the widespread rain is the second, and the influence of the drizzle is the least. The main reason is that the raindrop particles in a thunderstorm have a strong attenuation effect on ultraviolet light, so the received light power is small.

8.3.2 UV-Light Scattering Channel Characteristics on Foggy Days

The presence of fog reduces visibility. According to the standard of the World Meteorological Organization, weather with visibility less than 1 km is attributed to fog. When the visibility is between 1 and 10 km, it is light fog or haze; the term smog is commonly used when the horizontal visibility is less than 10 km. Haze is different from fog in that the water–vapor content and relative humidity of haze are small, while the water–vapor content in fog is basically saturated. The differences between foggy days and hazy days are shown in Table 8.3 [31].

The sizes of precipitation particles and fog particles have a certain influence on the UV-light scattering characteristics, and the size parameters depend on the particle shape, particle size, and concentration distribution. The radii and numerical densities of some atmospheric particles are given in Table 8.4 [30].

Two models are commonly used for the distribution of fog particles. The modified Γ function and the generalized gamma distribution are used to describe fog particles when they are in a steady state. The parameters of the modified Γ function must be determined by experimental data, so they have some fitting errors. The generalized gamma distribution has a wide application, not only to fog particles over continents and oceans, but also to aerosols in the stratosphere. The gamma spectrum model can be expressed as follows [31]:

Table 8.3 Foggy and hazy weather characteristics [31]

Weather phenomenon	Composition	Relative humidity (%)	Scale (μm)	Visibility (km)	Color	Concentration distribution
Fog	Tiny drops of water, ice crystals	≥ 90	1–10	< 1	Sky is milky white or cyan	Uneven
Haze	Fine dust particles, containing carbon particles and soluble salts	≤ 80	0.001–1	< 10	Sky is yellow or orange	Even and stable

Table. 8.4 Radius and concentration of atmospheric particles [31]

Type	Radius (μm)	Concentration (A/cm^3)
Air molecules	10^{-4}	10^{19}
Aitken particles	10^{-3}–10^{-2}	10^{-4}–10^{2}
Smog particles	10^{-2}–1	10–10^{3}
Fog particles	1–10	10–10^{2}
Rain particles	10^{2}–10^{4}	10^{-5}–10^{-2}

$$n(r_p) = a_d r_p^{\alpha_d} \exp(-b_d r_p^{\beta_d}), \tag{8.63}$$

where r_p is the radius of the fog particle. α_d and β_d are expressed as the gamma distribution index, where a_d and b_d are expressed as $a_d = (9.781/(V^6.W^5)) \times 10^{15}$ and $b_d = (1.304/(V.W)) \times 10^4$. V and W are the visibility and liquid water content, respectively.

Based on this spectral-distribution model, according to the empirical relationship between the water content and visibility, the fog droplet is divided into advection and radiation droplets, and the empirical formulas for the scale spectrum distribution of these two droplets are respectively [32]:

$$n(r) = 1.059 \times 10^7 V^{1.15} r^2 \exp(-0.8359 V^{0.43} r) \tag{8.64}$$

$$n(r) = 3.104 \times 10^{10} V^{1.7} r^2 \exp(-4.122 V^{0.54} r), \tag{8.65}$$

where V is the visibility in km, r is the radius of fog particles in μm, and $n(r)$ is in $m^{-1}\mu m^{-1}$.

Thick fog, with visibility $V = 0.05$ km, middle fog, with visibility $V = 0.2$ km, and mist, with visibility $V = 2$ km, are analyzed. From Eqs. (8.64) and (8.65), the size distributions of advection and radiation droplets are obtained, as shown in Fig. 8.17.

As can be seen from Fig. 8.17, at the same visibility, the diameter of the advection droplet is larger than that of the radiation droplet at the highest droplet concentration: the concentration of an advection droplet is two orders of magnitude lower than that of a radiation droplet. The lower the visibility, the larger the droplet radius under the highest droplet concentration. With the decrease of the visibility, the maximum droplet concentration gradually decreases.

In the performance analysis of an NLOS ultraviolet communication system, the attenuation of optical signals in an atmospheric channel has an important effect on

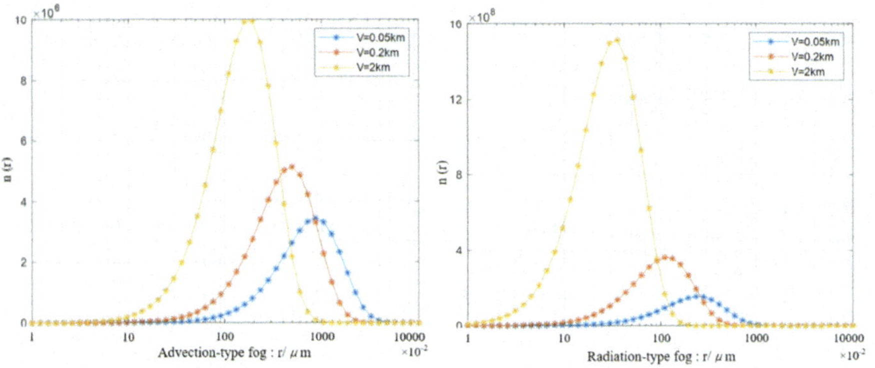

(a) Advection-type fog particles (b) Radiation-type fog particles

Fig. 8.17 Size distributions of different types of fog particles the communication performance. Analyzing the UV light attenuation accurately and easily is the key. In foggy weather, fog-induced light attenuation is usually analyzed by empirical attenuation models, such as the Kim, Kruse, and Al Nabousi models [33].

The empirical formula used by the Kim and Kruse models is [34]

$$\alpha = \frac{3.912}{V}\left(\frac{0.55}{\lambda}\right)^{q}, \tag{8.66}$$

where V is the visibility in km, λ is the wavelength of incident light in μm, and q is the constant related to V. The q values of the Kim and Kruse models are different, as follows [34]:

$$q_{Kim} = \begin{cases} 1.6 \text{ V >50 km} \\ 1.3 \text{6 km <V <50 km} \\ 1.6V + 0.34 \text{ V <6 km} \\ V - 0.5 \text{0.5 km <V <1 km} \\ 0 \text{ V <0.5 km} \end{cases} \tag{8.67}$$

$$q_{Kruse} = \begin{cases} 1.6 & V > 50\,\text{km} \\ 1.3 & 6\,\text{km} < V 50\,\text{km}\,. \\ 0.585V^{\frac{1}{3}} & V < 6\,\text{km}. \end{cases} \tag{8.68}$$

The Al Nabousi model divides fog into advection fog and radiation fog, and the attenuation models are [34]

$$\alpha_{adv} = \frac{0.11478\lambda + 3.8367}{V} \tag{8.69}$$

$$\alpha_{rad} = \frac{0.18126\lambda^2 + 0.13709\lambda + 3.7502}{V}, \tag{8.70}$$

where V is the visibility in km and λ is the wavelength of incident light in μm. The simulation of the decay coefficients of the three empirical formulas at the incident wavelength of 260 nm is shown in Fig. 8.18.

Overall, all three models showed a gradual decrease in attenuation with increasing visibility. As can be seen from Fig. 8.18, the attenuation curves of the Kim and Kruse models are almost identical at a 2-km visibility, while the magnitude of the attenuation coefficients of the Al Nabousi model and the first two are quite different. However, as the visibility increases, the difference between the three models gradually decreases.

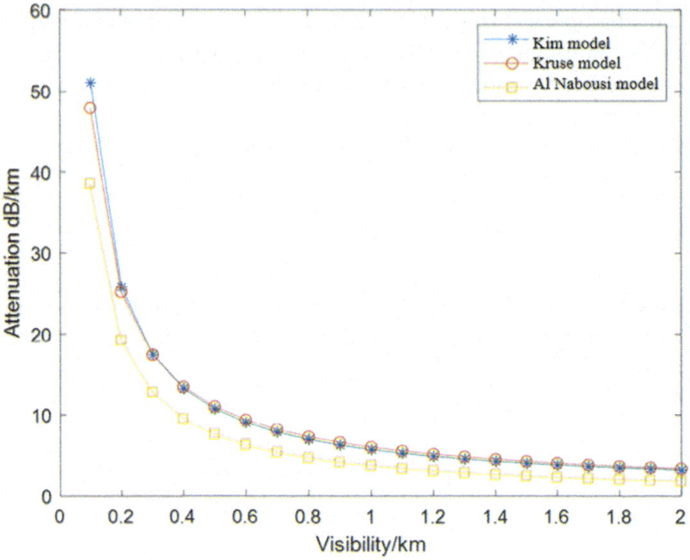

Fig. 8.18 Comparison of three empirical decay models [34]

8.3.3 Characteristics of a UV-Light Scattering Channel on a Dusty Day

Sandy and dusty weather is caused by strong winds that involve sand and soot particle into the atmosphere, and suspended in the atmosphere to form a stable form of dust particles, soot particles, aerosol, etc. As a result, the visibility and air quality are reduced.

The distribution of dust aerosol particles in the air differs in different grades of dusty weather. In dusty weather, the aerosol particles include dust, soot, and other particles that are difficult to dissolve in water. The size of aerosol particles in dusty weather follows a log-normal distribution, and its scale-spectrum function is [35]

$$n(r) = \frac{N_0}{r\sqrt{2\pi}\lg\sigma}\exp\left\{-\frac{1}{2}\left[\frac{\lg r - \lg r_m}{\lg\sigma}\right]^2\right\}, \tag{8.71}$$

where r is the radius of aerosol particles, r_m is the mean radius of the particles, σ is the standard deviation, and N_0 is the number density of the particles. In this case, the probability density function for the size spectrum of the aerosol is

$$p(r) = \frac{1}{r\sqrt{2\pi}\lg\sigma}\exp\left\{-\frac{1}{2}\left[\frac{\lg r - \lg r_m}{\lg\sigma}\right]^2\right\}. \tag{8.72}$$

Table. 8.5 Model parameters of the scale distribution of different types of aerosols

	Dust type	Soot type
r_m	0.5	0.0118
σ	2.99	2.00

Table 8.5 presents the relevant parameters for several types of aerosols commonly seen in dusty weather [36].

Because N_0 is a very difficult physical quantity to measure, the visibility V is typically used to describe the concentration of dust aerosols when studying the attenuation of UV light in dusty weather.

$$N_0 = \frac{15}{8.686 \times 10^3 \pi V \int_{r_1}^{r_2} r^2 p(r) dr} \tag{8.73}$$

Based on the parameters of different types of aerosols and the log-normal distribution function of the aerosol-scale spectrum, the scale-spectrum distribution of aerosol particles under dusty weather with different visibilities is obtained.

As can be seen from Fig. 8.19, the particle radii of different types of aerosols differ in dusty weather. The particle radius of dust-type aerosols is mostly around 10^{-2}–1 μm, while that of soot-type aerosols is mostly around 10^{-3}–10^{-1} μm.

Table 8.6 shows the particle-number densities of dust-type and soot-type particles at different visibilities. As can be seen from the table, the higher the visibility, the lower the number density of the aerosol particles. From Eq. (8.72), we find that the probability-density function of the same aerosol particle's scale-spectrum distribution is the same. Under the same scale-spectrum distribution probability-density function, visibility is the only variable that affects the number and density of aerosol particles; therefore, the visibility is inversely proportional to the number and density of aerosol particles.

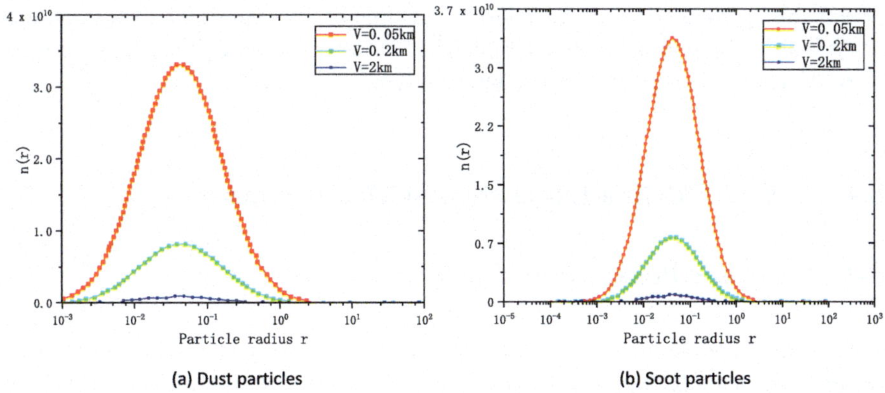

Fig. 8.19 Different particle-scale spectral distributions under dusty weather [36]

Table 8.6 Aerosol particle-number densities at different visibilities

		Number density of particles	
Visibility/km	0.05	0.2	2
Dust particles	1.1×10^7	2.75×10^6	2.64×10^5
Soot particles	1.1×10^{10}	2.75×10^9	2.67×10^8

In sandy/dusty weather, the visibility affects the number of aerosol particles and therefore, the light-attenuation coefficient. Therefore, visibility is considered as an independent variable of the empirical-attenuation model. According to the Mie scattering theory, the attenuation capacity of ultraviolet light per unit distance is

$$A = 4.343 \times 10^3 \int_{r_1}^{r_2} N_0 C_{ext}(m, x) P(r) dr, \tag{8.74}$$

where $x = 2\pi r/\lambda$ is the scale parameter of aerosol particles and A is in dB/km. The above equation reduces to

$$\begin{aligned} A &= 4.343 \times 10^3 \int_{r_1}^{r_2} N_0 C_{ext}(m, x) P(r) dr \\ &= \frac{15 \int_{r_1}^{r_2} C_{ext}(r) p(r) dr}{2\pi V \int_{r_1}^{r_2} r^2 p(r) dr}. \end{aligned} \tag{8.75}$$

At the incident wavelength of 260 nm, the attenuation coefficients of the above formulas in sandy/dusty weather and foggy weather are simulated. The results are shown in Fig. 8.20.

As can be seen from Fig. 8.20, in the UV band, the light attenuation of two types of aerosols on foggy days is much larger than that of two types of aerosols on dusty days; both are exponential attenuation. The UV attenuation abilities of three types of aerosols are fog type, dust type, and soot type.

8.4 Noise-Modeling Experimental Measurements

8.4.1 *Surface Background-Light Noise Model*

When sunlight passes through the atmosphere, the action of scattering elements, such as atmospheric molecules and aerosol particles, will cause the sunlight to scatter and form the sky background light. The complexity of the atmospheric channel causes the radiation intensity of the sky background light to change constantly, which increases

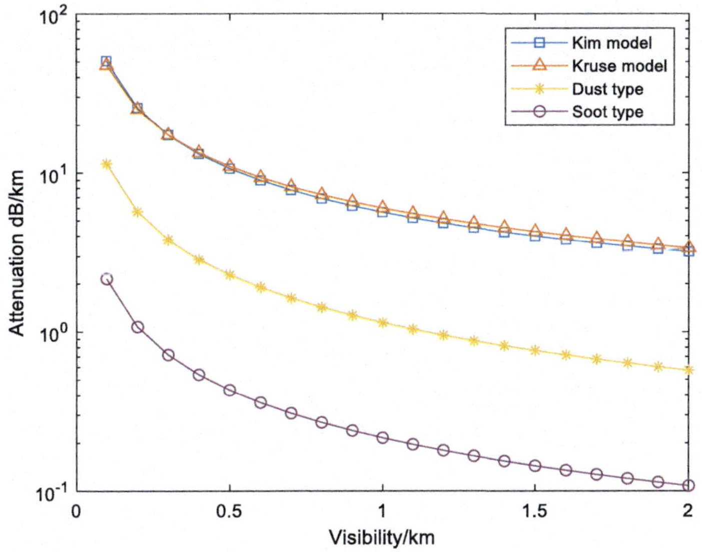

Fig. 8.20 Attenuation coefficients of the four attenuation models [36]

the difficulty of theoretical calculation. Measured and theoretical values vary for different types of weather.

8.4.1.1 Analysis of Clear-Weather Measurements

To visualize how the peaks change on different dates and to verify the accuracy of the theoretical model of Eq. (8.21), the measured and theoretical values of ultraviolet background noise in Xi'an in January and February of 2023 were compared. The measured conditions were the simulation conditions described above. The UV background noise in the 220 nm $\pm$ 10 nm and 300 nm $\pm$ 10 nm bands was measured and the results are shown in Fig. 8.21.

The peak value of the ultraviolet radiation on January 7 was small and the wave peak was at about 12:41. As time went on, the daytime became longer, the ultraviolet background radiation increased, and the peak value gradually increased. As the time of appearance approached 13:00, the peak shift reached its maximum on February 27, and the peak size continued to increase. The general trend of UV radiation in clear weather is consistent with the theoretical model.

As can be seen from Fig. 8.21, the intensity of the UV radiation varied with different dates and bands. On January 7, the daily maximum radiation value at 300 nm was 6.14×10^{-7} μW, and the daily average UV radiation value was 3.46×10^{-7} μW. The daily maximum radiation values increased as the date was delayed. The daily maximum radiation at 300 nm was 1.03×10^{-6} μW and the daily average ultraviolet radiation was 5.95×10^{-7} μW on February 26. The maximum measured

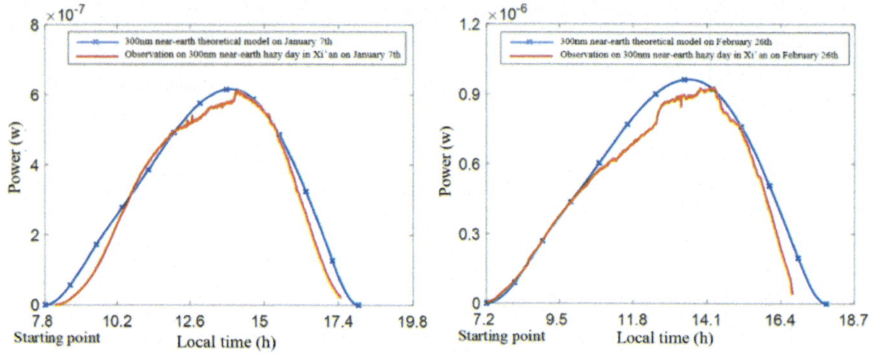

(a) Results at 300 nm on clear days at different times

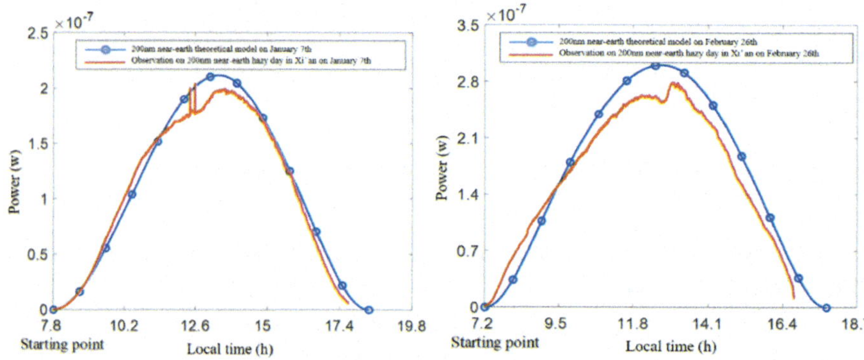

(b) Results at 220 nm on clear days at different times

Fig. 8.21 Theoretical results versus those observed on a clear day

on February 26 in the 300-nm band was 1.68 times higher than the maximum on January 7.

On January 7, the daily maximum radiation at 220 nm was $2.04 \times 10^{-7} \mu W$, and the daily average ultraviolet radiation was 1.13×10^{-7} μW. On February 26, the daily maximum radiation at 220 nm was 3.10×10^{-7} μW and the daily mean UV radiation was 1.87×10^{-7} μW. The maximum value of the 220-nm-band measurement data on February 26 increased by a factor of 1.52, compared to the maximum value on January 7.

8.4.1.2 Analysis of Field Measurements on Cloudy Days

In sunny weather, the changes of the theoretical value and the measured value are basically the same; however, in cloudy weather, because of the increase in the amount of clouds, the intensity of the ultraviolet radiation decreases seriously. Because of

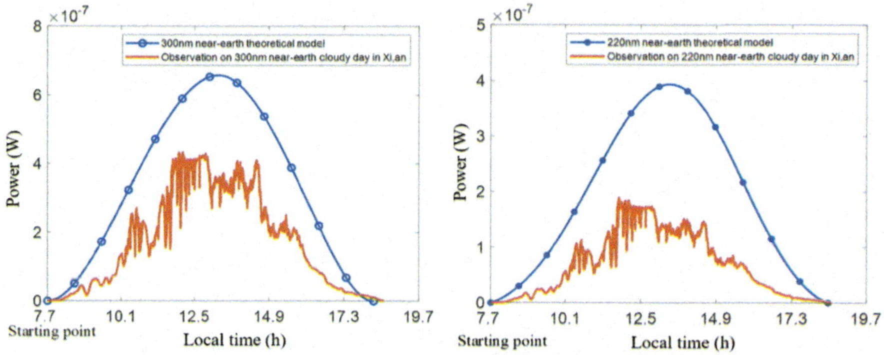

Fig. 8.22 Comparison between theoretical and field measurements on cloudy days

the influence of atmospheric turbulence, the cloud layer moves, the UV radiation fluctuates greatly, and the changing trend is irregular.

As can be seen from Fig. 8.22, the actual measured values are generally in line with the theoretical trend, but their values fluctuate greatly and have obvious attenuation. In cloudy weather, the daily maximum radiation value in the 300-nm band is 4.18×10^{-7} μW and the daily average ultraviolet radiation value is 2.48×10^{-7} μW. The daily maximum radiation value in the 220-nm band is 1.90×10^{-7} μW and the average daily UV radiation value is 6.84×10^{-8} μW. This differs from the theoretical value, which cannot be used to describe the changing trend of UV radiation in cloudy weather.

8.4.1.3 Analysis of Observed Data on Rainy Days

On rainy days, the clouds become denser, and the scattering and absorption of water vapor make the attenuation more severe and reduces the UV radiation amplitude. Compared with the severe wave band on overcast days, the fluctuation range is gentler and more concentrated on rainy days, and the influence of atmospheric turbulence on the cloud layer is not significant.

As can be seen from Fig. 8.23, the actual measured values are similar to the theoretical trend, reaching their maximum at noon; however, the amplitude difference is large. In cloudy weather, the daily maximum radiation at 300 nm is 1.55×10^{-7} μW and the daily average ultraviolet radiation is 5.44×10^{-8} μW. The daily maximum radiation at 220 nm is 7.00×10^{-8} μW and the average daily UV radiation value is 2.56×10^{-8} μW. This is different from the theoretical value and cannot be used to describe the changing trend of UV radiation in cloudy weather.

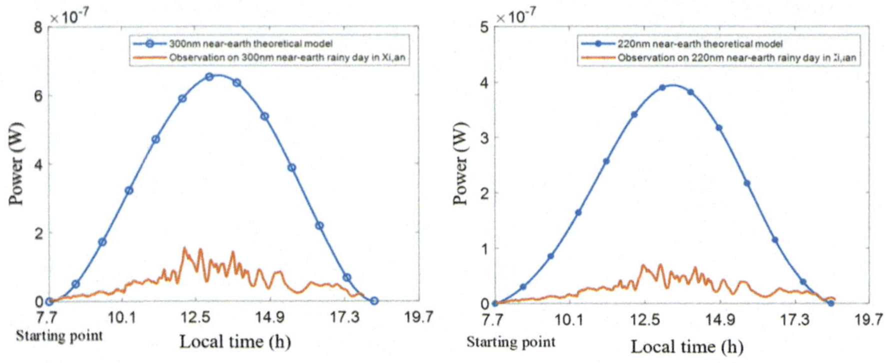

Fig. 8.23 Comparison between theoretical and rain measurements

8.4.1.4 Analysis of Measured Data on Hazy Days

The influence of the cloud layer and surface suspended particles in hazy weather makes the change of the UV background radiation more drastic. The UV background radiation does not simply change with the solar altitude angle, but is related to the degree of air turbidity. In the morning, the air is clearer and the UV radiation is higher. As the turbidity of the air increases, the UV radiation decreases continuously until the afternoon, when the UV radiation begins to rise. However, by this time the Sun's altitude angle has begun to decrease. There is a brief rise in the air turbidity recovery, and then the ultraviolet radiation begins to slowly decrease.

As can be seen from Fig. 8.24, the actual measured values do not conform to the theoretical trend and the amplitude difference is large. In cloudy weather, the daily maximum radiation at 300 nm is 1.35×10^{-7} μW and the daily average ultraviolet radiation is 4.44×10^{-8} μW. The daily maximum radiation at 220 nm is 4.00×10^{-8} μW and the daily mean UV radiation is 2.56×10^{-8} μW. This is different from the theoretical value, which cannot be used to describe the changing trend of UV radiation in foggy weather.

8.4.2 High-Altitude Background-Light Noise Model

In this study, the ultraviolet background noise at a high altitude was measured based on the existing surface-noise model to restore the high-altitude noise model. To ensure the accuracy of the high-altitude model, the ultraviolet background noise at 10 km in Urumqi, 8 km in Beijing, 8 km in Nanjing, and 7 km in Yantai was measured from Xi'an. The models were restored by cubic-spline interpolation.

The principle is to divide the interval [a, b] into n intervals $[((x_o, x_1),(x_1, x_2), \ldots,(x_{n-1}, x_n))]$, and there are $n + 1$ points. $x_o = a, x_n = b$, two of which are endpoints. A cubic-spline interpolation is a cubic equation with a curve in each

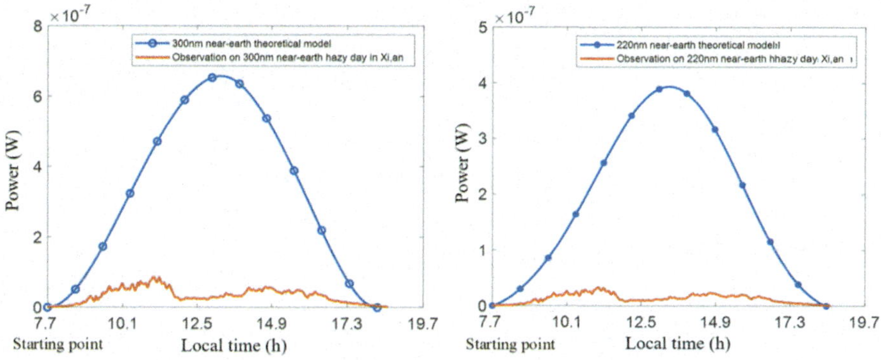

Fig. 8.24 Comparison between theoretical and actual measurements on a hazy day

interval. The cubic equation satisfies the following condition: in each interval $[x_i, x_{i+1}]$, $S(x) = S_i(x)$ is a cubic equation. It satisfies $S_i(x) = y_i$ $(i = 0, 1, ..., n)$, and $S(x)$, $S'(x)$, and $S''(x)$ are continuous. Then, these three conditions can be used to construct the equation $S_i(x) = a_i + b_i x + c_i x^2 + d_i x^3$; by solving a_i, b_i, c_i, and d_i, the complete interpolation function can be obtained.

8.4.2.1 Xi'an–Urumqi Altitude of 10 km

As shown in Fig. 8.25, the noise model of the Xi'an–Urumqi 10-km high-altitude route is restored. The round-trip flights to Urumqi are HU7898 and GS7525, which run from 12:25 to 16:10 on March 11, 2023 and from 7:45 to 11:15 on March 12. As shown in the figure, the high-altitude noise model also accords with the diurnal variation law of the solar-altitude angle.

The maximum value of the high-altitude ultraviolet background-noise model of 300 nm in Xi'an–Urumqi is 3.416 μW, and the peak appears at about 13.48 min,

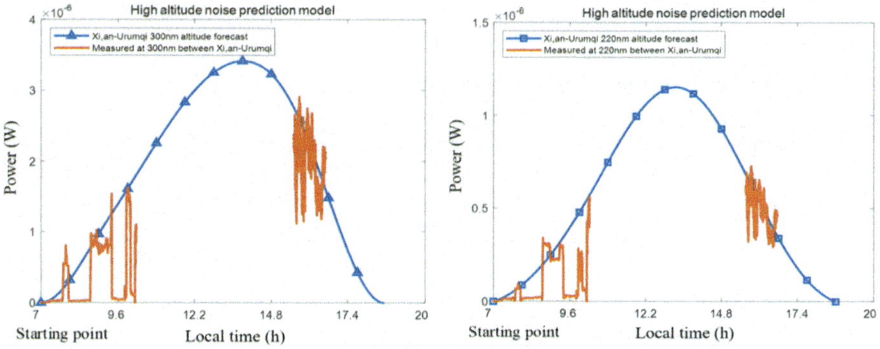

Fig. 8.25 Xi'an–Urumqi, at an altitude of 10 km on March 11

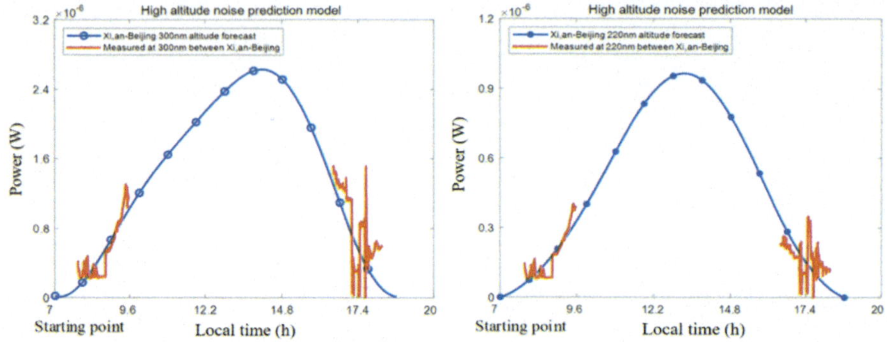

Fig. 8.26 Xi'an-Beijing at an altitude of 8 km on March 7

3.322 times as much as that at the surface at the same time. At 220 nm, the ozone absorption is more serious and the attenuation is more obvious. The maximum value of background noise is 1.152 μW, and the peak ozone time is 12:48, which is 3.152 times of that at the same time on the surface.

8.4.2.2 Xi'an-Beijing Altitude of 8 km

As shown in Fig. 8.26, the noise model of the 8-km high-altitude flight route between Xi'an and Beijing is restored. The flights to and from Beijing were HU7238 and HU7237, which ran between 15:30 and 18:00 on March 6 and 7:50 and 10:15 on March 7.

As the figure shows, with the decrease of altitude, the sunlight propagation path increases and the UV radiation decreases. The maximum value is 2.631 μW, and the peak appears at about 13:36, which is 2.623 times that at the same time on the surface. At 220 nm, the absorption of ozone is more serious and the attenuation is more obvious. The maximum value of background noise is 0.967 μW, and the peak ozone time is 12:36, which is 2.696 times that at the same time on the surface.

8.4.2.3 Xi'an-Nanjing Altitude of 8 km

As shown in Fig. 8.27, the noise model of the 8-km high-altitude flight route between Xi'an and Nanjing is restored. The flights to and from Nanjing were 3U3259 and MU2200 between 7:00 and 9:05 on April 15 and 15:55 and 17:50 on April 16.

As shown in the figure, the maximum of the 300-nm UV background noise in Xi'an–Nanjing is 2.551 μW, which is close to the UV radiation at an 8-km altitude in Xi'an–Beijing. The peak appears at about 13:42, 2.552 times as large as that of the surface at the same time. The 220-nm band is close to the Beijing route, and the maximum background noise is 0.951 μW. The peak time appears at about 13:00 h, which is 2.58 times that at the same time on the surface.

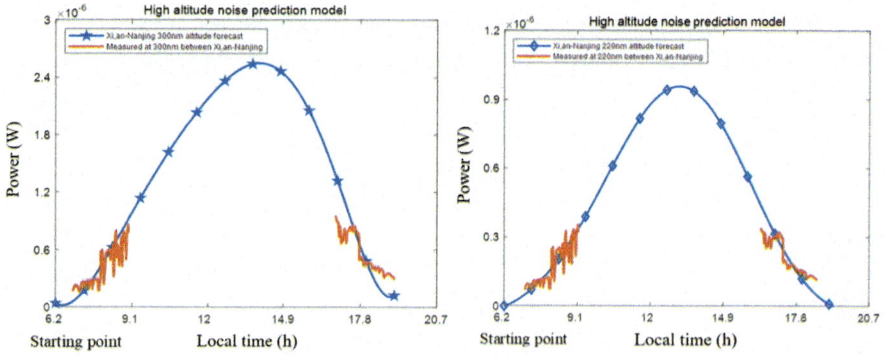

Fig. 8.27 Xi'an-Nanjing at an altitude of 8 km on April 15

8.4.2.4 Xi'an–Yantai Altitude of 7 km

As shown in Fig. 8.28, the noise model of the Xi'an-Yantai 7-km high-altitude flight route is restored. Flights 9H8397 and HU7506 to Yantai ran between 15:50 to 17:50 on March 31 and 9:40 to 12:00 on April 3.

As shown in the figure, the background noise of 300-nm ultraviolet light at a high altitude from Xi'an to Yantai increases with the decrease of altitude; the sunlight propagation path increases, and the ultraviolet radiation decreases. The maximum value is 2.324 μW, and the wave peak appears at about 14:00, 2.260 times that at the surface at the same time. The maximum value of UV background noise at 220 nm is 0.853 μW, and the peak time appears at about 13:00, which is 2.33 times that at the same time on the surface.

Table 8.7 shows the peak time, maximum peak, and multiples of the UV background-radiation intensity relative to the surface at different altitudes.

The experimental results show that the UV background noise at 220 nm is consistent with the variation of the solar-altitude angle, and reaches its maximum near

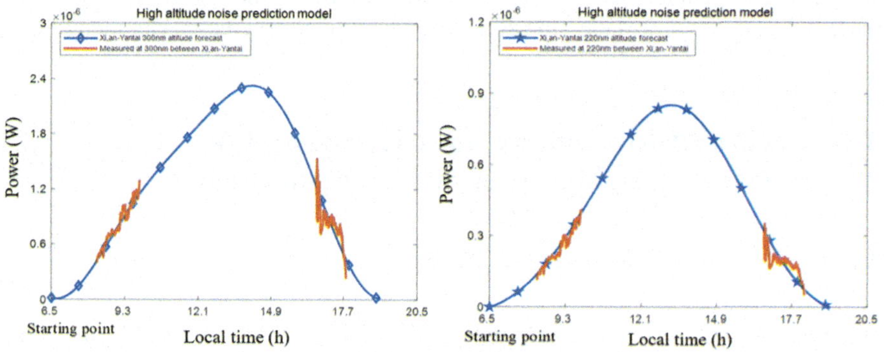

Fig. 8.28 Xi'an-Yantai at an altitude of 7 km on March 31

Table. 8.7 Model parameters of high-altitude background noise in different cities

Region/Wavelength (nm)	Altitude (km)	Peak time (h)	Peak amplitude (μW)	Multiple
Urumqi/300	10	13.8	3.416	3.322
Urumqi/220	10	12.8	1.152	3,0.152
Beijing/300	8	13.6	2.631	2.623
Beijing/220	8	12.6	0.967	2.696
Nanjing/300	8	13.7	2.551	2.552
Nanjing/220	8	13	0.951	2.58
Yantai/300	7	14	2.324	2.26
Yantai/220	7	13	0.853	2.33

1300. However, the peak UV background noise at 300 nm is slightly shifted, and the peak appears at about 1400.

The maximum ultraviolet background noise at 300 nm at the 7 km, 8 km, and 10 km altitudes were 2.324 μW, 2.631 μW, and 3.416 μW, respectively, which were 2.26 times, 2.623 times, and 3.322 times those of the surface noise, respectively. The UV background noise at 220 nm was 0.853 μW, 0.967 μW, and 1.152 μW, which were 2.33 times, 2.696 times, and 3.15 times those of the surface noise, respectively.

Above the clouds, the UV radiation increases by 20–23% for every 1000-m increase in altitude. For an all-weather ultraviolet communication system, the threshold of the filter can be adjusted flexibly to filter the background noise according to the time change in the clear weather, to improve the stability of the UV communication system.

8.4.2.5 Data Correlation Function Analysis

Figure 8.29 shows a correlation function of the 300-nm and 220-bands nm at the height of each city. The results show that the correlation between the two bands is strong. The correlation between the two bands is the strongest at the same time, and increases with the time delay; the correlation coefficient decreases exponentially.

8.5 Intensity-Distribution Characteristics of UV Non-line-of-Sight Scattering Communication

8.5.1 UV Non-line-of-Sight Model

When there is atmospheric turbulence in the UV communication link, the rough geometry of an NLOS UV single-scattering transmission is shown in Fig. 8.30. The NLOS transmission path is divided into two LOS links: the first is from T_X to the

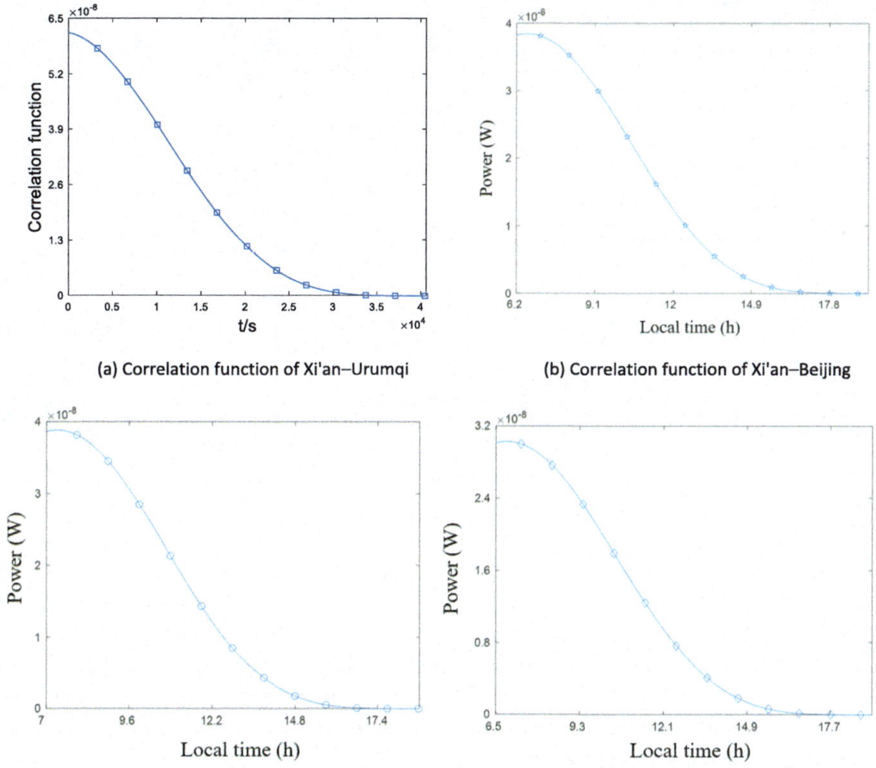

(a) Correlation function of Xi'an–Urumqi

(b) Correlation function of Xi'an–Beijing

(c) Correlation function of Xi'an–Nanjing

(d) Correlation function of Xi'an–Yantai

Fig. 8.29 Correlation functions of 300 and 220 nm bands at high altitudes

public volume V (the link length is L) and the other is from the scatter volume V to Rx (the link length is L). It is assumed that the emission plane of T_X and the angle of the T_X beam are small, so that the scintillation caused by turbulence in the scattering volume V is constant. For ease of processing, the single-scattering principle is used to calculate the received optical power.

When considering UV light attenuation, not only the light-power attenuation caused by scattering and absorption, but also the scintillation attenuation caused by turbulence is considered to obtain a more-accurate light-power probability-density function (PDF). The scintillation attenuation of the two LOS links is

$$\alpha_{turb\_l} = 2.\sqrt{23.17.C_n^2.(2\pi)^{7/6}.l^{11/6}} \tag{8.76}$$

$$\alpha_{turb\_L} = 2.\sqrt{23.17.C_n^2.(2\pi)^{7/6}.L^{11/6}}. \tag{8.77}$$

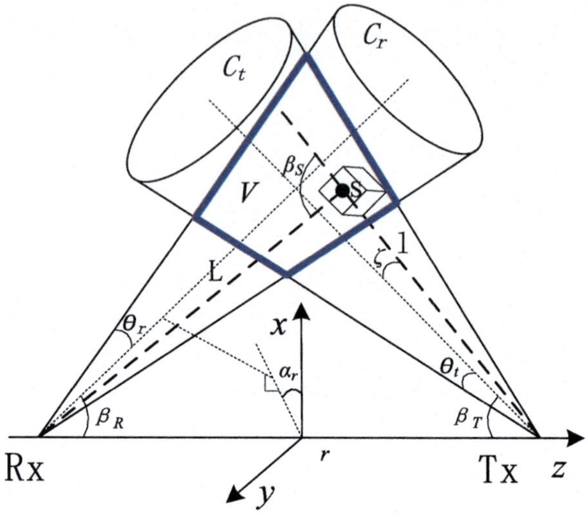

Fig. 8.30 Ultraviolet optical communication link in a turbulent atmosphere

Equation (8.76) is used to simulate the scintillation attenuation caused by turbulence, as shown in Fig. 8.31. It can be seen that the scintillation attenuation increases with the increase of communication distance or turbulence intensity. In addition, with the decrease of the light wavelength, the scintillation attenuation will be larger and the scintillation will be stronger. This means that, compared with the traditional free-space optical communication, UV light is more obviously affected by turbulence because of its smaller wavelength.

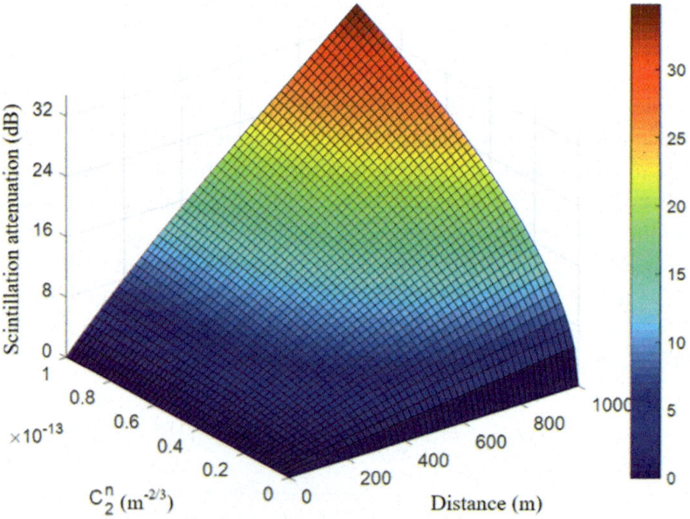

Fig. 8.31 Scintillation decay with turbulence intensity and distance

The entire UV communication link runs from the transmitting end to the scattering point and the scattering point to the receiving end. The flicker attenuation of the entire non-line-of-sight UV communication link is as follows:

$$\alpha_{turb} = \alpha_{turb\_l} + \alpha_{turb\_L}. \tag{8.78}$$

For the pre-scattering link, the log-normal distribution of the light power to the scattering volume V is the following PDF:

$$p(P_i) = \frac{1}{\sqrt{2\pi}\sigma_l P_l} \exp\left\{ -\frac{[\ln(P_l/P_{l0}) - \langle \ln(P_l/P_{l0}) \rangle]^2}{2\sigma_l^2} \right\}, \tag{8.79}$$

where P_i is the power at the scattering volume V and σ_l^2 is determined by a plane wave or a spherical wave. The average optical power when turbulence is not considered is $P_{l0} = P_T k_s V \exp(-k_s l)/(\Omega_T l^2)$, where P_T is the emitted optical power. $\langle P_l/P_{l0} \rangle$ can be expressed as

$$\langle P_l/P_{l0} \rangle = \exp\left[\langle \ln(P_l/P_{l0}) + \sigma_l^2/2 \rangle \right]. \tag{8.80}$$

α_{PL_l} is defined as path loss due to absorption and scattering:

$$\alpha_{PL\_l} = 10\lg(P_T/P_{l0}). \tag{8.81}$$

The total path loss when considering attenuation due to turbulent scintillation is

$$\alpha_{turb\_l} + \alpha_{turb\_L} = 10\lg(P_T/\langle P_l \rangle). \tag{8.82}$$

Applying a geometric transformation to the formula above, yields

$$\langle \ln(P_l/P_{l0}) \rangle = -\sigma_l^2/2 - (\alpha_{turb\_l} \cdot \ln 10)/10. \tag{8.83}$$

Introducing Eq. (8.83) into Eq. (8.79), yields

$$p(P_l) = \frac{\exp\left\{ -\frac{\left[\ln(P_l/P_{l0}) + \sigma_L^2/2 + (\alpha_{\mathrm{turb\_l}} \cdot \ln 10)/10\right]^2}{2\sigma_L^2} \right\}}{\sqrt{2\pi}\sigma_l P_l}. \tag{8.84}$$

Similarly, for scattered links, the condition PDF for the optical power received by RX is

$$p(P_R|P_l) = \frac{\exp\left\{ -\frac{\left[\ln(P_R/P_{L0}) + \sigma_L^2/2 + (\alpha_{\mathrm{turb\_L}} \cdot \ln 10)/10\right]^2}{2\sigma_L^2} \right\}}{\sqrt{2\pi}\sigma_L P_R}, \tag{8.85}$$

where p is the received light power and the variance is calculated from the turbulence intensity.

Excluding turbulence, the mean received light power is $P_{L0} = P_l P(\theta_s) A_R \cos(\xi) \exp(k_e L)/(4\pi L^2)$. Thus, the edge PDF is

$$p(P_R) = \int p(P_l, P_R) dP_l = \int p(P_R|P_l) p(P_l) dP_l, \tag{8.86}$$

8.5.2 UV Non-line-of-Sight PDF Experimental Measurements

Figure 8.32 shows the probability-density function of measured ultraviolet light intensity at different transmit–receive elevations.

According to Fig. 8.32, the received probability-density function conforms to the Gaussian model. After the received power is normalized, the variance of the PDF differs under different receiving and transmitting elevation angles. The path loss differs when the transmitted light intensity of the transmitter passes through the atmospheric channel.

When the elevation angle of the signal is increased from $\beta_R = \beta_T = 20^\circ$ to $\beta_R = \beta_T = 40^\circ$, the energy of the signal decreases gradually, and the variance of the PDF of the signal strength changes little. However, when the elevation angle of the signal

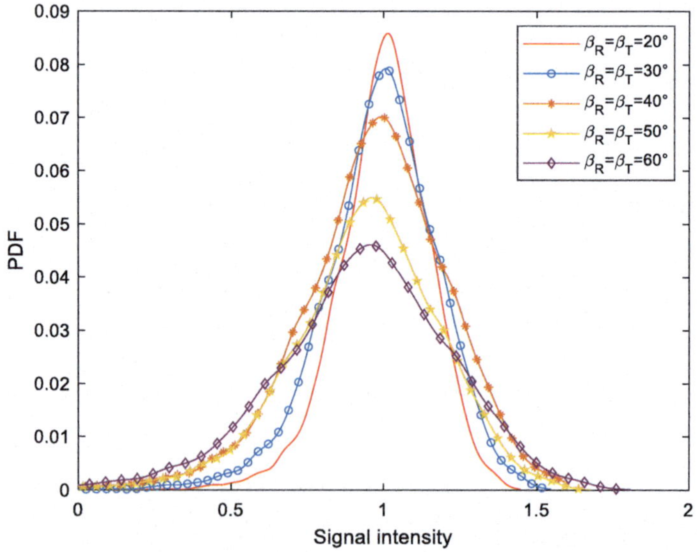

Fig. 8.32 Probability-density function of intensity at different elevations

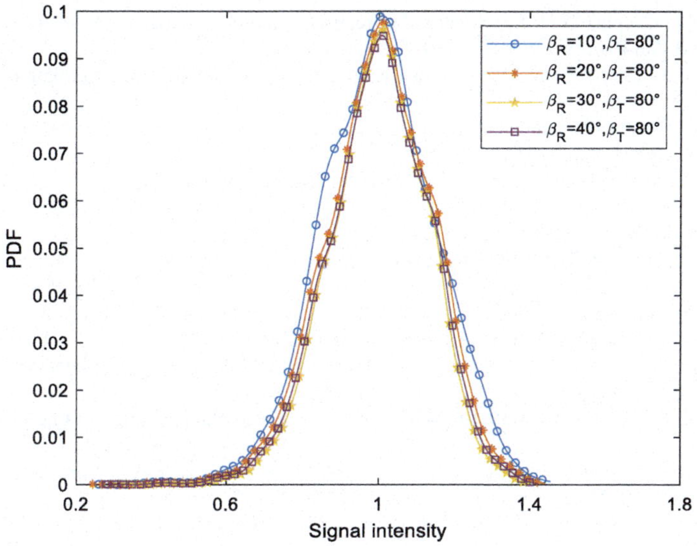

Fig. 8.33 Probability density function of light at different elevations

increases to $\beta_R = 60°$ and $\beta_T = 60°$, the attenuation of the signal energy is more serious; that is, with the synchronous increase of β_R and β_T, the variance of the signal strength PDF increases, and the energy attenuation gradually increases.

Figure 8.33 shows the probability-density function of the UV intensity measured by changing the elevation angle of the receiving end when the emitter elevation is fixed at $\beta_T = 80°$.

According to Fig. 8.33, the received PDF conforms to the Gaussian model. After the received power is normalized, the variance of the PDF differs under different receiving and transmitting elevation angles. The path loss is different when the transmitted light intensity of the transmitter passes through the atmospheric channel. With the increase of the receiving elevation angle β_R, the attenuation of the signal energy is very small, and the variance of signal strength probability density function is very small. This shows that the influence on the energy attenuation of the communication signal β_T is greater than β_R. Therefore, we should fix the receiving-elevation angle and reduce the transmitting-elevation angle to reduce the path loss when designing the ultraviolet transceiver system.

References

1. Chen G, Xu Z, Sadler BM (2010) Experimental demonstration of ultraviolet pulse broadening in short-range non-line-of-sight communication channels. Opt Express 18(10):10500–10509
2. Yong Z, Yanfeng L, Dengjia W et al (2022) Applicability analysis of daily total solar radiation calculation models in different climate zones in China and optimization of general calculation models. Solar Energy Sinica 43(09):1–7

3. Kasten F, Czeplak G (1980) Solar and terrestrial radiation dependent on the amount and type of cloud. Sol Energy 24(2):177–189
4. Prescott JA (1940) Evaporation from a water surface in relation to solar radiation. Trans Roy Soc S Aust 46(1):114–118
5. Chen S, Hinson JW, Lee J et al (2001) Branching fraction and photon energy spectrum for b→ s γ. Phys Rev Lett 87(25):13–15
6. Fu W (1984) Multiphoton process. Prog Laser Optoelectron 21(1):15
7. Zhixun H (2019) Several issues in the theory and application of single photon technology. J Commun Univ China (Nat Sci Ed) 26(02):1–18
8. Einstein A (1965) Concerning an heuristic point of view toward the emission and transformation of light. Am J Phys 33(5):367
9. Kirchhoff GI (1860) On the relation between the radiating and absorbing powers of different bodies for light and heat. Lond, Edinb, Dublin Philos Mag J Sci 20(130):1–21
10. Reda I, Andreas A (2004) Solar position algorithm for solar radiation applications. Sol Energy 76(5):577–589
11. Grena R (2008) An algorithm for the computation of the solar position. Sol Energy 82(5):462–470
12. Emilio M, Kuhn JR, Bush RI et al (2012) Measuring the solar radius from space during the 2003 and 2006 Mercury transits. Astrophys J 750(2):135
13. Ackermann ASE (1915) The utilisation of solar energy. J R Soc Arts 63(3258):538–565
14. Wong LT, Chow WK (2001) Solar radiation model. Appl Energy 69(3):191–224
15. Xianjin F, Duming W (1988) Time variation of atmospheric transparency coefficient in my country. Acta Geogr Sin 43(01):52–59
16. Yanli H, Yufang H (2004) Analysis on the causes of changes in ultraviolet radiation. J Heze Teach Colle 26(04):70–71
17. Jingmiao L, Yuguo D, Yongde H et al (2003) Correlation analysis of solar ultraviolet radiation intensity and meteorological elements. Plateau Meteorol 22(01):45–50
18. Xin W, Jun W, Rong L et al (2011) The influence of precipitation process on the near-surface solar radiation in Badain Jaran Desert. Arid Meteorol 29(4):427–432
19. Liequn H, Haiyan I (2008) Research on atmospheric transparency in Xinjiang. China Desert 28(2):332–337
20. Planck M (1901) On the law of distribution of energy in the normal spectrum. Ann Phys 4(553):1–2
21. Coskun C, Oktay Z, Dincer I (2011) Estimation of monthly solar radiation distribution for solar energy system analysis. Energy 36(2):1319–1323
22. Xu Z, Sadler BM (2008) Ultraviolet communications: potential and state-of-the-art. IEEE Commun Mag 46(5):67–73
23. Razavi M, Shapiro JH (2005) Optical-wireless communications via diversity reception and optical preamplification. IEEE Trans Wireless Commun 4(3):975–983
24. Yang C (1987) Twenty-four solar terms. Knowl Lit Hist 11:46–49
25. Luettgen MR, Shapiro JH, Reilly DM (1991) Non-line-of-sight single-scatter propagation model. JOSA A 8(12):1964–1972
26. Reilly DM, Warde C (1979) Temporal characteristics of single-scatter radiation. JOSA 69(3):464–470
27. Xizheng K (2011) Ultraviolet light self-organizing network theory. Science Press, Beijing, pp 40–42
28. Peng S, Chun L, Lei Z et al (2020) Analysis of single scattering path loss in wireless ultraviolet communication in mobile scenarios. Acta Opt Sin 40(04):28–37
29. Hale GM, Querry MR (1973) Optical constants of water in the 200-nm to 200-μm wavelength region. Appl Opt 12(3):555–563
30. Lei L, Hao L, Taichang G (2008) Approximate ellipsoidal model of raindrops and its near-infrared scattering characteristics. Meteorol Sci 28(3):271–275
31. Qian H, Xiaoyi L, Zhihao H et al (2011) The influence of different weather on the non-line-of-sight transmission of ultraviolet light. Commun Countermeasures 3:13–15

32. Zhengfang S (1990) Fundamentals of applied atmospheric optics. Meteorological Press, Beijing, pp 17–20
33. Yong L, Zhiyong X, Jingyuan W et al (2013) Study on non-line-of-sight atmospheric scattering transmission characteristics in fog environment. Acta Opt Sin 33(9):11–14
34. Bo L, Pengcheng L, Jin L et al (2019) Bit error rate performance of free space optical communication system under different fog concentrations. Electron Inf Countermeasure Technol 34(6):44–47
35. Zhong H, Zhou J, Du Z et al (2018) A laboratory experimental study on laser attenuations by dust/sand storms. J Aerosol Sci 121:31–37
36. Hongxia W, Youzhang Z, Tao T et al (2013) Research on the transmission characteristics of laser in different types of aerosols. Acta Phys Sin 62(2):324–333

Chapter 9
Example of a Noise-Model Analysis

This chapter begins with the mechanism and mathematical model for noise generation in the three parts of a communication system: laser, channel, and detector. Based on the optical-wireless communication system shown in Fig. 1.2, and the causal relationships affecting optical-wireless communication shown in Fig. 1.1, various factors affecting the noise model in a wireless laser communication system are analyzed. A unified noise model for optical-wireless communication systems is established by considering various noise factors. This chapter provides an example of applying the analysis theory presented in this book.

9.1 Noise Model

As shown in Fig. 1.2, the noise model of an optical-wireless communication system can be represented as [1–4]

$$P(t) = f_\beta f_{h_p} f_A f_d \Big(P_0(t) \Big) + \left(P_b + P_n \right) \tag{9.1}$$

where $P(t)$ represents the received optical power; f_β is the attenuation coefficient per unit distance β, related to light-intensity attenuation terms; f_{hp} is the aiming-error term; f_A is the light-intensity scintillation term caused by turbulence; f_d is the term for the distortion caused by the light source to the signal light; $P_0(t)$ is the ideal-signal optical power; P_b is the background-light noise power; and P_n is the sum of the detector noise power (which may include shot noise, generation–recombination noise, photon noise, etc.). Among them, f_β is a constant:

$$f_\beta = \exp(-\sigma L), \tag{9.2}$$

X. Ke and C. Ke, *Noise Models in Optical-Wireless Communication Systems*,
Optical Wireless Communication Theory and Technology,
https://doi.org/10.1007/978-981-97-7550-7_9

where σ is the attenuation coefficient and L is the propagation distance. f_{hp} is the probability of receiving the spot power, owing to the influence of the aiming error:

$$f_{h_p}(h_p) = \frac{\gamma^2}{A_0^{\gamma^2}} h_p^{\gamma^2 - 1}, \tag{9.3}$$

where γ^2 represents the ratio of the square of the equivalent radius of the received beam to the variance of the spot jitter, A_0 represents the maximum optical power that the antenna can receive when $r = 0$, and h_p is the received spot power. Taking a weakly-turbulent atmospheric channel as an example, f_A can be expressed as (lognormal model) [1]

$$f_A = f_I(I) = \frac{1}{2I} \frac{1}{\sqrt{2\pi\sigma_\chi^2}} \exp\left(- \frac{\left[\ln(I/I_0) + 2\sigma_\chi^2\right]^2}{8\sigma_\chi^2} \right), \tag{9.4}$$

where I is the received light intensity, I_0 is the received light intensity that has not been affected by turbulence, and ${\sigma_\chi}^2$ is the logarithmic variance of the received light intensity (i.e., Rytov variance ${\sigma_R}^2$).

9.2 Light-Source Noise Model

9.2.1 *Laser Static Nonlinear Distortion*

The static nonlinear distortion of a laser is caused by factors such as the intracavity loss, spatial hole-burning effect, and leakage current. For an ideal semiconductor laser, the relationship between the output optical power and the current is satisfied when the driving current is greater than the threshold current:

$$\frac{P - P_{th}}{I - I_{th}} = \frac{\mathrm{d}P}{\mathrm{d}I}, \tag{9.5}$$

where I_{th} is the threshold current and P_{th} is the output optical power of the laser when the driving current is the threshold current. The Taylor-series expansion of P near P_{th} can be written as

$$P = P_{th} + \sum_{k=1}^{\infty} \frac{1}{k!} \frac{\mathrm{d}^k P}{\mathrm{d}I^k} (I - I_{th})^k. \tag{9.6}$$

If the laser modulation is a subcarrier signal, the driving current can be expressed as

$$I(t) = I_{dc} + I_m \cos(2\pi f_{sc} + \theta_{sc}). \tag{9.7}$$

Substituting the above equation into Eq. (9.6) and expanding it to a summation form with the Peano remainder yields

$$P = P_{th} + \frac{\mathrm{d}P}{\mathrm{d}I}(I - I_{th}) + \frac{1}{2}\frac{\mathrm{d}^2 P}{\mathrm{d}I^2}(I - I_{th})^2 + o(I - I_{th})^3, \tag{9.8}$$

where $o\ (\cdot)$ is the Peano remainder. Because the P–I curve of the laser is not an ideal straight line, the light emitted from the laser will contain many higher-order harmonics, which will distort the modulation signal. This type of distortion typically occurs at points far from the emission threshold. The modulation index m can be introduced to measure the difference between the modulation signal and the emission threshold:

$$m = \frac{I_m}{P_{th}}\frac{\mathrm{d}P}{\mathrm{d}I}. \tag{9.9}$$

We rearrange Eq. (9.8) as

$$\begin{aligned} \frac{P}{P_{th}} &= 1 + m\cos(\omega t + \theta_n) \\ &+ \frac{1}{2}m^2 P_{th}\left[\frac{\mathrm{d}^2 P}{\mathrm{d}I^2} \Big/ \left(\frac{\mathrm{d}P}{\mathrm{d}I}\right)^2\right][\cos(\omega t + \theta_n)]^2 \\ &+ \frac{1}{6}m^3 P_{th}^2\left[\frac{\mathrm{d}^3 P}{\mathrm{d}I^3} \Big/ \left(\frac{\mathrm{d}P}{\mathrm{d}I}\right)^3\right][\cos(\omega t + \theta_n)]^3. \end{aligned} \tag{9.10}$$

Correspondingly, the ratio of the second harmonic to the fundamental component is

$$HD_2 \approx \frac{1}{4}mP_{th}\left[\frac{\mathrm{d}^2 p}{\mathrm{d}I^2} \Big/ \left(\frac{\mathrm{d}p}{\mathrm{d}I}\right)^2\right]. \tag{9.11}$$

Similarly, the direct ratio of the third harmonic to the fundamental component is

$$HD_3 \approx \frac{1}{24}m^2 P_{th}^2\left[\frac{\mathrm{d}^3 p}{\mathrm{d}I^3} \Big/ \left(\frac{\mathrm{d}p}{\mathrm{d}I}\right)^3\right]. \tag{9.12}$$

9.2.2 Generation–Recombination Noise

The difference between photoconductive and photovoltaic detectors lies in the random recombination process in the movement of electrons and holes within them. Therefore, the shot noise of photoconductive detectors is caused not only by the current fluctuations generated by carriers but also by the fluctuations generated by carrier recombination. Therefore, the shot noise of photoconductive detectors is twice that of photovoltaic detectors, and is called generation–recombination noise. The noise current and voltage are denoted by i_{g-r} and u_{g-r}, respectively, and the photoconductivity gain is denoted by M. The noise current and voltage are calculated as follows:

$$i_{g-r} = \sqrt{4ei_d \Delta f M^2} \tag{9.13}$$

$$u_{g-r} = \sqrt{4ei_d \Delta f M^2 R^2} \tag{9.14}$$

9.2.3 Dynamic Nonlinear Laser Distortion

When the modulation current is a sine wave, periodic changes in the laser's internal electro-optical coupling process can also cause nonlinear distortion of the laser-output light intensity. This nonlinear distortion is related to the modulation frequency; hence, it is called dynamic nonlinear noise.

Starting from semiconductor physics, the process of a photon emitted by the laser changing with time can be described by the following rate equations:

$$\frac{\mathrm{d}N}{\mathrm{d}t} = \frac{J}{qV} - g_0 \frac{(N - N_0)S}{1 + \varepsilon S} - \frac{N}{\tau_n} \tag{9.15}$$

$$\frac{\mathrm{d}S}{\mathrm{d}t} = \Gamma g_o \frac{N - N_0}{1 + \varepsilon S} S + \frac{\Gamma \beta N}{\tau_n} - \frac{S}{\tau_p}, \tag{9.16}$$

where N is the carrier concentration in the laser's active layer, N_0 is the carrier concentration at zero gain, S is the photon density in the laser cavity, J is the current injected into the active layer, q is the electronic charge, V is the volume of the active layer, g is the gain coefficient of the photon density, and g_0 is the derivative of g to N. ε is the gain-compression factor, τ_n is the carrier lifetime in the resonant cavity, τ_p is the lifetime of photons in the resonant cavity, Γ is the light-limiting factor, and β is the spontaneous-emission factor.

The perturbation method is used to solve the rate equation, and J, N, and S are respectively expressed as the sum of the mean value and the disturbance term:

$$\begin{cases} J = \overline{J} + j(t) \\ N = \overline{N} + n(t) \\ S = \overline{S} + s(t) \end{cases} \tag{9.17}$$

Substituting these into the rate equation yields

$$\begin{aligned} \frac{\mathrm{d}n}{\mathrm{d}t} = \frac{J}{qV} - \left(\frac{g_0\overline{S}}{1+\varepsilon\overline{S}} + \frac{1}{\tau_n}\right)n - \left[\frac{g_0\left(\overline{N}-N_0\right)}{1+\varepsilon\overline{S}} - \frac{g_0\left(\overline{N}-N_0\right)\varepsilon\overline{S}}{\left(1+\varepsilon\overline{S}\right)^2}\right]s \\ - \left[\frac{g_0}{1+\varepsilon\overline{S}} - \frac{g_0\varepsilon\overline{S}}{\left(1+\varepsilon\overline{S}\right)^2}\right]ns + \left[\frac{\varepsilon g_0\left(\overline{N}-N_0\right)}{\left(1+\varepsilon\overline{S}\right)^2}\right]s^2 \end{aligned} \tag{9.18}$$

$$\begin{aligned} \frac{\mathrm{d}s}{\mathrm{d}t} = \left(\frac{\Gamma g_0\overline{S}}{1+\varepsilon\overline{S}} + \frac{\beta\Gamma}{\tau_n}\right) + \left[\frac{\Gamma g_0\left(\overline{N}-N_0\right)}{1+\varepsilon\overline{S}} - \frac{\Gamma g_0\left(\overline{N}-N_0\right)}{\left(1+\varepsilon\overline{S}\right)^2} - \frac{1}{\tau_p}\right]s \\ + \left[\frac{\Gamma g_0}{1+\varepsilon\overline{S}} - \frac{\Gamma\varepsilon\overline{S}}{\left(1+\varepsilon\overline{S}\right)^2}\right]n - \left[\frac{\Gamma\varepsilon g_0\left(\overline{N}-N_0\right)}{\left(1+\varepsilon\overline{S}\right)^2}\right]s^2 \end{aligned} \tag{9.19}$$

The undetermined-coefficient method was used to solve the above differential equation, assuming that the perturbation term in Eq. (9.17) is expressed as follows [4]:

$$\begin{cases} j(t) = j_1 e^{j\omega t} \\ n(t) = n_1 e^{j\omega t} \\ s(t) = s_1 e^{j\omega t} \end{cases} \tag{9.20}$$

The rate equation containing only the fundamental wave can be obtained:

$$j\omega n_1 = \frac{j_1}{qV} - An_1 - Bs_1 \tag{9.21}$$

$$j\omega s_1 = Cn_1 + Ds_1 \tag{9.22}$$

In this equation, parameters A, B, C, and D are represented as [4]

$$\begin{cases} A = \frac{g_0\overline{S}}{1+\varepsilon\overline{S}} + \frac{1}{\tau_n} \\ B = \frac{g_0\left(\overline{N}-N_0\right)}{1+\varepsilon\overline{S}} - \frac{g_0\left(\overline{N}-N_0\right)\varepsilon\overline{S}}{\left(1+\varepsilon\overline{S}\right)^2} \\ C = \frac{\Gamma g_0\overline{S}}{1+\varepsilon\overline{S}} + \frac{\beta\Gamma}{\tau_n} \\ D = \frac{\Gamma g_0\left(\overline{N}-N_0\right)}{1+\varepsilon\overline{S}} - \frac{\Gamma g_0\left(\overline{N}-N_0\right)}{\left(1+\varepsilon\overline{S}\right)^2} - \frac{1}{\tau_p} \end{cases} \tag{9.23}$$

When the perturbation term includes the second harmonic, Eq. (9.20) becomes

$$\begin{cases} j(t) = j_1 e^{j\omega t} \\ n(t) = n_1 e^{j\omega t} + n_2 e^{j2\omega t} \\ s(t) = s_1 e^{j\omega t} + s_2 e^{j2\omega t} \end{cases} \tag{9.24}$$

At this point, the rate equation containing the first and second harmonics can be obtained:

$$j2\omega n_2 = -An_2 - Bs_2 - En_1 s_1 + Fs_1^2 \tag{9.25}$$

$$j2\omega s_2 = Cn_2 + Ds_2 + \Gamma En_1 s_1 - \Gamma Fs_1^2 \tag{9.26}$$

In the equation, parameters E and F are represented as

$$\begin{cases} E = \frac{g_0}{1+\varepsilon\overline{S}} - \frac{g_0\varepsilon\overline{S}}{\left(1+\varepsilon\overline{S}\right)^2} \\ F = \frac{\varepsilon g_0\left(\overline{N}-N_0\right)}{\left(1+\varepsilon\overline{S}\right)^2} \end{cases} \tag{9.27}$$

By solving the rate equation, the following can be obtained [4]:

$$s_1 = \frac{j_1 C/qV}{g(\omega)} \tag{9.28}$$

$$s_2 = s_1^2 \frac{\left[-\left(\frac{ED}{C} + F\right)(\Gamma A - C) + \left[\frac{E(\Gamma A - C)}{C} - 2\Gamma\left(\frac{ED}{C} + F\right)\right]j\omega + \frac{2\Gamma E}{C}(j\omega)^2\right]}{\left[(BC - AD) + 2(A - D)j\omega + 4(j\omega)^2\right]} \tag{9.29}$$

The ratio of the two is the second harmonic distortion (HD_2):

$$2HD = \frac{s_2}{s_1} = \frac{\left[-\left(\frac{ED}{C} + F\right)(\Gamma A - C) + \left[\frac{E(\Gamma A - C)}{C} - 2\Gamma\left(\frac{ED}{C} + F\right)j\omega\right] + \frac{2\Gamma E}{C}(j\omega)^2\right]}{(BC - AD) + 2(A - D)j\omega + 4(j\omega)^2}\overline{S}m. \tag{9.30}$$

9.2.4 Laser-Clipping Distortion

Owing to the problem of a high peak-to-average power ratio (PAPR) in orthogonal frequency-division multiplexing (OFDM) signals, there may be high or low driving currents when using OFDM signals to drive lasers. At this point, the output light intensity of the laser does not accurately reflect the size of the driving signal. This distortion is caused by the threshold of the laser-driving current and is called clipping distortion. The current signal driving the laser can be expressed as

$$x(t) = \sum_{k=1}^{N} I_k \sin(2\pi f_k + \theta_k), \tag{9.31}$$

where I_k, f_k, and θ_k represent the subcarrier amplitude, frequency, and phase, respectively.

When $x(t)$ is composed of a very large number of sine waves with different frequencies ($N > 10$), $x(t)$ can be considered as a zero mean and $\sigma^2 = N_I^2/2$ variance Gaussian random process. The autocorrelation function R_o of the corresponding laser output light intensity can be expressed as the autocorrelation function R_i of the input current $x(t)$:

$$R_o(\tau) = \sum_{k=1}^{\infty} \frac{1}{k!} h_k^2 \left[\frac{R_i(\tau)}{\sigma^2} \right]^k \tag{9.32}$$

In the equation, R_i and h_k can be calculated by the following equations:

$$R_i(\tau) = \frac{1}{2} I^2 \sum_{k=1}^{N} \cos(2\pi f_k \tau) \tag{9.33}$$

$$h_k = \frac{1}{\sigma} \int_{-\infty}^{\infty} P(x) H_k\left(\frac{x}{\sigma}\right) \frac{1}{\sqrt{2\pi}} \exp\left(-\frac{x^2}{2\sigma^2}\right) \mathrm{d}x, \tag{9.34}$$

where H_k is a kth-order Hermite polynomial. The autocorrelation function of $x(t)$ is

$$R_i(\tau) = \frac{1}{2} I^2 \sum_{n=1}^{N} \cos(2\pi f_n \tau) \tag{9.35}$$

We assume that the optical-power curve output near the laser threshold can be expressed as

$$P(x) = \begin{cases} S(x + I_b - I_{th}), & x + I_b \geq I_{th} \\ 0, & x + I_b < I_{th} \end{cases} \tag{9.36}$$

By substituting Eq. (9.36) into (9.34), we obtain the components of each order of h_k, as follows:

$$h_0 = s\sigma \left\{ \frac{1}{\sqrt{2\pi}} e^{-\frac{1}{2\mu^2}} + \frac{1}{2\mu} \left[1 + \mathrm{erf}\left(\frac{1}{\sqrt{2}\mu} \right) \right] \right\} \tag{9.37}$$

$$h_1 = s\sigma\left[1 - \frac{1}{2}\mathrm{erfc}\left(\frac{1}{\sqrt{2}\mu}\right)\right], \tag{9.38}$$

where $\mu = N_m{}^2/2$ is the root-mean-square modulation index, while erf and erfc are error functions and error-compensation functions, respectively. By substituting each order of h_k into the autocorrelation function, the power expression for each order can be obtained, and the ratio is the harmonic distortion ratio.

9.3 Atmospheric-Channel Noise Model

The model of light propagating in the atmosphere divides the energy of the light field into three parts, as shown in Fig. 9.1. The beam propagates in the atmosphere, and the energy is divided into the line-of-sight direct-propagation component U_L, line-of-sight coupled forward-scattering propagation component $U_S{}^C$, and scattering component $U_S{}^G$. Here, the direct-propagation component U_L is the direct component of the beam, and the forward-scattering component $U_S{}^C$ and the scattering-component $U_S{}^G$ of the line-of-sight coupling are both caused by scattering. The difference is that $U_S{}^C$ is the collection of energy scattered forward by the dielectric particles on a direct optical path (near the optical axis), whereas $U_S{}^G$ is the collection of scattering energy from other dielectric particles outside the optical axis.

The light field is divided into the above three parts, and the light field U can be expressed as

$$U = \left(U_L + U_S^C + U_S^G\right)\exp(\chi + jS) \tag{9.39}$$

The direct-propagation component U_L can be expressed as

$$U_L = \sqrt{G}\sqrt{\Omega}\exp(j\phi_A) \tag{9.40}$$

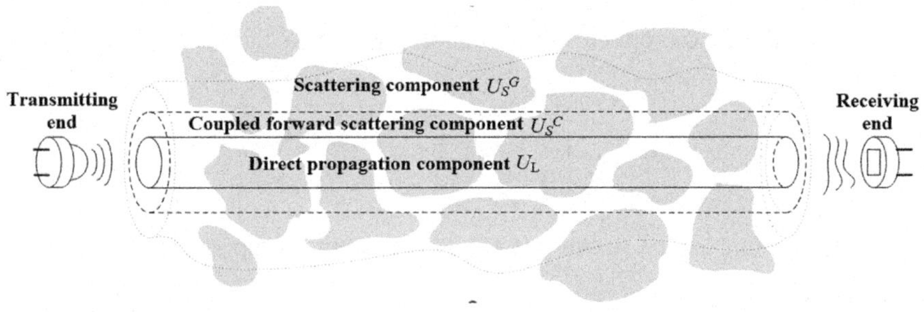

Fig. 9.1 Schematic diagram of a beam-propagation model in the atmosphere

The coupled forward-scattering component ${U_S}^C$ can be expressed as

$$U_S^C = \sqrt{\rho}\sqrt{G}\sqrt{2b_0}\exp(j\phi_B) \tag{9.41}$$

The scattering component ${U_S}^G$ can be expressed as

$$U_S^G = \sqrt{(1-\rho)}U_S' \tag{9.42}$$

U_L, ${U_S}^C$, and ${U_S}^G$ are random processes with different distributions, where U_L and ${U_S}^G$ are independent of each other, while ${U_S}^C$ and ${U_S}^G$ are independent of each other. The parameter G represents the slow fluctuation of the direct component U_L and ${U_S}^C$, satisfying $E[G] = 1$.

Parameter $\Omega = E[|U_L|^2]$ is the average power of the direct component of the beam; parameter b_0 is the average power of scattered light, which satisfies $2b_0 = E\left[\left|U_S^C\right|^2 + \left|U_S^G\right|^2\right]$; ϕ_A and ϕ_B represent the phases of the direct component U_L and the coupled forward-scattering component ${U_S}^C$ of the light field, respectively.

Parameter ρ represents the percentage of the coupled forward-scattering component in the total scattering energy, satisfying $0 \le \rho \le 1$. The magnitude of ρ is related to a number of conditions, such as the propagation distance, wavelength, beam diameter, and beam extension produced in propagation. ${U_S}'$ is a complex random number that characterizes the ${U_S}^G$ phase of the scattering component; χ and S represent the logarithmic amplitude and phase of the light field, respectively. Therefore, light intensity I can be expressed as the sum of the three aforementioned terms:

$$\begin{aligned} I &= \left|U_L + U_S^C + U_S^G\right|^2 \exp(2\chi) \\ &= \left|\sqrt{G}\sqrt{\Omega}\exp(j\phi_A) + \sqrt{\rho}\sqrt{G}\sqrt{2b_0}\exp(j\phi_B) + \sqrt{(1-\rho)}U_S'\right|^2 \exp(2\chi). \end{aligned} \tag{9.43}$$

The fluctuation in the light intensity can be expressed as the product of the large-scale fluctuation X and small-scale fluctuation Y:

$$I = XY \tag{9.44}$$

In the equation, X and Y can be represented as

$$\begin{cases} X \triangleq \exp(2\chi) \\ Y \triangleq \left|U_L + U_S^C + U_S^G\right|^2 \end{cases} \tag{9.45}$$

For the stochastic process X, which describes large-scale fluctuations in light intensity, Rytov's theory states that X follows a lognormal distribution in a weakly turbulent environment; for more general cases, X approximately follows the gamma distribution. In the derivation in the following sections, the gamma distribution can be transformed into a lognormal distribution under specific conditions through a parametric transformation.

Here is an expression for X, following a gamma distribution:

$$f_X(x) = \frac{\alpha^{\alpha}}{\Gamma(\alpha)} x^{\alpha-1} \exp(-\alpha x), \tag{9.46}$$

where parameter α is used to describe the intensity of large-scale fluctuations in atmospheric turbulence.

For the stochastic process Y describing the small-scale light-intensity fluctuation,

$$\begin{aligned} Y &= \left|U_L + U_S^C + U_S^G\right|^2 \\ &= \left|\sqrt{G}\left[\sqrt{\Omega}\exp(j\phi_A) + \sqrt{\rho}\sqrt{2b_0}\exp(j\phi_B)\right] + \sqrt{(1-\rho)}U_S'\right|^2, \end{aligned} \tag{9.47}$$

where Y includes two random variables G and $U_S{}'$, while the other parameters Ω, b_0, ϕ_A, ϕ_B, and ρ are all constants. Parameter G, characterizing the direct components U_L and $U_S{}^C$, follows a Nakagami distribution. Parameter $U_S{}'$, characterizing the scattering component $U_S{}^G$, can be expressed as the sum of a series of Rayleigh random vectors. Therefore, Y follows the Nakagami distribution, which contains Rayleigh components:

$$f_Y(y) = \frac{1}{\gamma}\left(\frac{\gamma\beta}{\gamma\beta + \Omega'}\right)^{\beta} \exp\left(-\frac{y}{\gamma}\right) {}_1F_1\left(\beta; 1; \frac{1}{\gamma}\frac{\Omega'}{(\gamma\beta + \Omega')}y\right) \tag{9.48}$$

In the equation, β is the ratio of the mean square to the variance of the random variable G; Ω' is the vector sum of the direct-light component and the forward-scattering component. The relationship between Ω' and Ω, b_0, ϕ_A, ϕ_B, and ρ is

$$\Omega' = \Omega + \rho 2b_0 + 2\sqrt{2b_0\Omega\rho}\cos(\phi_A - \phi_B) \tag{9.49}$$

$\gamma = 2b_0(1-\rho)$ is a parameter in the equation; function ${}_1\mathrm{F}_1(a; c; x)$ is the first type of confluent hypergeometric function (Kummer function):

$${}_1F_1(a; c; z) = \sum_{k=1}^{\infty} \frac{(a)_{k-1}}{(c)_{k-1}} \frac{z^{k-1}}{(k-1)!} \tag{9.50}$$

In the equation, $(a)_k$ is the Pochhammer symbol:

$$(a)_k = \frac{\Gamma(a+k)}{\Gamma(a)} \tag{9.51}$$

The random process $I = XY$, product of the above distribution, satisfies the generalized M distribution (Málaga distribution), and is recorded as $I \sim M\ (G)\ (\alpha, \beta, \gamma, \rho, \Omega')$:

$$f_I(I) = A^{(G)} \sum_{k=1}^{\infty} a_k^{(G)} I^{\frac{\alpha+k}{2}-1} K_{\alpha-k}\left(2\sqrt{\frac{\alpha I}{\gamma}}\right) \tag{9.52}$$

In this equation, parameters $A^{(G)}$ and $a_k{}^{(G)}$ are calculated using the following equations:

$$\begin{cases} A^{(G)} = \frac{2\alpha^{\frac{\alpha}{2}}}{\gamma^{1+\frac{\alpha}{2}}\Gamma(\alpha)}\left(\frac{\gamma\beta}{\gamma\beta+\Omega'}\right)^{\beta} \\ a_k^{(G)} = \frac{(\beta)_{k-1}(\alpha\gamma)^{\frac{\alpha}{2}}}{[(k-1)!]^2\gamma^{k-1}(\Omega'+\gamma\beta)^{k-1}}, \end{cases} \tag{9.53}$$

where $K_v(\cdot)$ is the second type of modified Bessel function of order v and $\Gamma\ (\cdot)$ is the gamma function. Finding $f_I(I)$ in the above equation requires calculating the sum of the infinite series $a_k{}^{(G)}$, which complicates the calculation. Therefore, by performing mathematical transformations, it can be concluded that

$$f_I(I) = A \sum_{k=1}^{\beta} a_k I^{\frac{\alpha+k}{2}-1} K_{\alpha-k}\left(2\sqrt{\frac{\alpha\beta I}{\gamma\beta+\Omega'}}\right) \tag{9.54}$$

Parameters A and a_k are calculated using the following equations:

$$\begin{cases} A = A^{(G)}\left(\frac{\gamma\beta}{\gamma\beta+\Omega'}\right)^{\frac{\alpha}{2}} \\ a_k = C_{\beta-1}^{k-1}\frac{1}{(k-1)!}\left(\frac{\Omega'}{\gamma}\right)^{k-1}\left(\frac{\alpha}{\beta}\right)^{\frac{k}{2}}\left(\gamma\beta+\Omega'\right)^{1-\frac{k}{2}}, \end{cases} \tag{9.55}$$

where C_m^n is the number of combinations and β is a natural number. When $\rho = 0.5$, $\gamma = 0.5$, and $\Omega' = 0.5$, the probability-density function of the M distribution for different values of α and β is shown in Fig. 9.2.

When the following values are fixed, $\alpha = 5$, $\beta = 10$, $\gamma = 0.5$, and $\Omega' = 0.5$, the probability-density functions of the M distribution for different ρ values are shown in Fig. 9.3.

9.3.1 Evolution of Atmospheric-Turbulence Noise Models

In the generalized M distribution, the line-of-sight direct propagation component U_L, the line-of-sight coupled forward-scattering propagation component $U_S{}^C$, and the scattering component $U_S{}^G$ are simultaneously considered. The model can be

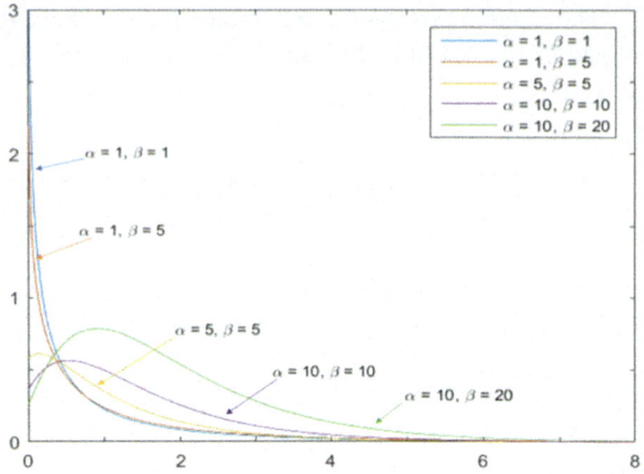

Fig. 9.2 Diagram of different α and β M-distribution probability-density function [4]

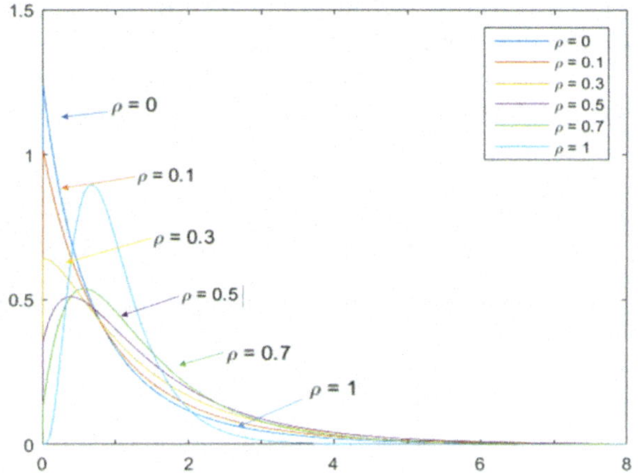

Fig. 9.3 Different ρ M-distribution probability-density functions [4]

simplified for different specific applications. When $\rho = 0$, the variance of $|U_L|$ is zero and $E[G] = 1$; the following can be obtained:

$$E[|U_L|] = \sqrt{\Omega} \tag{9.56}$$

At this point, the three terms in light intensity I degenerate into two terms, U_L and $U_S{}^G$ (first-order Born approximation):

$$I = \left| \sqrt{G}\sqrt{\Omega}\exp(j\phi_A) + U_S' \right|^2 \tag{9.57}$$

At this point, light intensity I follows the modified Rice–Nakagami distribution:

$$f_I(I) = \frac{1}{\gamma}\exp\left[-\frac{(A_0^2 + I)}{\gamma}\right] I_0\left(\frac{2A_0}{\gamma}\sqrt{I}\right), \tag{9.58}$$

where $A_0 = (G\Omega)^{1/2}$, $\gamma = 2b_0$, and $I_0(\cdot)$ is a zero-order modified Bessel function of the first kind.

Considering the case where the scattered light power is infinitely small, when $A_0^2/\gamma \to \infty$, according to the first-order Rytov approximation, $I = I_0 \exp(2\chi)$, $I_0 = |A_0|^2$, and $\sigma_\chi^{\ 2}$ is the variance of χ. At this point, the three terms of light intensity I degenerate into only one term, U_L:

$$I = |U_L|^2 \exp(2\chi) = \left|\sqrt{G}\sqrt{\Omega}\exp(j\phi_A)\right|^2 \exp(2\chi) \tag{9.59}$$

Therefore, the modified Rice–Nakagami distribution described above can be transformed into a lognormal distribution, as follows:

$$f_I(I) = \frac{1}{2I}\frac{1}{\sqrt{2\pi\sigma_\chi^2}}\exp\left(-\frac{\left[\ln(I/I_0) + 2\sigma_\chi^2\right]^2}{8\sigma_\chi^2}\right) \tag{9.60}$$

Consider another extreme case where $\Omega = 0$ and $\rho = 0$. The receiving end cannot receive direct light from the transmitting end at all; it can only receive scattered light. At this time, Y degenerates into an exponential distribution:

$$f_Y(y) = \frac{1}{\gamma}\exp\left(-\frac{y}{\gamma}\right) \tag{9.61}$$

Therefore, $I = XY$ transforms into the product of the random processes of the exponential and gamma distributions. At this point, I follows the K distribution

$$f_I(I) = \frac{2\alpha}{\Gamma(\alpha)}(\alpha I)^{\frac{\alpha-1}{2}} K_{\alpha-1}\left(2\sqrt{\alpha I}\right) \tag{9.62}$$

Furthermore, if large-scale turbulence parameter α approaches infinity, $\alpha \to \infty$, the K distribution will approach the negative-exponential distribution. When the scattering component $U_S^{\ G}$ in light intensity I is approximately zero, $\rho = 1$, $\gamma = 0$, and $\Omega' = 1$. At this point, X and Y follow independent gamma distributions, including

$$\begin{cases} f_X(x) = \frac{\alpha(\alpha x)^{\alpha-1}}{\Gamma(\alpha)}\exp(-\alpha x) \\ f_Y(y) = \frac{\beta(\beta y)^{\beta-1}}{\Gamma(\beta)}\exp(-\beta y) \end{cases} \tag{9.63}$$

The light intensity $I = XY$ follows a gamma–gamma distribution:

$$f_I(I) = \int_0^\infty f_Y(I|x) f_X(x) dx$$
$$= \frac{2(\alpha\beta)^{\frac{\alpha+\beta}{2}}}{\Gamma(\alpha)\Gamma(\beta)} I^{\frac{\alpha+\beta}{2}-1} K_{\alpha-\beta}\left(2\sqrt{\alpha\beta I}\right) \tag{9.64}$$

Assuming $\beta \to \infty$, the probability-density distribution function of Y can be reduced to

$$f_Y(y) = \frac{1}{\gamma} \exp\left(-\frac{y+\Omega'}{\gamma}\right) I_0\left(\frac{2\sqrt{\Omega' y}}{\gamma}\right) \tag{9.65}$$

With everything else unchanged, X follows the gamma distribution, and then the light intensity I follows the gamma–Rician distribution:

$$\begin{aligned} f_I(I) &= \int_0^\infty f_Y(I|x) f_X(x) \mathrm{d}x \\ &= \int_0^\infty \frac{1}{\gamma} \frac{\alpha^\alpha}{\Gamma(\alpha)} \exp\left(-\frac{\Omega'}{\gamma}\right) I_0\left(\frac{2\sqrt{\Omega' I}}{\gamma\sqrt{x}}\right) x^{\alpha-1} \exp\left(-\frac{I}{x\gamma} - \alpha x\right) \mathrm{d}x \\ &= \frac{1}{\gamma} \frac{\alpha^\alpha}{\Gamma(\alpha)} \exp\left(-\frac{\Omega'}{\gamma}\right) \sum_{k=0}^\infty \frac{(-1)^k (\Omega' I)^k}{k!\Gamma(k+1)\gamma^{2k}} \int_0^\infty x^{\alpha-2-k} \exp\left(-\frac{I}{x\gamma} - \alpha x\right) \mathrm{d}x \\ &= \frac{1}{\gamma} \frac{\alpha^\alpha}{\Gamma(\alpha)} \exp\left(-\frac{\Omega'}{\gamma}\right) \sum_{k=1}^\infty \frac{(-1)^{k-1} (\Omega' I)^{k-1}}{(k-1)!\Gamma(k)\gamma^{2(k-1)}} \left(\frac{I}{\alpha\gamma}\right)^{\frac{\alpha-k}{2}} K_{\alpha-k}\left(2\sqrt{\frac{\alpha I}{\gamma}}\right). \end{aligned} \tag{9.66}$$

Comparing the obtained gamma–Rician distribution with the generalized M distribution, it is easy to obtain the following corresponding relationship:

$$f_I(I) = \hat{A} \sum_{k=1}^\infty \hat{a}_k I^{\frac{a+k}{2}-1} K_{\alpha-k}\left(2\sqrt{\frac{\alpha I}{\gamma}}\right) \tag{9.67}$$

In the equation, $\hat{A}$ and $\hat{a}_k$ are

$$\begin{cases} \hat{A} = \frac{2\alpha^{\frac{\alpha}{2}}}{\gamma^{1+\frac{\alpha}{2}}\Gamma(\alpha)} \exp\left(-\frac{\Omega'}{\gamma}\right) \\ \hat{a}_k = \frac{(-1)^{k-1}\Omega'^{k-1}(\alpha\gamma)^{\frac{k}{2}}}{(k-1)!\Gamma(k)\gamma^{2(k-1)}} \end{cases} \tag{9.68}$$

When $\beta \to \infty$, the generalized M distribution can evolve into a gamma–Rician distribution. Based on the generalized M-distribution model, common models, such as the lognormal, gamma–gamma, K, Rice–Nakagami, and gamma–Rician distributions, can be developed. Thus, in practical applications, it is only necessary to adjust the parameters of the generalized M distribution model to simulate the flickering of light intensity under various channel-noise conditions.

Taking as examples the typical light-intensity distribution in weak turbulence (lognormal distribution) and the typical light-intensity distribution in medium–strong turbulence (gamma–gamma distribution), the signal variance and system signal-to-noise ratio are calculated when the system receives additive noise.

The probability-density distribution functions of the lognormal and gamma–gamma distributions are shown in Eqs. (9.60) and (9.64), respectively. Using parameters α and β of the gamma–gamma distribution defined by Eq. (3.41), the simplified system model is

$$P(t) = f_A P_0(t) + P_n \tag{9.69}$$

In this formula, f_A is simulated by the lognormal distribution model and gamma–gamma model, respectively; P_n is the power spectrum of the electrical noise e_n, which follows a normal distribution.

The variance and signal-to-noise ratio (SNR) of the simulated received signals were separately calculated. To obtain a relatively smooth simulation surface, the average values of 500,000 samples were calculated for each grid node in the figure. The variance and signal-to-noise ratio results are shown in Figs. 9.4 and 9.5, respectively. The coordinates of the x- and y-axes in the figure are the Rytov exponent and electrical-noise intensity in dB, respectively. It can be seen that the variance of the received signal increases with the increase of the Rytov exponent and electrical noise, and the corresponding received signal-to-noise ratio gradually deteriorates.

From Figs. 9.4 and 9.5, it can be seen that the logarithmic-normal noise model agrees well with the theoretical results of the first-order Rytov approximation. When the electrical noise is small (the circuit signal-to-noise ratio is greater than 30 dB), the simulation results are basically consistent with the theoretical value of Eq. (4.18), and the signal variance simulated using the gamma–gamma noise model is significantly greater than the theoretical value of Eq. (4.18).

9.3.2 Background-Light Noise

The receiving field of view in optical-wireless communication is usually narrow; therefore, the background light source in the field of view can be approximated as a surface light source that fills the entire field of view, and is regarded as a uniformly-emitting surface light source. However, when the receiving field of view is relaxed, certain light sources, such as the sun and other light, can also be considered point light sources.

The radiation power-spectrum function $W(\lambda)$ can be used to measure the intensity of the background light radiation. It represents the power radiated by each unit area

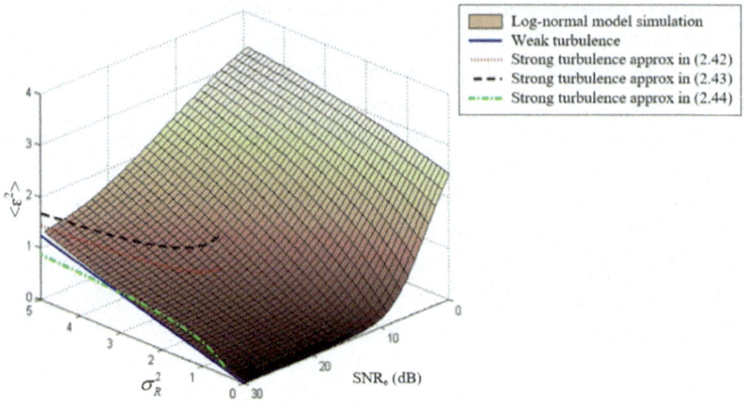

(a)

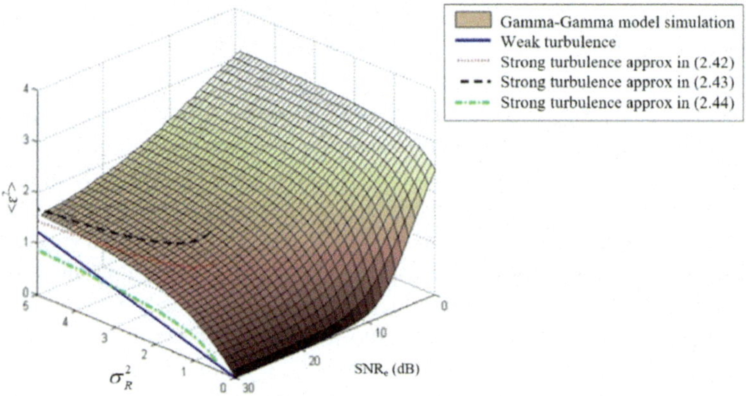

(b)

Fig. 9.4 Received signal variance of two noise models under different electrical noises [4]. **a** Lognormal noise-model simulation of the received signal variance, **b** Gamma–gamma noise-model simulation of the received signal variance

of a light source per unit bandwidth to a unit solid angle at wavelength λ. The unit of the radiation spectrum is W/m^2 sr μm. If the receiving antenna area is A, and the background light-source distance is Z, the solid angle is A/Z^2. If the area of the background light source is A_s and the field of view of the receiver is Ω_{fv}, then the background optical power collected by the receiver in the range of wavelength λ, for receiving bandwidth $\Delta\lambda$ is

$$\begin{aligned} P_b &= W(\lambda)(\Delta\lambda)\left(\Omega_{\mathrm{fv}}Z^2\right)\left(A/Z^2\right) \quad A_s > \Omega_{\mathrm{fv}}Z^2 \text{ Surface light source} \\ &= W(\lambda)(\Delta\lambda)A_s\left(A/Z^2\right) \qquad\qquad A_s < \Omega_{\mathrm{fv}}Z^2 \text{ Point light source} \end{aligned} \tag{9.70}$$

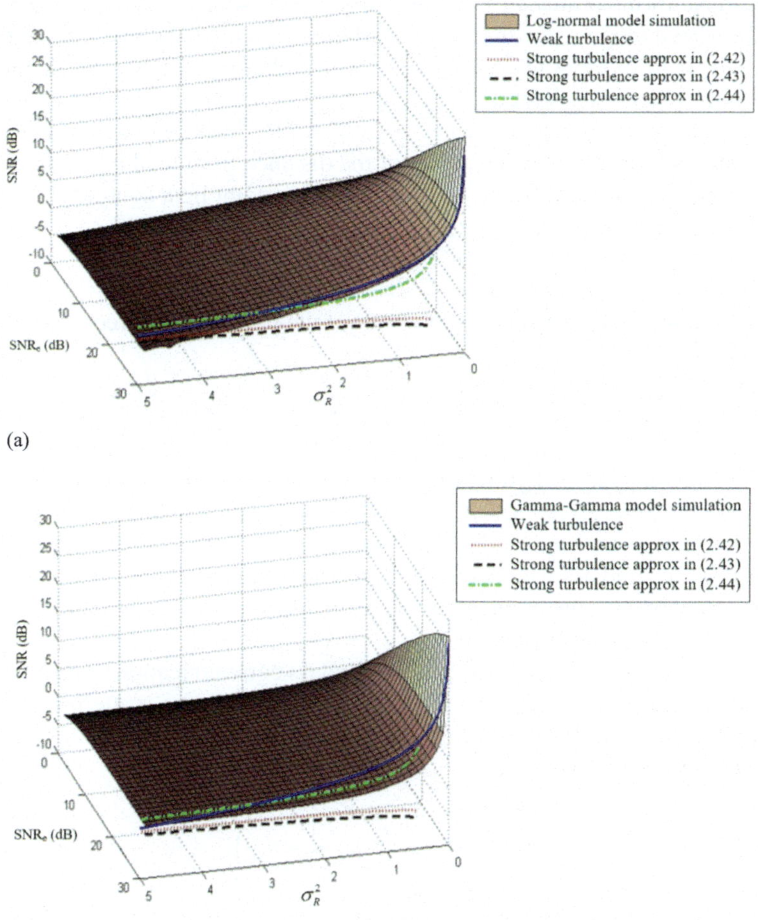

(a)

(b)

Fig. 9.5 Signal-to-noise ratio of received signals for two noise models under different electrical-noise levels. **a** Lognormal noise-model simulation of the received signal's signal-to-noise ratio, **b** Gamma–gamma noise-model simulation of received signal variance [4]

The solid angle of the background light source to the receiver is expressed as Ω_s; then $\Omega_s \approx A_s/Z^2$, and the above equation becomes

$$\begin{aligned} P_b &= W(\lambda)(\Delta\lambda)\Omega_{\mathrm{fv}}A \quad \Omega_s > \Omega_{\mathrm{fv}} \text{ Surface light source} \\ &= W(\lambda)(\Delta\lambda)\Omega_s A \quad \Omega_s < \Omega_{\mathrm{fv}} \text{ Point light source} \end{aligned} \tag{9.71}$$

The above formula indicates that for surface light sources, the background light power is related to the receiving area A, and the receiving bandwidth $\Delta\lambda$ is proportional to the field-of-view angle Ω_{fv} of the receiver. For point light sources, the background light power is related to the receiving area A and the receiving bandwidth $\Delta\lambda$ is proportional to the field-of-view angle Ω_s of the background light source. The background light power is independent of the distance from the light source.

We define the light irradiance $\omega(\lambda)$ to describe a point light source:

$$\omega(\lambda) = W(\lambda)\Omega_s \tag{9.72}$$

At this point, the background light power of the point light source (in W/m^2 μm) can be rewritten as

$$P_b = \omega(\lambda)(\Delta\lambda)A \tag{9.73}$$

When modeling the background-light noise in an optical-wireless communication system, the actual received background-light noise can be considered a superposition of the sky background-light noise and various point light-source noises. The sky background-light radiation model can be described by the blackbody radiation model. Various actual measurement junctions are available to fit the background-light noise models of various point light sources.

The sky background light in an optical-wireless communication system is caused by the radiation of the background light source itself or the scattering or diffuse reflection of sunlight. The sky background-light radiation spectrum can be given by the blackbody radiation model:

$$W(\lambda) = \frac{c^2 h}{\lambda^5}\left(\frac{1}{e^{hc/\kappa\lambda T} - 1}\right), \tag{9.74}$$

where c is the speed of light, h is Planck's constant, κ is Boltzmann constant, and T is the blackbody radiation temperature (unit: K). By integrating the above equation, the total radiant flux can be obtained as follows:

$$W = 5.67 \times 10^{-8} T^4 \tag{9.75}$$

Blackbody radiation is proportional to the fourth power of the absolute temperature; this is the Stefan–Boltzmann law, where 5.67×10^{-8} is called the Stefan constant (in W/m^2 K^4).

The radiation wavelength and absolute temperature also satisfy an inversely proportional relationship, that is, Wien's displacement law:

$$\lambda T = 2897.8 \tag{9.76}$$

The 2897.8 in the equation is referred to as the Venn constant (unit: μM K). Below, we discuss typical background light sources used in the collection.

9.3.2.1 Point Light-Source Background Light: Sunlight

Sunlight is the main background light source for daytime optical-wireless communication. When the receiver's field-of-view angle is greater than the angle of the sun, it can be considered a point light source. The national standards involving the solar spectrum in many industries have detailed solar-spectrum data. For example, through data fitting, the national standard SJ/T 10174 is close to the blackbody radiation curve with an absolute temperature of 6000 K, as shown in Fig. 9.6. Solar-radiation energy is mainly concentrated in the visible region at a wavelength of about 0.5 μm.

When calculating the background power of sunlight, we substitute the $W(\lambda)$, receiving area A, and receiver-filter bandwidth $\Delta\lambda$ in the figure above into Eq. (9.70) to find P_b.

For example, for an 850-nm optical-wireless receiver with a 100-mm aperture and a 10-nm filter bandwidth, if the receiver faces directly at the sun, the received optical power is $960 \times \pi \times (50 \times 10^{-3})^2 \times 10 \times 10^{-3} \approx 0.075$ W is approximately 18.75 dBm, which is far greater than the power of a general signal light. Therefore, optical-wireless system receivers must avoid direct exposure to sunlight. Generally, optical-wireless communication systems with horizontal transmissions are more susceptible to sunlight interference during sunrise and sunset.

When the background light source occupies a small solid angle in the receiving field of view, it can be regarded as a point light source. For example, when a large field-of-view receiver is directly illuminated by sunlight, when the light is far from the receiver at night, or when starlight enters the receiver's field of view, it can be considered a point light source in background noise. The contribution of these point sources to the background-noise power of the receiver is

$$P_b = T_a T_r \Delta\lambda A_r E_\lambda, \tag{9.77}$$

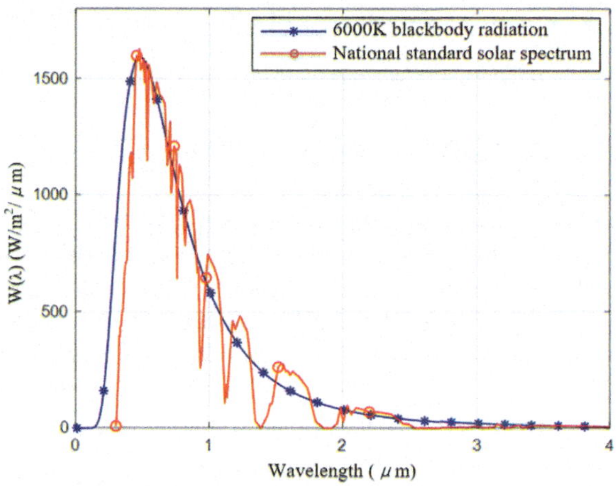

Fig. 9.6 Solar-radiation spectrum is approximately 6000 K blackbody radiation spectrum [4]

where T_a is the atmospheric transmittance, T_r is the transmittance of the receiving system, $\Delta\lambda$ is the bandwidth of the filter or detector, and A_r is the equivalent receiving area of the detector. If the equivalent aperture of the receiver is d_r, then $A_r = \pi\, d_r^2/4$. If there are multiple background-light point sources in the field-of-view of the receiver, the radiation power of each point source is calculated separately and then summed.

9.3.2.2 Spherical Background Light Source

If the solid angle occupied by the background light source in the receiving field of view is not infinitely small and does not occupy the entire receiving field of view, the background light source can be treated as spherical, according to the equivalent area.

When the diameter d of the spherical light source is R and the distance from the receiver is R, the hemisphere area of the light source is $\pi d^2/2$, and the total power radiated by the spherical light source into the large sphere with a radius R is $\pi d^2/2 \cdot M_\lambda$. For receivers located on a spherical surface, the background power of the spherical light source is $\pi d^2/2\ M_\lambda/2\ \pi R^2$. Considering that the aperture area of the receiver is $A_r = \pi d_r^2/4$, the background power received by the detector from the spherical light source can be expressed as

$$P_b = \frac{\pi T_a T_r \Delta\lambda d^2 d_r^2}{16R^2} M_\lambda \tag{9.78}$$

If the spherical light source is a Lambert source with the same brightness in all directions, the power received by the detector can be expressed as

$$P_b = \frac{\pi^2 T_a T_r \Delta\lambda d^2 d_r^2}{16R^2} L_\lambda \tag{9.79}$$

9.3.2.3 Expanded Background Sources

If the background light source occupies the field of view of the receiver, it is called an extended background light source. At this time, the angle of view of the receiver only accounts for a part of the solid angle of the extended source radiation, and $\Omega_r = \pi\theta_r^2/4$ represents the solid angle of the receiver field of view, θ_r is the plane angle of the receiver's field of view, and $\Omega = \pi d^2/(4R^2)$ represents the solid angle that the spread source opens to the receiver. Then, the background radiation power received by the receiver from the spread source is

$$P_b = \frac{\pi T_a T_r \Delta\lambda \theta_r^2 d_r^2}{16R^2} M_\lambda \tag{9.80}$$

If the extended background light source is a Lambert source, the background radiation power can be expressed as

$$P_b = \frac{\pi^2 T_a T_r \Delta\lambda \theta_r^2 d_r^2}{16R^2} L_\lambda \tag{9.81}$$

9.3.2.4 Reflected of Sunlight by the Target

When the optical-wireless system operates in the daytime, it receives strong interference from sunlight. It not only reflected direct sunlight but also reflected light in the receiver's field of view and atmospheric scattering of sunlight (Fig. 9.7).

For the former, assuming that the light reflected from the reflector in the field of view is a uniform diffuse reflection (Lambert diffuse reflection), the diffuse reflection coefficient is ρ, the solid angle of the sun opening is Ω_s, and the spectral illuminance of the sun to the ground is $E_\lambda = \Omega_s L_\lambda$. When the angle between the sunlight and the target normal is θ, the radiance M of the target $M_\lambda{}'$ and brightness $L_\lambda{}'$ are represented as

$$M_\lambda' = \rho L_\lambda \Omega_s \cos\theta \tag{9.82}$$

$$L_\lambda' = \frac{\rho L_\lambda \Omega_s \cos\theta}{\pi} \tag{9.83}$$

The angle between the receiver optical axis and the target normal is φ; therefore, according to Lambert's cosine law, it can be concluded that

$$M_\lambda' = \rho L_\lambda \Omega_s \cos\theta \cos\varphi \tag{9.84}$$

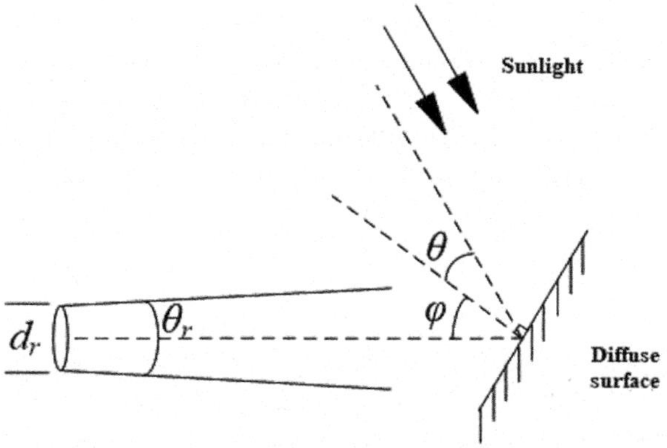

Fig. 9.7 Diffuse reflection of sunlight by the target

$$L'_\lambda = \frac{\rho L_\lambda \Omega_s \cos\theta \cos\varphi}{\pi} \tag{9.85}$$

At this time, the radiation power of the reflected surface received within the receiver's field-of-view angle is

$$\begin{aligned} P_b &= \frac{\pi T_a T_r \Delta\lambda \theta_r^2 d_r^2}{16} M'_\lambda = \frac{\pi^2 T_a T_r \Delta\lambda \theta_r^2 d_r^2}{16} L'_\lambda \\ &= \frac{\pi}{16} T_a T_r \Delta\lambda \theta_r^2 d_r^2 \rho L_\lambda \Omega_s \cos\theta \cos\varphi \end{aligned}, \tag{9.86}$$

where cos θ and cos φ indicate the relative positions between the sunlight, target, and receiver.

In practice, there are many cases where φ is zero; thus, the background optical power of sunlight entering the receiver by reflection is

$$P_b = \frac{\pi}{16} T_a T_r \Delta\lambda \theta_r^2 d_r^2 \rho L_\lambda \Omega_s \cos\theta \tag{9.87}$$

For example, for general diffuse-reflection targets on the ground (such as vertical walls), ρ is 0.4. The sun can be approximated as a 6000-K blackbody radiation source, and the sun can be approximated as a parallel light. When the operating wavelength λ is 0.5-μm red light, it can be obtained as $L_{0.5} = 1.5 \times 10^3$ W/(cm^2 sr μm).

The variation range of the solar zenith angle in Xi'an, China on the summer solstice was 33–80° (80° during the summer solstice and 33° during the winter solstice). Taking the summer solstice as an example, without considering the transmittance of the atmosphere and optical system, the receiving-antenna aperture is 100 mm, the filter bandwidth is 10 nm, and the divergence angle is 3 mrad. At this time, the power of the diffuse reflected light received by the receiver facing a wall directly under the sun is approximately $\pi/16 \times 10 \times 10^{-3} \times (3 \times 10^{-3})^2 \times 10^2 \times 0.4 \times 1.5 \times 10^3 \times \cos 80° \approx 1.84^{-4}$ W is approximately − 7.4 dBm. If the power of the signal light is low at this time, the diffuse reflection of sunlight on the wall affects the system's signal-to-noise ratio.

9.3.2.5 Atmospheric Sunlight Scattering

Assuming a uniform atmosphere, the scattering of sunlight by atmospheric molecules is isotropic. The volume element of the atmosphere in the receiver's field of view can be expressed as follows:

$$\Delta V = \frac{\pi}{4} (\theta_r x)^2 \mathrm{d}x, \tag{9.88}$$

where x is the distance from the volume element to the receiver, as shown in Fig. 9.8. The illumination of sunlight on the atmosphere is $E_\lambda = \Omega_s L_\lambda$, the scattering coefficient of the volume element is β, and its scattered light intensity is $E_\lambda \beta \Delta V / 4\pi$. The

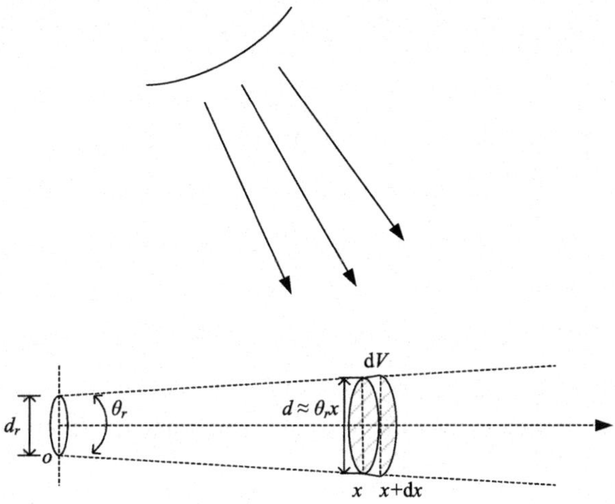

Fig. 9.8 Schematic diagram of atmospheric sunlight scattering received by the detector

solid angle occupied by the receiver aperture in the volume element is

$$\frac{A_r}{x^2} = \frac{\pi}{4}\frac{d_r^2}{x^2} \tag{9.89}$$

Therefore, the contribution of the scattered light from this volume element to the background-light power of the receiver is.

$$\mathrm{d}P_b = \frac{T_r \Delta\lambda \Omega_s L_\lambda}{4\pi}\frac{\beta\pi}{4}(\theta_r x)^2 \mathrm{d}x \exp(-ax)\frac{A_\tau}{x^2}, \tag{9.90}$$

where a is the atmospheric attenuation coefficient, which can be obtained from a table. By integrating Eq. (9.90), the sum of the radiation powers of the volume elements in the field of view of all receivers can be obtained as follows:

$$P_b = \int \mathrm{d}P_b = \frac{\pi}{64}\Omega_s L_\lambda \Delta\lambda \beta \theta_r^2 d_r^2 \frac{1 - T_a}{a} \tag{9.91}$$

When the atmospheric transmittance is uniform, $T_a = \exp(-ax)$. By combining the atmospheric-scattered sunlight with the reflected light of the target, the main background-light noise power of the wireless photosystem during daytime operation can be obtained as follows:

$$P_b = \frac{\pi}{16} T_r \Delta\lambda \theta_r^2 d_r^2 \Omega_s L_\lambda \left[\rho T_a + \frac{\beta}{4\alpha}(1 - T_a)\right] \tag{9.92}$$

9.3.3 Scattering Causes Light-Intensity Attenuation in Rain

When a light beam propagates in rain, its intensity is attenuated by the scattering of raindrops. The ratio of the received intensity to the emitted intensity satisfies the Beer–Lambert law:

$$\frac{I_t}{I_0} = \frac{I_c + I_i}{I_0} = \exp(-\rho \sigma_a L), \tag{9.93}$$

where ρ is the particle number density, σ_a is the absorption cross section of the medium, and L is the propagation distance. σ_a is the difference between the total particle cross section σ_t and the scattering cross section σ_s. According to the Mie scattering theory, the total and scattering cross sections of a single spherical particle can be expressed as follows:

$$\sigma_a = \sigma_t - \sigma_s \tag{9.94}$$

$$\sigma_t = 2\pi \sum_{n=1}^{\infty} (2n+1)\mathrm{Re}(a_n + b_n) \tag{9.95}$$

$$\sigma_s = 2\pi \sum_{n=1}^{\infty} (2n+1)\left(|a_n|^2 + |b_n|^2\right), \tag{9.96}$$

where a_n and b_n are the Mie scattering coefficients, which are the *n*th terms of two infinite series and can be calculated by the following equations:

$$a_n = \frac{\psi_n(\alpha)\psi_n'(m\alpha) - m\psi_n'(\alpha)\psi_n(m\alpha)}{\zeta_n(\alpha)\psi_n'(m\alpha) - m\zeta_n'(\alpha)\psi_n(m\alpha)} \tag{9.97}$$

$$b_n = \frac{m\psi_n(\alpha)\psi_n'(m\alpha) - \psi_n'(\alpha)\psi_n(m\alpha)}{m\zeta_n(\alpha)\psi_n'(m\alpha) - \zeta_n'(\alpha)\psi_n(m\alpha)} \tag{9.98}$$

In the equation, α is the scale parameter of the particle, $\psi_n(\alpha)$ and $\zeta_n(\alpha)$ are functions related to the Bessel function, and *m* is the complex refractive index of the medium. The calculation methods are as follows:

The value of α is the product of the particle radius and the wave number:

$$\alpha = kr = \frac{2\pi r}{\lambda}, \tag{9.99}$$

where *r* is the particle radius, *k* is the wavenumber, and λ is the wavelength.

$\psi_n(\alpha)$ and $\zeta_n(\alpha)$ are obtained recursively from the following equations:

$$\psi_{n+1}(\alpha) = \sqrt{\frac{\pi\alpha}{2}} J_{n+\frac{1}{2}}(\alpha) \tag{9.100}$$

$$\zeta_{n+1}(\alpha) = \sqrt{\frac{\pi\alpha}{2}} \left(J_{n+\frac{1}{2}}(\alpha) + (-1)^n i J_{-n-\frac{1}{2}}(\alpha) \right), \tag{9.101}$$

where $J_n(\cdot)$ is the order-n Bessel function of the first kind.

The corresponding $\psi'_n(\alpha)$ and $\zeta'_n(\alpha)$ are the derivatives of $\psi_n(\alpha)$ and $\zeta_n(\alpha)$. The recursive higher order $\psi_n(\alpha)$ and $\zeta_n(\alpha)$ need to be calculated from the zero order of the series, and the corresponding initial values of $\psi_0(\alpha)$ and $\zeta_0(\alpha)$ are respectively

$$\psi_0(\alpha) = \sin\alpha \tag{9.102}$$

$$\zeta_0(\alpha) = \sin\alpha + i\cos\alpha \tag{9.103}$$

In the calculation, the series order n needs to be calculated in relation to the scale parameter α of the particle. When α is very small, only the first few terms of a_n and b_n need to be calculated to obtain more accurate results. When α is larger, more terms of a_n and b_n need to be calculated accordingly to get more accurate results. Otherwise, the calculated value of the scattering cross section will oscillate and incorrect results will be obtained. Specifically, the order of the series n and the scale parameter α must be calculated to meet the relationship between the relationship:

$$n = \begin{cases} 3; & \alpha < 0.02 \\ \lceil \alpha + 4\sqrt[3]{\alpha} + 1 \rceil; & 0.02 \le n < 8 \\ \lceil \alpha + 4.05\sqrt[3]{\alpha} + 2 \rceil; & 8 \le n < 4200 \\ \lceil \alpha + 4\sqrt[3]{\alpha} + 2 \rceil; & 4200 \le n < 20,000 \end{cases} \tag{9.104}$$

In the formula, $\lceil\cdot\rceil$ means rounded up. m is the complex refractive index of the medium corresponding to each wavelength. In the calculation of the light-intensity rain attenuation, m is the complex refractive index of water (the imaginary part of the complex refractive index is defined as the opposite of the absorption index). The water refractive and absorption indices of commonly used wavelengths of light are listed in Table 9.1 for reference.

We calculate the optical cross section σ_t, scattering cross section σ_s, and absorption cross section σ_a of the water-droplet particle according to Eq. (9.94). The ratio of the result to the cross-section area of the particle is recorded as the corresponding scattering coefficient of the particle. We calculate the scattering cross-section coefficient a_s, the extinction cross-section coefficient a_t, and the absorption cross-section coefficient a_a of the particle. The relationship between each coefficient and the particle-scale parameter α is shown in Fig. 9.9.

The scattering parameters of each particle of different sizes are calculated by several recursive sequences, which is time-consuming. To simplify the calculation,

Table 9.1 Optical refractive index and absorption index of water

λ (nm)	n (λ)	k (λ)	λ (nm)	n (λ)	k (λ)
200	1.396	1.1×10^{-7}	675	1.331	2.23×10^{-8}
225	1.373	4.9×10^{-8}	700	1.331	3.35×10^{-8}
250	1.362	3.35×10^{-8}	725	1.330	9.15×10^{-8}
275	1.354	2.35×10^{-8}	750	1.330	1.56×10^{-7}
300	1.349	1.6×10^{-8}	800	1.329	1.25×10^{-7}
325	1.346	1.08×10^{-8}	825	1.329	1.82×10^{-7}
350	1.343	6.5×10^{-9}	850	1.329	2.93×10^{-7}
375	1.341	3.5×10^{-9}	875	1.328	3.91×10^{-7}
400	1.339	1.86×10^{-9}	900	1.328	4.86×10^{-7}
425	1.338	1.3×10^{-9}	925	1.328	1.06×10^{-6}
450	1.337	1.02×10^{-9}	950	1.327	2.93×10^{-6}
475	1.336	9.35×10^{-10}	975	1.327	3.48×10^{-6}
500	1.335	1.00×10^{-9}	1000	1.327	2.89×10^{-6}
525	1.334	1.32×10^{-9}	1200	1.324	9.89×10^{-6}
550	1.333	1.96×10^{-9}	1400	1.321	1.38×10^{-4}
575	1.333	3.60×10^{-9}	1600	1.317	8.55×10^{-5}
600	1.332	1.09×10^{-8}	1800	1.312	1.15×10^{-4}
625	1.332	1.39×10^{-8}	2000	1.306	1.1×10^{-3}
650	1.331	1.64×10^{-8}	2200	1.296	2.89×10^{-4}

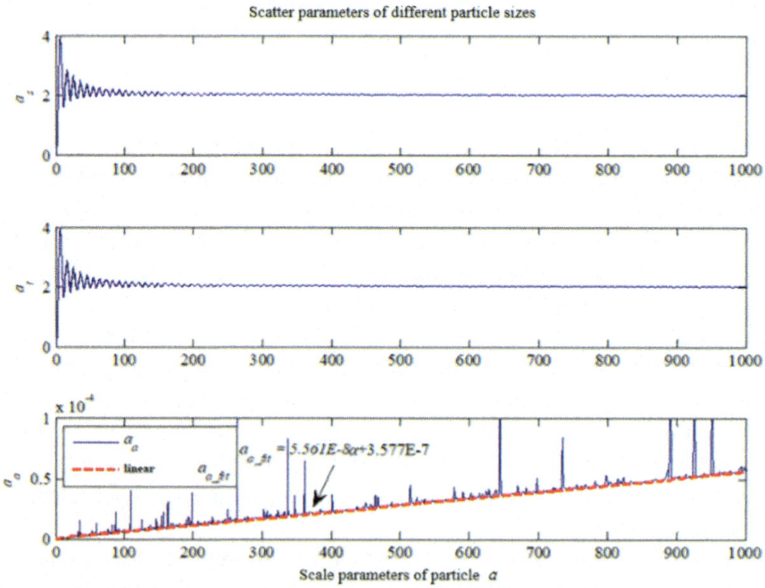

Fig. 9.9 Particle-scattering parameters at a 650-nm wavelength [4]

linear fitting of the particle-absorption parameters is used instead of calculation. The particle absorption coefficient obtained by linear fitting is

$$a_{a\_\mathrm{fit}}^{650} = 5.564 \times 10^{-8}\alpha + 3.577 \times 10^{-7} \tag{9.105}$$

Similarly, the absorption coefficient of water particles at other wavelengths can be calculated; for example, the absorption coefficient of particles at 550 nm is approximately equal to

$$\sigma_{a\_\mathrm{fit}}^{550} = 6.655 \times 10^{-9}\alpha + 1.161 \times 10^{-7} \tag{9.106}$$

The ratio of I_t to I_0 is converted into the attenuation coefficient of the light intensity per unit length in dB/km as follows:

$$\frac{I_t}{I_0}(\mathrm{dB}) = 4.343 \times 10^3 \times \frac{\pi}{4} \int_{D_{\min}}^{D_{\max}} a_a(\alpha, m) D^2 N(D) dD \tag{9.107}$$

In the equation, α is the particle-size parameter, m is the complex refractive index of the medium, N is the particle-size distribution function, and D is the particle diameter.

Many fitting formulas conform to the actual local situation of the raindrop particle-size distribution function under various rainfall potentials. The common raindrop particle-size distribution is an exponential-type distribution, specifically the Marshall–Palmer (MP) and gamma distributions.

Chinese meteorologists have conducted long-term measurements of rainfall spectra in various regions, accumulated a large amount of data, and fitted them to obtain different empirical functions for raindrop spectra. At present, raindrop scale distributions in the literature include Guangzhou, Qingdao, Shenyang, Changchun, Xi'an, and some areas in Ningxia Province, China [3–8]. The raindrop spectrum-distribution function of each area is listed in Table 9.2.

Table 9.2 Raindrop spectral-distribution functions in various regions of China [4]

Distribution	Expression (unit: individual/m^3/mm)
Marshall–Palmer distribution	$8000\exp(-4.1p^{-0.21}D)$
Guangzhou City	$230{,}600p^{0.364}D^{-0.274}\exp(-7.411p^{-0.0527}D^{0.452})$
Changchun City	$2020p^{-0.2}\exp(-3.18\mathrm{p}^{-0.28}D)$
Shenyang City	$13{,}829.7p^{0.1}\exp(-4.56\mathrm{p}^{-0.16}D)$
Qingdao City	$1387p^{0.4052}\exp(-2.38p^{-0.0877}D)$
Part of Ningxia Province	$5688.2p^{0.3996}\exp(-4.05p^{-0.1213}D)$
Xi'an City	$257p0.315(8315p^{-0.202})2D/2K_0(8315p^{-0.202}D/2)$

The various raindrop spectra in Table 9.2 are substituted into Eq. (9.105) to calculate the total attenuation of light propagation under different rainfall levels, as shown in Figs. 9.10 and 9.11.

Figures 9.10 and 9.11 show that the attenuation of light intensity caused by raindrop absorption increases with an increase in rainfall. Compared with the MP distribution, the attenuation of the rainfall light intensity in different regions was divided into two groups. The attenuation of light intensity per unit distance in Guangzhou and Qingdao was greater than that in the MP distribution, and the attenuation of light intensity per unit distance in other regions was smaller than that in the MP distribution.

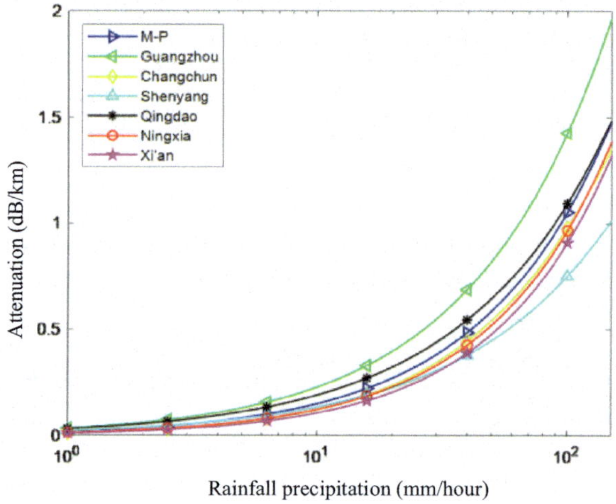

Fig. 9.10 Total attenuation of light transmission at a 550-nm wavelength varies with local rainfall rates [4]

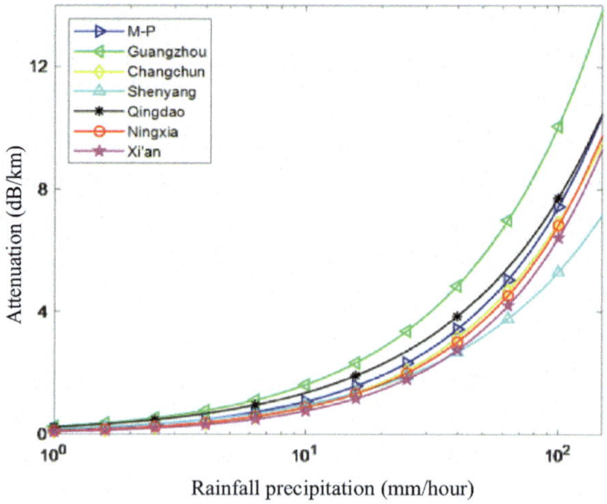

Fig. 9.11 Total attenuation of light transmission at a 650-nm wavelength varies with local rainfall rates [4]

This indicates that raindrops with the same amount of rainfall in Guangzhou and Qingdao had a stronger ability to absorb light. Comparing Figs. 9.10 and 9.11, owing to the increase in the wavelength of the imaginary part of the complex refractive index of water in the visible-light band, 650-nm red light is absorbed more by the water medium than 550-nm green light, and the attenuation per unit distance is greater.

Coherent fields can be considered as proportional attenuation caused by the optical thickness of the medium along the light-propagation path, according to the Beer–Lambert theorem:

$$\frac{I_c}{I_0} = \exp(-\rho\sigma_t L) \tag{9.108}$$

Similar to Eq. (9.107), converting the ratio of I_c to I_O into the attenuation coefficient of the light intensity per unit length in dB/km, we obtain

$$\frac{I_c}{I_0}(\text{dB}) = 4.343 \times 10^3 \times \frac{\pi}{4} \int_{D_{\min}}^{D_{\max}} a_t(\alpha, m) D^2 N(D) dD \tag{9.109}$$

σ_t is calculated based on the Mie scattering theory. By substituting the raindrop spectra from different regions, the attenuation of light transmission under different rainfall conditions was determined, as shown in Fig. 9.12.

Figure 9.12 shows that the overall attenuation of light intensity increases with increasing rainfall. For visible light, the scale parameters of raindrop particles are generally large, and the extinction cross-section coefficient of most particles is approximately 2. Therefore, the wavelength has little effect on the attenuation of the coherent field.

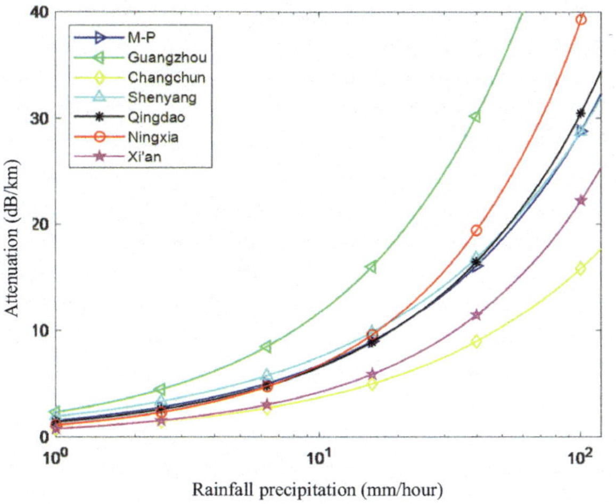

Fig. 9.12 Attenuation of the coherent field of a 650-nm-wavelength light transmission with local rainfall rates [4]

The coherent-field attenuation of a 550-nm optical transmission is almost identical to that shown in Fig. 9.12, which compares the raindrop spectra and MP distribution in different regions. The extinction rate and MP distribution in the rain in the Qingdao and Shenyang regions are very similar, and the extinction rate in rain in Guangzhou is greater than the MP distribution. When the rainfall is greater than 11 mm/h, the extinction rate in some areas of Ningxia Province is also greater than the MP distribution. When the rainfall is less than 11 mm/h, the extinction rate is slightly lower than the MP distribution, and the extinction rate in rain in the Xi'an and Changchun regions is significantly lower than the MP distribution.

Figure 9.12 also shows that, under the same rainfall conditions, to achieve an equal received signal-to-noise ratio, a higher transmission power is required in Guangzhou than in Changchun.

Owing to the positive nature of the light-intensity signal, the fluctuation of light intensity caused by scattering also has a mean value of [1]

$$\frac{I_i}{I_0} = \exp(-\rho\sigma_a L) - \exp(-\rho\sigma_t L) \tag{9.110}$$

The above equation is the difference between Eqs. (9.107) and (9.109). The non-coherent field attenuation of 650-nm red light in rain is calculated as shown in Fig. 9.13. The average scattered light intensity initially increases and subsequently decreases with increasing rainfall. This may be because when the rain changes from light to moderate, the radii of the raindrop particles in the raindrop distribution spectrum increase, while the number density increases slightly. At this time, the forward-scattering intensity increases with the increase in particles. When the rainfall continues to increase, the radius of the raindrop particles increases inexplicably, and the number density increases sharply. The total optical thickness of the raindrop particles increases, resulting in a decrease in the average scattered light intensity.

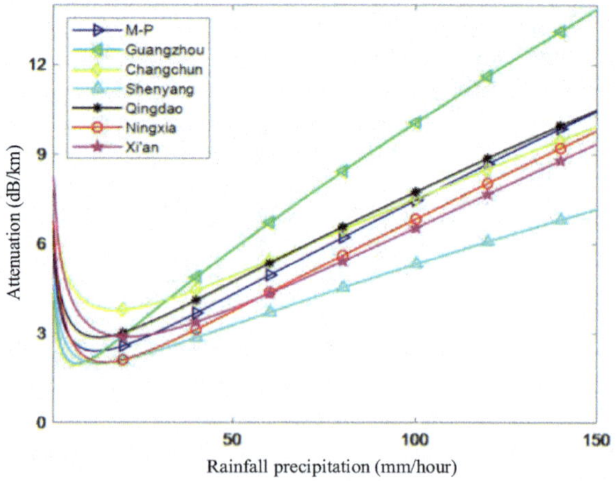

Fig. 9.13 Incoherent field attenuation of 650-nm light transmissions with local rainfall rates [4]

The average scattered light intensity first increases and then decreases, resulting in an extreme value, which is related to the wavelength and raindrop spectrum. The complex refractive indices of raindrops with different wavelengths are different, and the influence of the different raindrop spectra is shown in Fig. 9.13.

For 650-nm red light, the maximum scattered light intensity in the MP-distribution rain appears at 12 mm/h, whereas the rainfall at the extreme point of the raindrop-spectrum distribution in other regions is 7 mm/h in Guangzhou, 13 mm/h in Shenyang and Qingdao, 15 mm/h in Ningxia Province, 18 mm/h in Changchun, and 21 mm/h in Xi'an.

Similarly, the noncoherent field attenuation of 550-nm green light was calculated, as shown in Fig. 9.14. Owing to the different complex refractive indices, the absorption ability of the medium for different wavelengths of light varies, and the corresponding extreme points also change. Comparing Figs. 9.13 and 9.14, it can be seen that the scattered light intensity of green light attenuates less than that of red light in rain. The results also indicate that, during a rainfall, the received scattered light intensity first increases and then decreases with an increase in rainfall. The specific extreme points are related to the transmission wavelength and different raindrop-spectral distributions.

Although the incoherent field caused by scattering transfers energy from the transmitting end to the receiving end, the random motion of the scattering medium can cause random changes in the incoherent field. The intermittent extinction and enhancement of light-intensity signals in the scattering media measured in actual environments can be attributed to the scattering of the moving media. The scattered light intensity in an actual environment is not constant, and the incoherent field intensity varies randomly with time.

The time-varying scattering field is generated by a moving medium, and the differential interface in a stationary state is σ_d. When a single particle d moves at velocity V, its time-dependent differential scattering cross section can be written as [3]

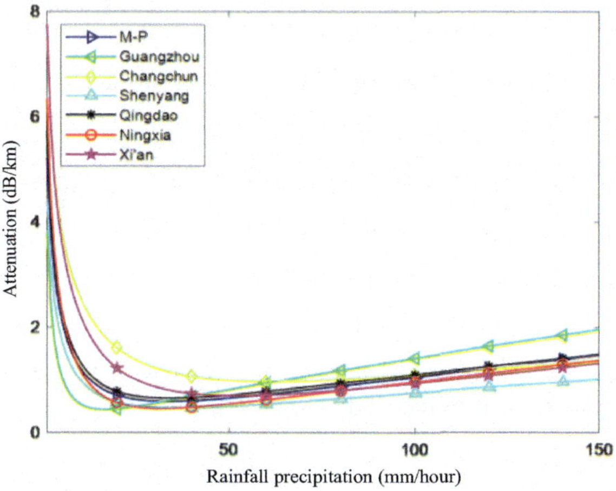

Fig. 9.14 Incoherent field attenuation of 550-nm light transmissions with local rainfall rates [4]

$$\sigma_d(\tau) = \sigma_d \langle \exp(i\mathbf{k}_s \cdot \mathbf{V}\tau) \rangle \tag{9.111}$$

We decompose the velocity into V and divide it by the sum of the average velocity U and velocity fluctuation V_f:

$$\begin{aligned} \sigma_d(\tau) &= \sigma_d \langle \exp(i\mathbf{k}_s \cdot (\mathbf{U} + \mathbf{V_f})\tau) \rangle \\ &= \sigma_d \langle \exp(i\mathbf{k}_s \cdot \mathbf{V_f}\tau) \rangle \exp(i\mathbf{k}_s \cdot \mathbf{U}\tau) \end{aligned} \tag{9.112}$$

In the formula, k_s is the scattering wave vector and τ is the time difference. The relationship between the differential scattering cross section and the total scattering cross section is

$$\sigma_s = \int_{4\pi} \sigma_d \, d\omega, \tag{9.113}$$

where ω is the solid angle of space. The time-correlation function of the scattering-field strength of moving particles is

$$\begin{aligned} B_u(\tau) &= \langle u_f(\mathbf{r}, t_1) u_f^*(\mathbf{r}, t_2) \rangle \\ &= \int_V \sigma_d \langle \exp(i\mathbf{k}_s \cdot \mathbf{V}_f\tau) \rangle \exp(i\mathbf{k}_s \cdot \mathbf{U}\tau) \exp(-\gamma_0 - \gamma') \rho dV \end{aligned} \tag{9.114}$$

The power spectrum of the scattered-field strength is obtained by doing a Fourier transform of $B_u(\tau)$ (Wiener–Khinchin's theorem):

$$\begin{aligned} W_u(\omega) &= 2 \int_{-\infty}^{\infty} B_u(\tau) e^{j\omega\tau} d\tau \\ &= (4\rho\sigma_s L) \left(\frac{\pi \alpha_p}{k^2 U^2} \right)^{\frac{1}{2}} \exp\left(-\gamma - \frac{\alpha_p \omega^2}{k^2 U^2} \right) \end{aligned} \tag{9.115}$$

Substituting the raindrop size distribution above into Eq. (9.115), we obtain

$$W_u(\omega) = \int_{D_{\min}}^{D_{\max}} N(D)\, dD 4\sigma_s L \left(\frac{\pi \alpha_p(D)}{k^2 U^2(D)} \right)^{\frac{1}{2}} \exp\left(-\int_{D_{\min}}^{D_{\max}} N(D) \sigma_t L dD - \frac{\alpha_p(D) \omega^2}{k^2 U^2(D)} \right), \tag{9.116}$$

where $\alpha_p(D) = 2.77/(1.02\lambda/D)^2$.

The following describes the approximate relationship between the raindrop velocity and raindrop diameter [3]:

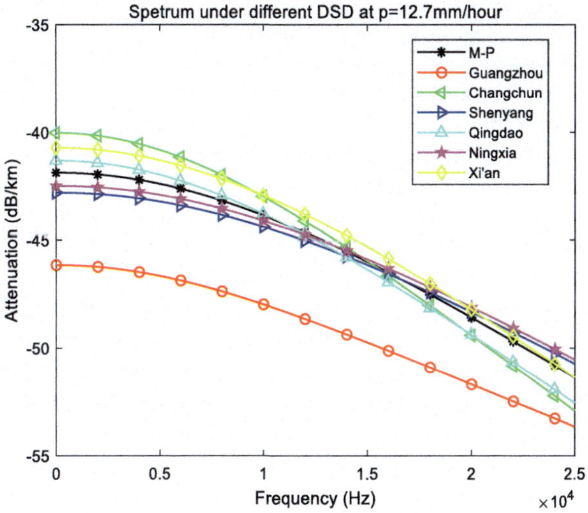

Fig. 9.15 Scattered light-intensity spectra at a 650-nm wavelength in rainy environments in various regions [4]

$$U(D) = 200.8\left(\frac{D}{2}\right)^{\frac{1}{2}} \tag{9.117}$$

For example, when the rainfall $p = 12.7$ mm/hour, various raindrop spectra were substituted into the above equation to calculate the scattering-field intensity spectra under different rainfall conditions, as shown in Fig. 9.15. The spectral bandwidth under the MP distribution is 12.43 kHz, and models with smaller bandwidth than the MP distribution include Changchun: 10.1 kHz, Qingdao: 11.16 kHz, and Xi'an: 11.8 kHz. The models with larger bandwidths than the MP distribution include Guangzhou: 13.38 kHz, Ningxia: 13.9 kHz, and Shenyang: 14 kHz. Different models estimate the number of raindrop particles, resulting in different bandwidth calculations. The main factor affecting the spectral broadening during rainfall is the speed of heavy particles in the raindrop spectrum.

The field-strength variance of the scattered field can be obtained from the spectrum. When $\omega = 0$, Eq. (9.116) yields

$$\begin{aligned} W_u(0) &= 2\int_{-\infty}^{\infty} B_u(\tau)d\tau \\ &= 2\int_{-\infty}^{\infty} \langle u_{\mathrm{f}}(\mathbf{r}, t)u_{\mathrm{f}}^*(\mathbf{r}, t)\rangle d\tau \end{aligned} \tag{9.118}$$

For the discrete light-intensity signal collected by the receiver, Eq. (9.118) is represented as

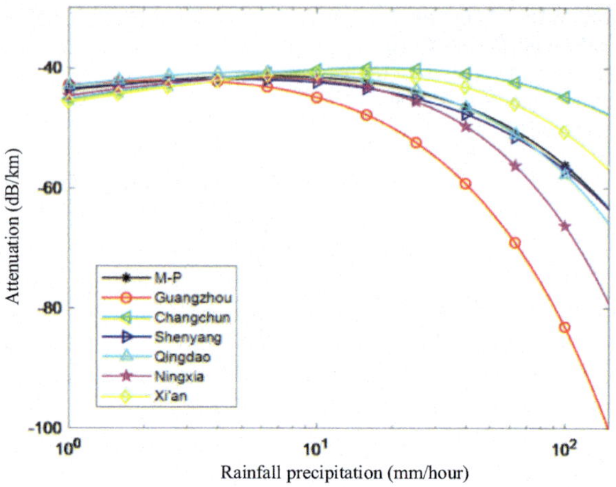

Fig. 9.16 Zero-frequency attenuation of 650-nm wavelength under various rainfall conditions in various regions [4]

$$W_u(0) = 2\sum u_{\mathrm{f}}^2 \tag{9.119}$$

The field-strength variance of the scattered field can be expressed as

$$D\left(u_f\right) = E\left(u_{\mathrm{f}}^2 - E^2(u_{\mathrm{f}})\right) \tag{9.120}$$

In a weakly turbulent environment, the mean field strength can be considered zero, and the variance of the field strength is proportional to $W_u(0)$:

$$D(u_{\mathrm{f}}) \propto W_u(0) \tag{9.121}$$

We calculate the variance of the scattering-field intensity of 650-nm red light under different rainfall conditions using Eq. (9.121), as shown in Fig. 9.16. The figure shows that the variance in the scattering field first increases and then decreases with an increase in rainfall.

From the perspective of the raindrop-spectrum model, with an increase in rainfall, the increase in small raindrops with a significant weight leads to the enhancement of scattering. Then, the number of raindrops will not increase until their size increases to a certain extent. At this time, the number of raindrops increases to increase the rainfall. An increase in the number of particles results in an increase in the optical thickness, which leads to a decrease in the scattering-field variance.

Combining Figs. 9.12 and 9.16, the relationship between the mean received light intensity, field-intensity variance, and rainfall can be obtained. For free-space optical (FSO) receiving light intensity without recording the external rainfall, the received light intensity can be associated with the field-intensity variance. This chapter cites the measured data of 650-nm red-light FSO under partial rainfall conditions in a certain area of Xi'an from July 2007 to September 2014 to verify the relationship between the mean light intensity and variance, as shown in Fig. 9.17 [2].

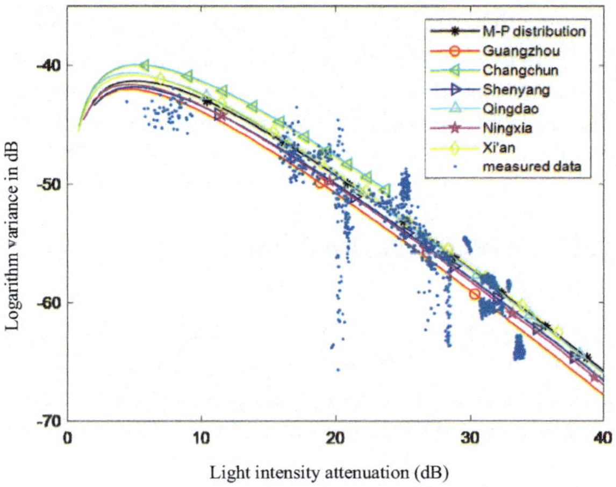

Fig. 9.17 Relationship between the measured light-intensity attenuation and variance at a 650-nm wavelength [4]

Because most of the variance values in the figure are small, they are converted into logarithms according to the definition of $W_u(0)$. Limited experimental data indicate that the variance in light intensity decreases with an increase in the mean light intensity, and this trend is consistent with the MP and regional distribution. These experimental data can verify the relationship between the mean light intensity and variance in various models. However, because of factors, such as circuit noise and alignment errors, in the experimental data, the results are somewhat scattered. Therefore, further research is needed on how to distinguish the differences between different models.

In addition, the variance values of the measured data in the figure show significant changes when the mean light intensity remains unchanged, because the received light experiences short-term fading or enhancement in a rainy environment. Owing to its short duration, the occurrence has little impact on the statistical mean; however, it can cause the variance to deviate from the median, either large or small. This reflects the time-varying non-stationary characteristics of light propagation in turbulent media.

In this section, the mean values of the coherent and incoherent fields in the received light field are calculated by substituting various raindrop spectral distributions into the Mie scattering model with a focus on comparing the differences in raindrop spectra in some regions of China.

The results showed that the mean scattering field first weakened and then increased with increasing rainfall, and the extreme points were related to the wavelength and different raindrop distributions. The spectrum of the received light intensity signal was calculated using the correlation function of the scattering field, and the spectral broadening of the received light intensity was obtained by substituting the raindrop-particle velocity.

We qualitatively analyzed the variance in the scattering-field strength by calculating the zero-frequency attenuation under different rainfall levels. The results showed that the variance of the scattering field also depended on different raindrop spectral distributions, and as the rainfall increased, the receiving variance first increased and then decreased. Finally, the model was validated using experimental data to verify the relationship between the mean and variance of light intensity.

9.4 Detector Noise Model

The working process of a photodetector can introduce noise into the system. According to the classification of the physical causes of noise, detector noise generated by semiconductor photodetectors can generally be divided into four categories: shot noise, generation-compound noise, thermal noise, and low-frequency noise.

9.4.1 Shot Noise

Shot noise refers to the noise of fluctuating random electrons generated by thermal excitation in the absence of light. It is called "shot" noise because the fluctuating electric unit is a single electron charge. The shot-noise model is shown in Fig. 9.18 below. When an electron in the medium moves to x between the plates, the induced charges Q_1 and Q_2 on plates 1 and 2 are, respectively,

$$Q_1 = e\left(1 - \frac{x}{d}\right) \tag{9.122}$$

$$Q_2 = e\frac{x}{d}, \tag{9.123}$$

where d is the distance between the plates. The current in the external circuit between the plates is

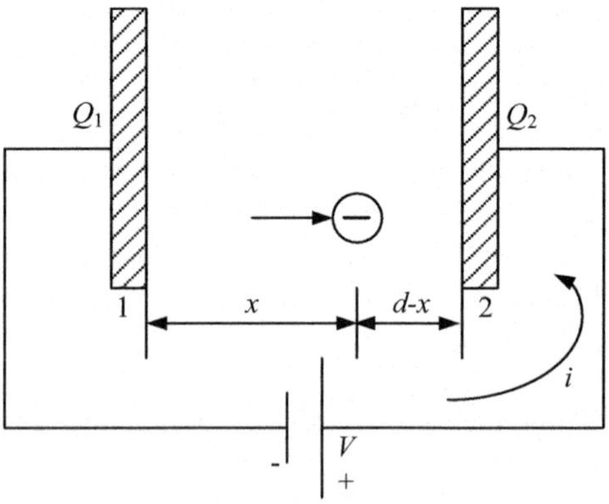

Fig. 9.18 Schematic diagram of the shot-noise model [4]

$$i(t) = \frac{\mathrm{d}Q_2}{\mathrm{d}t} = \frac{ev_x}{d} \tag{9.124}$$

From t to $t + T$, there are always N_T excited electrons. The total current is

$$i_T(t) = \sum_{i=1}^{N_T} ef(t - t_i) \quad 0 < \mathrm{t} < \mathrm{T} \tag{9.125}$$

By applying the Fourier transform to the above equation, the following is obtained:

$$\begin{aligned} F_i(\omega) &= \int_{-\infty}^{+\infty} f(t - t_i)\exp(-j\omega t)\mathrm{d}t \\ &\overset{u=t-t_i}{=} \exp(-j\omega t_i)\int_{-\infty}^{+\infty} f(u)\exp(-j\omega u)\mathrm{d}u \\ &= \exp(-j\omega t_i)F(\omega) \end{aligned} \tag{9.126}$$

In the formula, F(ω) is the Fourier transform of the single electron-pulse current function $F(u)$. Therefore,

$$i_T(\omega) = \sum_{i=1}^{N_T} eF(\omega)\exp(-j\omega t_i) \tag{9.127}$$

$$\begin{aligned} |i_T(\omega)|^2 &= i_T(\omega)i_T^*(\omega) \\ &= e^2|F(\omega)|^2 \sum_{i=1}^{N_T}\sum_{j=1}^{N_T} \exp\left[-j\omega\left(t_i - t_j\right)\right] \end{aligned} \tag{9.128}$$

We calculate the mean of the above equation, and obtain

$$\overline{|i_T(\omega)|^2} = e^2\overline{|F(\omega)|N_T} \tag{9.129}$$

The mean-square value of the noise current per unit time is

$$\overline{|i_n(\omega)|^2} = e^2\overline{N}\,\overline{|F(\omega)|^2} = ei_d\overline{|F(\omega)|^2} \tag{9.130}$$

where $i_d = e\overline{N}$ is the mean thermal-excitation current (dark current). Considering the average electron-transit time τ, $\overline{f(t)} = \overline{v_x}/d = 1/\tau$; moreover,

$$\overline{F(\omega)} = \int_{-\infty}^{+\infty} \frac{\overline{v_x}}{d} \exp(-j\omega t)\mathrm{d}t = \int_{-\tau/2}^{\tau/2} \frac{1}{\tau} \exp(-j\omega t)\mathrm{d}t$$
$$= \left| \frac{\sin(\omega\tau/2)}{\omega\tau/2} \right|_{t\to 0} = 1 \tag{9.131}$$

Therefore, the current power spectrum of shot noise can be obtained:

$$g(\omega) = \overline{|i_n(\omega)|^2} = ei_d \tag{9.132}$$

For a system bandwidth of Δf, the mean-square value of the shot-noise current is

$$\overline{i_n^2} = 2ei_d\Delta f \tag{9.133}$$

The effective value of the noise current is

$$i_n = \sqrt{2ei_d\Delta f} \tag{9.134}$$

The noise voltage on a load with an internal resistance of R is

$$u_n = \sqrt{2ei_d\Delta f R^2} \tag{9.135}$$

For photodetectors with an internal gain M (photomultiplier tube, avalanche photodetector, etc.), the average noise current and voltage must be multiplied by M.

9.4.2 Photon Noise

The function of a photodetector is to detect optical signals. The optical power of the signal and background light are the statistical means of the number of photons that reach the detector; a random number of photons reach the detector at each moment. Therefore, the number of charge carriers excited by photons is also random, and the resulting fluctuation noise is a type of shot noise, called photon noise, because it is excited by photons.

The noise photocurrents excited by the fluctuation of the signal light and background light are i_s and i_b, respectively, and the photon noise current can be expressed as

$$i_{nb} = \sqrt{2ei_b\Delta f} \tag{9.136}$$

$$i_{ns} = \sqrt{2ei_s\Delta f} \tag{9.137}$$

Considering the internal gain M of the photoconductive detector, the generation of the composite photon noise in the photoconductive detector can be expressed as

$$i_{bg-r} = \sqrt{4ei_b M^2 \Delta f} \quad (9.138)$$

$$i_{sg-r} = \sqrt{4ei_s M^2 \Delta f} \quad (9.139)$$

Considering the quantum efficiency of detectors η, i_s and i_b can be represented by signal-light power P_s and background-light power P_b:

$$i_b = \frac{e\eta}{hv} P_b \quad (9.140)$$

$$i_s = \frac{e\eta}{hv} P_s \quad (9.141)$$

Therefore, the total shot noise caused by i_d, i_b, and i_s of the photodetector can be uniformly expressed as

$$i_n = \sqrt{Se(i_d + i_b + i_s)M^2 B}, \quad (9.142)$$

where S is the detector-type parameter. The photovoltaic detector $S = 2$, and the photoconductivity detector $S = 4$. M is the internal gain of the detector and B is the detector bandwidth.

9.4.3 Thermal Noise

The thermal noise of the photodetector is caused by the random thermal motion of free electrons in the medium. At thermal equilibrium, the average kinetic energy of a single electron is

$$E_k = \frac{3}{2} K_B T, \quad (9.143)$$

where K_B is Boltzmann's constant, T is the absolute temperature, and m is the electron mass. The rate of motion of the electrons is

$$\overline{v_x^2} = \overline{v_y^2} = \overline{v_z^2} = \frac{K_B T}{m} \quad (9.144)$$

The detector conductivity σ is

$$\sigma = \frac{ne^2 \tau_0}{m}, \quad (9.145)$$

where n is the electron concentration in the medium, e is the electron charge, and τ_0 is the average electron collision time. The internal resistance R of the medium is

$$R = \frac{d}{\sigma A} = \frac{md}{ne^2\tau_0 A}, \tag{9.146}$$

where d is the length of the medium and A is its cross-sectional area.

The current produced by a single electron in the interval τ between two adjacent collisions is

$$i_1(t) = \frac{ev_x}{d} \quad 0 < t < \tau, \tag{9.147}$$

where v_x and τ are both independent variables. By applying the Fourier transform to $i_1(t)$, we obtain

$$|i_1(\omega, v_x, \tau)|^2 = \frac{e^2 v_x^2}{\omega^2 d^2}\left[2 - \exp(j\omega\tau) - \exp(-j\omega t)\right] \tag{9.148}$$

Because the collision probability of electrons is

$$P(\tau) = \frac{1}{\tau_0}\exp\left(-\frac{\tau}{\tau_0}\right), \tag{9.149}$$

and the collision time is extremely short, with $\omega^2\tau_0{}^2 \ll 1$, averaged over τ and v_x, respectively,

$$|i_1(\omega)|^2 = \frac{2e^2\tau_0^2 K_B T}{md^2} \tag{9.150}$$

For a medium with an electron concentration n and volume $V = Ad$, there are nV electrons inside. the average number of collisions is

$$\overline{N} = \frac{nV}{\tau_0} \tag{9.151}$$

Therefore, the power spectrum $g(\omega)$ of the total thermal-noise current $i_n(t)$ is

$$g(\omega) = \overline{|i_n(\omega)|^2} = \overline{N}\,\overline{|i_1(\omega)|^2} = \frac{2K_B T}{R} \tag{9.152}$$

Consider detector bandwidth Δf, the mean-square values of the thermal-noise current and voltage of the detector with an internal resistance of R can be obtained as

$$\begin{cases} \overline{i_n^2} = \dfrac{4K_B T \Delta f}{R} \\ \overline{u_n^2} = 4K_B TR\Delta f \end{cases} \tag{9.153}$$

Correspondingly, the effective values of the thermal-noise current and voltage of the detector are

$$\begin{cases} i_n = \sqrt{\dfrac{4K_B T \Delta f}{R}} \\ u_n = \sqrt{4K_B TR\Delta f} \end{cases} \tag{9.154}$$

9.4.4 1/f Low-Frequency Noise

In addition to the aforementioned detector shot noise and thermal noise, any electrical device has $1/f$ noise, and photodetectors are no exception. The $1/f$ noise energy of the photodetector is mainly concentrated in the frequency band below 1 kHz, and is inversely proportional to the modulation frequency of the incident light signal.

At present, the $1/f$ noise of photodetectors is not very detailed, and its generation mechanism is not yet understood. The detection of surface-processing techniques (structural defects, surface unevenness, etc.) has a significant impact on $1/f$ noise. Therefore, $1/f$ noise can also become surface or excess noise. The $1/f$-noise empirical formula is

$$\overline{i_f^2} = \frac{Ai^\alpha \Delta f}{f^\beta}, \tag{9.155}$$

where A, α, and β are specific coefficients, and i is the detector's DC current. For most photodetectors, $\alpha \approx 2$ and $\beta \approx 1$. The effective current value of its $1/f$ noise is

$$i_f = \sqrt{\frac{Ai^2 \Delta f}{f}} \tag{9.156}$$

When designing optoelectronic systems, increasing the modulation frequency of optical signals can effectively avoid the impact of $1/f$ noise.

9.5 Summary

This chapter provided examples of noise models in optical-wireless communication systems, including relative-intensity noise and modulation distortion in light-source noise; background noise, turbulence noise, and scattering attenuation in atmospheric channels; and shot noise, thermal noise, and $1/f$ noise in optical receivers. Finally, the possible impact of various types of noise on the received signal-to-noise ratio of the system was discussed.

References

1. Li Z, Liao Z, Liang J et al (2023) Research and prospects of atmospheric turbulence models and atmospheric channel models. Opt Commun Technol 47(03):9–17
2. Ke X, Wu J, Yang S (2021) Research progress and prospects of atmospheric turbulence for optical-wireless communication. J Radio Wave Sci 36(03):323–339
3. Ishimaru A (1978) Wave propagation and scattering in random media. IEEE Press
4. Xie M (2017) A study on the noise model of optical-wireless communication system directly detected by subcarrier modulation, Xi'an University of Technology, Xi'an
5. Zhao Z (1995) Raindrop size distribution model and rain attenuation prediction in Guangzhou Region. J Radio Wave Sci 10(4):33–37
6. Chen D, Gu S (1989) Study on the average spectrum of raindrops in heavy rainstorm. J Meteorol 47(1):124–127
7. Zheng J, Chen B (2007) Selection of raindrop spectral distribution function: a comparative study of M-P and gamma distributions. Meteorol Sci 27(1):17–25
8. Huang J, Hu D (1991) Size distribution model in Qingdao region. J Radio Wave Sci 6(Z1):177–180

The manufacturer's authorised representative in the EU is Springer Nature Customer Service Centre GmbH, Europaplatz 3, 69115 Heidelberg, Germany. If you have any concerns regarding our products, please contact ProductSafety@springernature.com

Printed and bound by CPI Group (UK) Ltd, Croydon, CR0 4YY
01/10/2026
02225907-0014